STATISTICAL QUALITY CONTROL METHODS

STATISTICS

Textbooks and Monographs

A SERIES EDITED BY

D. B. OWEN, Coordinating Editor

Department of Statistics
Southern Methodist University
Dallas, Texas

PAUL D. MINTON

Virginia Commonwealth University
Richmond, Virginia

JOHN W. PRATT

Harvard University
Boston, Massachusetts

VOLUME 1: The Generalized Jackknife Statistic
H. L. Gray and W. R. Schucany

VOLUME 2: Multivariate Analysis
Anant M. Kshirsagar

VOLUME 3: Statistics and Society
Walter T. Federer

VOLUME 4: Multivariate Analysis:
A Selected and Abstracted Bibliography, 1957-1972 *(out of print)*
Kocherlakota Subrahmaniam and Kathleen Subrahmaniam

VOLUME 5: Design of Experiments: A Realistic Approach
Virgil L. Anderson and Robert A. McLean

VOLUME 6: Statistical and Mathematical Aspects of Pollution Problems
edited by John W. Pratt

VOLUME 7: Introduction to Probability and Statistics (in two parts).
Part I: Probability. Part II: Statistics
Narayan C. Giri

VOLUME 8: Statistical Theory of the Analysis of Experimental Designs
J. Ogawa

VOLUME 9: Statistical Techniques in Simulation (in two parts)
Jack P. C. Kleijnen

VOLUME 10: Data Quality Control and Editing *(out of print)*
Joseph I. Naus

VOLUME 11: Cost of Living Index Numbers: Practice, Precision, and Theory
Kali S. Banerjee

VOLUME 12: Weighing Designs. For Chemistry, Medicine, Economics, Operations Research, Statistics
Kali S. Banerjee

VOLUME 13: The Search for Oil: Some Statistical Methods and Techniques
edited by D. B. Owen

VOLUME 14: Sample Size Choice: Charts for Experiments with Linear Models
Robert E. Odeh and Martin Fox

VOLUME 15: Statistical Methods for Engineers and Scientists
Robert M. Bethea, Benjamin S. Duran, and Thomas L. Boullion

VOLUME 16: Statistical Quality Control Methods
Irving W. Burr

OTHER VOLUMES IN PREPARATION

STATISTICAL QUALITY CONTROL METHODS

Irving W. Burr

Professor Emeritus
Statistics Department
Purdue University
West Lafayette, Indiana

MARCEL DEKKER, INC. New York and Basel

MARCEL DEKKER, INC.
270 Madison Avenue, New York, New York 10016

LIBRARY OF CONGRESS CATALOG CARD NUMBER: 75-17457
ISBN: 0-8247-6344-0
Current printing (last digit):
10 9

PRINTED IN THE UNITED STATES OF AMERICA

To Terry, Mary Kate, Peter, and John

PREFACE

The basic aim of this book is to give a clear, sound and natural exposition of statistical methods, especially useful in quality control in industry. The criteria used in choosing topics and their presentation are usefulness and wide applicability, rather than elegance. The emphasis is on data-analysis and decision-making. The statistical techniques discussed are also of great use in areas not specifically quality control, such as laboratory analyses and scientific research.

Methods of statistical quality control are becoming of greater importance for many reasons: (a) the heightening of competition, both domestic and international (quality being one of the key ingredients), (b) the increasing need to avoid loss of material and to save man-hours of time, (c) the intense profit squeeze, (d) the rise of consumerism, (e) the rapid increase in number of legal liability cases which emphasizes the need for greater reliability of the product and adequate documentary records for legal defense, (f) the need to know one's process capabilities, (g) new and stricter quality related laws (e.g., on weights and purity), (h) the proliferation of mandatory industrial standards, and (i) the growth of international standards for international trade. This book is therefore of direct interest to the over 20,000 members of the American Society for Quality Control.

The book is written for junior, senior and graduate students in engineering, industrial management, statistics and science, for industrial courses and also for self-study by those in industry. It is assumed that the student has at least a little background in statistics. Chapter 2 reviews the basics of probability and statistics for those with some background. Moreover, this chapter can almost stand on its own for the reader with very little or no previous statistical background. Some use is made of calculus, but not to a great extent as the book deals mostly in concepts.

Nearly all of the problems and illustrative examples are from actual cases in the author's extensive collection, gathered through the years from industry and from research at Purdue University and elsewhere. They thus represent actual practical problems. The aim of the book is, however, unity and cohesiveness, to emphasize principles rather than to present a host of case histories. The latter can easily leave the student with a feeling of frustration and disunity. Presentation of too many competing techniques and too many details and exceptions is avoided.

Some attention is given to the growing tendency toward automation in industry. Some calculational aids are presented. Very little is included in the direction of "decision theory," because of the large amount of inputs required which are often unobtainable, and the inability to codify such approaches so as to avoid starting from "scratch" every time. Nevertheless, elements of this viewpoint often appear in the book, subjectively guiding the choice of critical levels and assumed risks.

The present book is to some extent an outgrowth of the author's "Engineering Statistics and Quality Control," McGraw-Hill, 1953. Some paragraphs, data and examples taken from this volume are used in the present book without specific referencing.

For notations, the author has chosen to use Greek letters for population charactersitics and ordinary (Latin) letters for sample characteristics for measurement data, as is common practice in most statistical areas. Prime marks are used for population characteristics for counted data. In connection with notations, the reader will find the glossary to be of much use.

The author owes much to the pioneers in statistical quality control at the Bell Telephone Laboratories: George D. Edwards, Walter A. Shewhart, Harry G. Romig, and especially to Harold F. Dodge, with whom the author worked on the Standards Committee of the American Society for Quality Control for over 20 years. The author has had the privilege of working with many other people from industry, of whom we may particularly mention Arthur Bender, Jr., Richard Bingham, Walter Breisch, Matt Dalton, Richard Ede, Ernest Fay, Hugh Ferguson, Aldis Hayes, Carl Hoover, John Malmstrom, Harry Marsh, George McDermott, Paul Peach, Carl Noble, Kort Pfabe, Keith Ross and Wade Weaver.

It has also been most rewarding to have taught in many intensive courses with Professors Cecil Craig, John Henry, Charles Hicks, Lloyd Knowler, Gayle McElrath, Edwin Olds, Mason Wescott and Holbrook Working, as well as with some of those from industry mentioned previously.

The author would like to heartily thank Professor Shanti S. Gupta, Head of the Statistics Department at Purdue University, for encouragement and generous use of facilities. Warm thanks are also due to the department's secretaries for expert typing of the manuscript: Norma Lucas, Edna Hicks, Sandy Emory and Rebecca Fagan.

Finally, the author must heartily thank his wife, Elsie, whose interest, encouragement and enthusiasm have been so helpful.

Irving W. Burr

CONTENTS

CHAPTER 1

INTRODUCTION

1.1. What is Statistics? Statistics is the science concerned with the drawing of inferences and making decisions from data which are subject to variation. Wherever results vary from trial to trial, time to time or person to person, you have a statistical problem, whether you like it or not, whether you know it or not. Since random variation is so universally present in industry, engineering and science in general, the question becomes not whether you are going to perform the functions of a statistician; you are, each time you make a decision or draw an inference! The only question is as to how well you are going to perform that function.

Sometimes, of course, especially where the random variation is relatively minor, the decision is perfectly "obvious." In such cases no statistical methods are needed. But often enough what appears obvious, proves to be unjustifiable or false. Moreover there are many cases where the decision is not at all clear, and statistical methods are invaluable in estimating the risks and providing objective decisions. Statistical methodology is also concerned with the setting up of experiments, the drawing of samples and the proper analysis of results, for maximum efficiency and objectivity.

1.2. Why Statistical Quality Control? Quality control in various forms has been practiced for thousands of years, certainly dating back to the building of the pyramids in Egypt. But the use of statistical quality control, that is, the use of statistical methods in the control and improvement of quality in industrial production, is a quite recent development. The real beginnings were made in the 1920's by men in the Bell Telephone Laboratories, among whom we may mention Walter A. Shewhart, Harold F. Dodge, Harry G. Romig, George D. Edwards, Thornton C. Fry and E. C. Molina.

A basic reason for the need for statistical quality control is that industry is continually trying to work to ever closer tolerances. It is not difficult to make parts all "exactly" alike to some fairly gross degree of precision. But when we are asked to have a certain dimension of a part to always lie within an extremely narrow tolerance range (two limits), we may be forced to the utmost with our production and measurement capabilities, and statistical control becomes basic. For example, years ago, a diesel engine plant had to make plunger rods for forcing fuel through small holes. The diameter of the rods was to meet a total tolerance range of only .00004". Even with guages they had constructed, measuring to the nearest .00001", the problem was very difficult. They could of course produce rods all exactly alike to the nearest .001" or even .0001" perhaps. But to the nearest .00001" there were varying diameters.

Another form of variation to be controlled is the incidence of non-conforming parts in industrial production. These may have one or another dimension out of tolerance, or have some defect of slight or substantial importance. Production completely free of all such non-conforming parts is the ideal of course. But it may be unnecessary, or uneconomic, or even impossible. In the latter case is the production of millions of parts each of which can have any one or more of 20 different defects of varying importance. Another common example in which we cannot have absolute assurance of all parts meeting standards, is that in which the test destroys the part. Some examples of such tests are tensile strengths of rivets, life tests of electronic parts, and muzzle velocity of ammunition. If we test every part and perhaps find that all met the standard in such a test, we have none left to use!

In all these cases statistical quality control methods are of basic importance.

1.3. Aim of Book. In the last, say, twenty years, there has been a great proliferation of methods of statistical quality control. In this book we cannot begin to cover this broad a field. What we can do is (a) to cover a basic set of methods of statistical quality control of wide usage and flexibility, (b) to emphasize general principles and philosophy, and (c) to provide the reader with background upon which to grow in his use and application of statistical quality control. To make full use in a company of just the techniques herein presented could take fifty years.

Furthermore the aim is not to present statistical methods in general, however interesting this would be, but instead to concentrate on those of particular use in statistical quality control.

1.4. Prerequisites. It will be assumed that the student is somewhat familiar with basic statistical methods, as covered in a beginning course. For those without such background, however, as well as a refresher for those with this background we present a review in Chapter 2. Also some knowledge of elementary calculus is assumed. However, in order to make the book readable to those who are weak in these two assumed prerequisites we shall try to make the early parts of sections and chapters as readable as possible and take up more technical aspects further along. The answers to problems can also be a considerable aid to those having difficulties.

1.5. Suggestions to the Student. In statistics it seems to be impossible to avoid some difficulties with notations. This is perhaps especially true in statistical quality control. Hence the student is urged to make full use of the glossary.

A great many of the problems in the book use data, in most cases, actually obtained in industry. Care in calculation is important as always. A slide rule is usually sufficient. (But please, not like some people's idea of "slide-rule accuracy" permitting about a 3% error!) Of course a desk calculator, logarithms and tables of square roots are also useful. In substantial calculations, retaining more places than fully justified, and then rounding off at the end, is good practice.

In this book many different industrial parts and products occur in examples and problems. If you are not fully (or at all) acquainted with one, do not let this bother you, the author may not be either. The important thing is that the part or product has one or more measurable characteristics whose distribution we may study. Or there may be some properly defined "defects" we wish to study the incidence of. In both cases we can use statistical distributions without intimate knowledge of the product in question.

The student is urged to try to think of examples and illustrations in areas where he does have some background. This can considerably strengthen his grasp of the methods and make them more interesting.

A healthy skepticism of the obvious conclusion and an active seeking for all possible errors of tabulation, calculation, logic, nonapplicability of a technique or interpretation should be practiced and developed. Until one has made several serious blunders, it is difficult to realize the diabolical variety of traps which lie in wait for the unwary statistical worker. Only by eternal vigilance can they be avoided.

CHAPTER 2

BRIEF REVIEW OF STATISTICAL BACKGROUND

2.1. Population and Sample. In statistical work we are concerned with populations of objects, parts, people, trials or measurements. Such a population may be finite as in the first three cases or conceptually infinite as in the last two. Each individual in the population can be characterized by (a) a number such as length, weight or strength, or (b) can possess one or more attributes, such as, being defect-free or having one or more defects. These two ways of characterizing the individuals of a population gives rise to numerical or measurement data, and to attribute data respectively. Rather different statistical population models are used for the two different cases. The student should often be asking: Is this collection measurement or attribute data?

Probably the most basic question in all of statistics is the relation between a sample and the population from which it comes. How does the population determine what to expect of samples from it, and what does the sample tell about the population? Of course neither question has any answer unless the samples are drawn in an approved, unbiased manner, usually and hopefully at random, such as, by using a table of random numbers. (See Table IV in the back of the book.)

In general, no matter what the population is like, it will determine, in some manner, the behavior of the samples, whether or not we can fully characterize this relation.

2.2. Parameter and Statistic. In general a population is characterized by one or more "parameters," and a mathematical model, particularized by the given parameters in question. Thus a stable industrial process producing parts whose diameter is of interest, will have a population of some form with a mean diameter and standard deviation (typical departure from the mean). The "form" may be "normal" or not. Or a stable industrial process may be producing light bulbs, a fixed proportion of which fail to light when first used. This proportion is the parameter characterizing this "binomial" population. It is binomial because, from this viewpoint, there are but two kinds of bulbs produced: those lighting and not lighting.

As distinct from population parameters, we also have the characteristics of a sample described by a sample statistic or several. Thus for a sample of diameters from the population, we may find the sample mean and the sample standard deviation. These statistics are supposed to be, and hopefully are, close to the respective population parameters. (How close? See Section 2.5.) Such sample statistics are often used as estimates of the respective population parameters.

In the other example cited, we might count the light bulbs in a sample of 500 which fail to light initially. If this is five, then we would find the sample proportion to be 5/500 = .010. This would be regarded as an estimate of the population parameter: the long-run proportion not lighting, under this process.

A sample containing a substantial collection of observations is often studied by making a frequency tabulation (of the observations into numerical classes), and then picturing the distribution by some frequency graph such as a histogram (blocks of height proportional to the frequency in each numerical class).

2.3. Notations. We now come to the troublesome problem of notations. Unfortunately there is not just one notational system for statistical measures. Fortunately there are now basically only two such systems. The most widely used one reserves Greek letters for population parameters, and ordinary letters for sample statistics. Corresponding letters in the two alphabets are used insofar as convenient or feasible. On the other hand a notational system, widely used in the quality control area [1]* and in industrial standards groups around the world, is to use for a sample statistic any convenient symbol, whether ordinary letter or Greek, and then to put a prime mark on it to designate the corresponding population characteristic. This is a flexible system and makes good sense. Moreover it introduces fewer Greek letters for the layman to learn to use. But unfortunately it is not the system toward which most fields of statistics have gradually evolved.

The author was greviously torn between using one or the other. Fortunately at least some progress has been made toward uniformity, in the standard [1] of the American Society for Quality Control. After considerable soul-searching, the author decided to use the notation, ordinary letter — Greek letter, for the sample statistics versus population characteristics, but use the prime notation for attribute data. Parentheses will also be used to include alternative notations.

We shall use the following notations. An observed measurement will be denoted by x. Then for a finite population of N numbers x, the population mean will be[†]

$$\frac{\Sigma_{i=1}^{N} x_i}{N} = \mu = \mu_x \quad (\text{or } \bar{x}'). \tag{2.1}$$

*Such bracketed numbers refer to references at the ends of chapters.

† $\Sigma_{i=1}^{N} x_i = x_1 + x_2 + \ldots + x_N$.

If the population is infinite, then the definition is strictly analogous, but defined by an integral instead of a summation.

The sample mean for n observed x values is analogous

$$\frac{\Sigma_{i=1}^{n} x_i}{n} = \bar{x}. \tag{2.2}$$

For the "variance" of a finite population of N x's:

$$\frac{\Sigma_{i=1}^{N} (x_i - \mu)^2}{N} = \sigma^2 = \sigma_x^2 \quad (\text{or } \sigma'^2). \tag{2.3}$$

This is the mean squared deviation from the population mean. For a sample of n x's we use

$$\frac{\Sigma_{i=1}^{n} (x_i - \bar{x})^2}{n - 1} = s^2 = s_x^2. \tag{2.4}$$

The n - 1 in the denominator (instead of n) is the number of degrees of freedom for the sample. It is used so that if we were to take all possible samples from the population, and for each, find s^2, then the mean of all of these very many s^2's will be exactly σ^2. We may summarize this relation by using the expectation symbol E, and say that

$$E(s^2) = \sigma^2 \quad (\text{or } \sigma'^2). \tag{2.5}$$

(This is true for all populations having a σ^2, and all samples of two or more.) Likewise it may be proven that

$$E(\bar{x}) = \mu = \mu_x \quad (\text{or } \bar{x}'). \tag{2.6}$$

Next we have the respective standard deviations. They are merely the square roots of (2.3) and (2.4):

$$\sqrt{\frac{\Sigma_{i=1}^{N} (x_i - \mu)^2}{N}} = \sigma = \sigma_x \quad (\text{or } \sigma'), \tag{2.7}$$

$$\sqrt{\frac{\Sigma_{i=1}^{n} (x_i - \bar{x})^2}{n - 1}} = s. \tag{2.8}$$

Each is a type of average departure of x's from their mean.

For discrete or counted data, we shall be concerned with two types: We may have a sample of n parts and count how many of these n are non-conforming or defective, that is, possess one or more "defects", suitably defined. In this way, each part or piece is either a good one or a defective. If we let d be the number of defective or non-conforming parts in a sample of n, then we use for the sample "fraction defective"

$$p = d/n = \text{sample fraction defective}. \tag{2.9}$$

We note that p is a sample statistic. For a <u>finite</u> lot of N parts, we suppose that there are D defectives. Then for the lot or population fraction defective

$$p' = D/N = \text{population fraction defective}. \tag{2.10}$$

Here p' is a parameter. (In the Greek system π or ϕ might be used. neither of which is too happy a choice.) Another usage of p' is for a stable process, for which the probability of a defective is constantly p', as each part is produced. This is analogous to an infinite lot.

The second type of discrete data is that in which we count the number of defects in a sample of one or more parts. Thus for example we can count the number of "seeds" (small bubbles) on say, 24 glass bottles. Then

$$c = \text{number of defects}, \tag{2.11}$$

Here c is a statistic. The parameter is the long-run average number of defects on such samples.

$$c' = \text{population average number of defects}. \tag{2.12}$$

(In the Greek system, letters used are λ, γ and μ.)

2.4. Probability Definitions and Laws. In general we are concerned with experiments or trials in which the outcome is uncertain. We can describe the "sample space" of possible individual outcomes. For example, we may have a lot of 100 parts and a trial consists of drawing two parts without replacement. Regarding the 100 parts as distinguishable somehow, perhaps by a serial number, we then have a sample space consisting of all possible samples of two from the lot. Counting unordered samples we have by combinations C(100, 2) = 100(99)/2 = 4950 possible outcomes or points in the sample space. If drawing is at random, these are all equally likely and the probability for each is 1/4950. In general an "event" is a collection or set of possible outcomes. For example if 10 of the 100 parts are defective, then an event might be that the sample of two contain no defectives. How many such samples are there? The answer is C(90, 2) = 90(89)/2 = 4005, and these too are equally likely.

With finite numbers of equally likely outcomes the definition of the probability of an event A is

$$P(A) = \frac{\text{no. of ways A can occur}}{\text{total no. of possible outcomes}} = \frac{N(A)}{N(S)} \tag{2.13}$$

where S is the sample space (an event too). In this equally likely case we see that

1. $0 \leq P(A) \leq 1$
2. $P(S) = 1$
3. If events A and B have no outcome in common $P(A \text{ or } B) = P(A) + P(B)$

In our example, we have

$$P(\text{no defectives}) = \frac{N(\text{no defectives})}{N(S)} = \frac{4005}{4950} = .809.$$

If we extend the last to a countable sequence of sets, these three properties may be used as defining postulates for probabilities of events.

For finite sample spaces with equally likely individual outcomes, there is much need for counting techniques such as permutations and combinations, because the numbers are so huge.

In probability we often have use for "conditional probabilities." We may ask what is the probability of event A, if it is known that event B did occur, designated

$$P(A|B) = \text{probability of A given B did occur.} \quad (2.14)$$

For this we limit our space from S down to just those outcomes of S in which B does occur. Then among these, how many are also favorable to A?

$$P(A|B) = \frac{N(A \text{ and } B)}{N(B)}. \quad (2.15)$$

In the example suppose that part number 1 is a good one and we define event B to be the event that the sample of two contains part number 1. Now N(B) = 99, because there are 99 other parts with which to complete the sample of two. Next let A be the event that the sample contains no defective. Then

$$N(A|B) = 89$$

because there are 89 good parts other than number 1. Hence

$$P(A|B) = 89/99 = .899.$$

Note that this is higher than the .809 figure for A by itself. P(A|B) can be higher, lower or equal to P(A). If equal then we call events A and B "independent," because the occurrence of B has no effect upon A.

Thus:

$$P(A|B) = P(A) \qquad \text{A, B independent.} \quad (2.16)$$

Dividing numerator and denominator of the fraction in (2.15) by N(S) we have

$$P(A|B) = \frac{P(A \text{ and } B)}{P(B)}, \quad (2.17)$$

$$P(A \text{ and } B) = P(B)P(A|B). \quad (2.18)$$

From (2.16) and (2.18) we have an alternative definition of independence:

$$P(A \text{ and } B) = P(A)P(B) \qquad \text{A, B independent} \quad (2.19)$$

We are now in a position to state four useful laws of probability:

1. Multiplicative laws

   a. $P(A \text{ and } B) = P(B)P(A|B) = P(A)P(B|A)$ (2.20)

   b. $P(A \text{ and } B) = P(A)P(B)$ A, B independent (2.19)

2. Additive laws

   a. P(A and/or B) = P(A) + P(B) − P(A and B) (2.21)

   b. P(A and/or B) = P(A) + P(B) A, B mutually exclusive (2.22)

2.4.1 Interpretation of a Probability. Suppose that on trial after trial the probability of an event remains constant at p'. The observed ratio of occurrence of the event in question is p = d/n. How are p and p' related? Intuition and logic both indicate that p approaches p' somehow, as n increases. The manner of approach is not, however, absolute as in calculus. Instead it is an approach "in probability." Suppose that p' = .5; then we might wish to continue trials until p is between .48 and .52. There is however, no absolute guarantee that p will ever lie between these limits no matter how large an n is taken. But we can determine an n, sufficiently large so that we can have, say, a .99 probability that p lies between .48 and .52. Such an approach is called "stochastic," or we say p approaches p' in a statistical sense.

2.5. Distribution of Sample Statistics. We shall briefly review the behavior of sample statistics for samples drawn at random from a more or less completely specified population.

The population commonly assumed for measurements x is the normal distribution. It is often at least a reasonable first approximation to observed distributions of data. Given that the mean is μ and standard deviation σ, then the probability density function is

$$f(x) = \frac{1}{\sigma\sqrt{2\pi}} \exp\left[-\frac{(x-\mu)^2}{2\sigma^2}\right] \qquad -\infty < x < \infty. \tag{2.23}$$

In order to use tables of normal curve probabilities it is necessary to standardize the variable. For this let

$$z = \frac{x - \mu}{\sigma} \quad \text{or } x = \mu + z\sigma. \tag{2.24}$$

Then for this standardized variable, z, the density function becomes

$$\phi(z) = \frac{1}{\sqrt{2\pi}} \exp\left[-(z^2/2)\right] \qquad -\infty < z < \infty. \tag{2.25}$$

Tables of probabilities for (2.25) are areas under curves between two limits. Many tables, such as our Table I, give the probability for the random variable Z to lie between $-\infty$ and z in the table. This is a cumulative probability.

2.5.1. Samples from a Normal Population. Of particular interest for such samples is the behavior of the sample mean, $\bar{x}$, and the variance s^2. If the population has mean μ_x and standard deviation σ_x, then we have

Distribution of Sample Means from Popn. $N(\mu_x, \sigma_x^2)$

1. Mean $\bar{x}$'s $= E(\bar{x}) = \mu_x$ (2.26)
2. Standard deviation $\bar{x}$'s $= \sigma_{\bar{x}} = \dfrac{\sigma_x}{\sqrt{n}}$ (2.27)
3. The $\bar{x}$'s are normally distributed

Note that $\bar{x}$'s center at the same value μ_x as the x's do, but that the variability $\sigma_{\bar{x}}$ is less than that for the x's, σ_x, by a factor $\sqrt{n}$. Thus to cut σ_x in half we need n = 4 observations for $\bar{x}$, etc. This might be called the "inverse square-root law of statistics."

Next consider the distribution of sample variances, s^2.

Distribution of Sample Variances from Popn. $N(\mu_x, \sigma_x^2)$

1. Mean s^2's $= E(s^2) = \sigma^2$ (2.28)
2. Standard deviation of s^2's $= \sigma_{s^2} = \sigma^2\sqrt{\dfrac{2}{n-1}}$ (2.29)
3. The variable $\dfrac{(n-1)s^2}{\sigma^2}$ follows the chi-square distribution with n - 1 degrees of freedom.

The chi-square distribution is tabulated in Table II. The chi-square distribution is extremely skewed, being J-shaped for one or two degrees of freedom, and becomes less unsymmetrical as the degrees of freedom increase, but the longer tail is always to the right side, that is, for high χ^2 values.

2.5.2. Non-normal Continuous Populations. If a population is non-normal, but with mean μ and standard deviation σ, then (2.26), (2.27) and (2.28) still hold exactly. Moreover the distribution of $\bar{x}$'s approaches normality as n increases, by the Central Limit Theorem. Formula (2.29) is approximately true, and we will commonly assume the chi-square distribution for $(n - 1)s^2/\sigma^2$, knowing that it is only approximate.

2.5.3. Binomial Data. When we have the following conditions we have a binomial sampling distribution:

Conditions for Binomial Distribution

1. Two possible outcomes per trial, e.g. good vs. non-conforming, or success vs. failure.
2. Constant probability p' for one particular type of outcome, for example, a non-conforming part.
3. Independence of trials: each trial independent of previous outcomes.
4. Take n trials, with interest in how many yield the type of occurrence in question.

Given these conditions, then

$$P(d \text{ occ. in } n) = C(n,d)\, p'^{d}(1-p')^{n-d} \qquad d = 0, 1, 2, \ldots, n. \tag{2.30}$$

This is essentially a sampling distribution consisting of a sample of n trials for each of which we note whether we have an occurrence or non-occurrence.

For the binomial we can show

$$E(d) = \mu_d = np' \tag{2.31}$$

$$\sigma_d = \sqrt{np'(1 - p')}. \tag{2.32}$$

The shape of the distribution is always at least somewhat unsymmetrical if $p' < .5$, with the longer tail for d values above np'. As n increases, with p' fixed, the distribution becomes "normal."

Likewise, for $p = d/n$, we have

$$E(p) = \mu_p = p' \tag{2.33}$$

$$\sigma_p = \sqrt{p'(1 - p')/n}. \tag{2.34}$$

The distribution shape is like that for d of course.

2.5.4. Poisson Data. Another sampling distribution much used in statistical quality control is the Poisson distribution. It is the typical model for analyzing data on the number of defects on samples of material. For example, we may count visual defects on 24 bottles, specks and other blemishes on test areas of paper, typographical errors on a page, defects in each of a series of sub-assemblies, or breakdowns of insulation in 1000 meters of insulated wire being tested.

Practical Conditions for Poisson Distribution

1. The number, c, of "defects" suitably defined to be counted over "equal areas of opportunity."
2. The defects occur independently, that is, the occurrence of one defect does not make it more, or less, likely that another will occur in the sample.
3. The maximum possible number of defects on a sample is vastly greater than the average number c'.

Under these conditions (made more rigorous) we have

$$P(c \text{ defects}) = \frac{e^{-c'}(c')^c}{c!} \qquad c = 0, 1, 2, \ldots, \tag{2.35}$$

where $e = 2.71828\cdots$.

Thus knowing the parameter c', we can find the probability for any number of defects c to occur. Moreover we can prove

$$E(c) = \mu_c = c' \tag{2.36}$$

$$\sigma_c = \sqrt{c'}. \tag{2.37}$$

The Poisson distribution is also unsymmetrical with the long tail toward high values of c, for all $c' > 0$, but becomes more and more symmetrical as c' increases.

2.6. Statistical Estimation. One of the main uses for drawing samples and analyzing them is to estimate what the population is like. More specifically we commonly assume a type of population, for example, normal, binomial or Poisson, and then try to estimate the parameter(s) from the sample or samples at hand. (In quality control, as we shall see, we often have a series of samples.) There are two forms of estimation: point and interval.

2.6.1. Point Estimation. In point estimation the objective is to obtain a one-number guess at the parameter, from the sample. In general there are more than just one such estimator and we want to choose the "best." Several criteria of "best" are: (a) The theoretical average value of the estimator should be the parameter. We then have an unbiased estimator of the parameter. (b) The estimator should be subject to as little error as possible, that is, among unbiased estimators we would like that estimator with minimum standard deviation. (c) The estimator should use every bit of relevant information in the sample. The following are estimators, each possessing all three of these desirable properties:

$\bar{x}$ for μ_x (normal population of x's) (2.38)

s_x^2 for σ_x^2 (normal population of x's) (2.39)

p for p' (binomial population) (2.40)

c for c' (Poisson population). (2.41)

Hence we use them.

2.6.2. Interval Estimation. Since it is not to be expected that we can hit a parameter with a single guess or estimate we commonly use interval estimation. For this we choose (preferably in advance of sample-drawing) a desired degree of confidence or probability that the interval which we determine from the sample will contain the true parameter. This is often 90% or 95%. Then from the sample data we find two limits, and assert that the parameter lies between, having, say, 90% confidence that we are telling the truth.

Slightly less rigorously we say that we are 90% confident that the parameter lies between the two limits from the sample. This seems entirely satisfactory for practical purposes.

For a normal population of measurements we have μ_x and σ_x^2 to be estimated. In general, elementary statistics courses discuss limits giving $1 - \alpha$ confidence. For μ_x, when σ_x is somehow known, we have the following limits

$$\bar{x} \pm z_{(\alpha/2 \text{ above})} \frac{\sigma_x}{\sqrt{n}} \tag{2.42}$$

where z is the standard normal variable and the probability (area) above is $\alpha/2$. Such limits are very nearly correct, even when the x's are not normal and n is small.

However, again assuming a normal population of x's, if we do not know σ_x, then we are forced to use the sample standard deviation s, and the so-called t distribution as follows:

$$\bar{x} \pm t_{(n - 1 \text{ degr. of freedom, } \alpha/2 \text{ above})} \frac{s}{\sqrt{n}} . \tag{2.43}$$

Then we use Table III in the Appendix.

For $1 - \alpha$ confidence limits for σ_x, we use the sample standard deviation and the chi-square distribution of Table II. Limits for σ, derivable from the distribution of s^2 (Section 2.5.1) are

$$s \cdot \sqrt{\frac{n-1}{\chi^2_{(n-1 \text{ d.f., } 1-\alpha/2 \text{ below})}}} , \quad s \cdot \sqrt{\frac{n-1}{\chi^2_{(n-1 \text{ d.f., } \alpha/2 \text{ below})}}} . \tag{2.44}$$

For discrete data populations, there are many approaches possible, the easiest of which is to use such a table as [2], and for given d and n merely look up the limits. However, there are rather approximate limits available which are useful. For binomial data, with $p \leq .5$ and np at least 10, say, the following approximate limits may be used:

$$\text{Limits for } p': p \pm z_{(\alpha/2 \text{ above})} \cdot \sqrt{\frac{p(1-p)}{n}} . \tag{2.45}$$

These stem from (2.33), (2.34) and the approach of the distribution of p to normality.

Likewise for a Poisson sample we observe c defects. Then if c is, say, at least 10, then we may use the approximate limits:

$$\text{Limits for } c': \quad c \pm z_{(\alpha/2 \text{ above})} \cdot \sqrt{c} . \tag{2.46}$$

These came from (2.36), (2.37) and the approach toward normality of the Poisson distribution as c' increases.

Many other cases of interval estimation are usually given in beginning statistics courses, for example, limits for functions of parameters for two populations, such as $\mu_1 - \mu_2$.

2.7. Testing Hypotheses. Complementary to the technique of estimation is that of testing hypotheses. The basic question is as to whether the observed sample could reasonably have come from a population more or less completely specified by the hypothesis. The hypothesis may be concerned with a parameter or with the <u>type</u> of population. For the latter we might for example test the hypothesis that the population is normal. Tests of goodness of fit are used for such hypotheses, such as the chi-square test. More often we are concerned with one (or more) parameters, and compare the observed sample statistics with the hypothesized parameter.

Elements of Testing a Hypothesis on One Parameter, for Example, μ.

1. Basic <u>assumptions</u> are made which are assumed true and not open to question in the test. Commonly the <u>type</u> of population is assumed, for example, that it is normal. The value of parameters not being tested may also be assumed, for example, $\sigma = 5$.
2. Although in a sense a <u>null hypothesis</u> is an assumption, it is not like the assumptions in 1. Instead the hypothesis is under test

and may be rejected, whereas we never use our test to reject an assumption as in 1. An example of a simple null hypothesis is $\mu = 100$. (We could also have a composite hypothesis $\mu \leq 100$.)

3. Rejecting a hypothesis when it is actually true is committing an <u>error of the first kind</u>.
4. Because of variability it is in general impossible or infeasible to make a test for which the probability α of an error of the first kind is zero. Nevertheless we do want to keep the <u>risk</u> α at some specified low value, perhaps .01.
5. A <u>sample statistic</u> (an estimator) for the parameter in question is chosen. For μ we could use the mean $\bar{x}$ or the median, commonly choosing the former because of its smaller standard deviation. A sample size n is usually decided upon also at this point.
6. An <u>alternative hypothesis</u> is next chosen, containing other values of the parameter considered possible, or of economic or scientific interest. For example, as alternatives to $\mu = 100$ we could have $\mu < 100$, $\mu > 100$ or $\mu \neq 100$.
7. The chosen risk α in 4, the type of alternative hypothesis in 6 and knowledge of the distribution of the statistic in 5 enables us to set a critical region or rejection region for the statistic. Then whenever the statistic lies in this region we <u>reject</u> the basic or null hypothesis and accept the alternative hypothesis. But if the statistic does not lie in the rejection region we "accept" the null hypothesis, meaning that it perfectly well could be true. A critical region can be stated for the sample statistic itself, e.g. $\bar{x}$ or s, or for a standardized transformation of it, for example, $z = \dfrac{\bar{x} - \mu_0}{\sigma/\sqrt{n}}$ or

$\chi^2 = \frac{(n - 1)s^2}{\sigma_0^2}$. The critical region will have two parts for alternatives such as $\mu \neq 100$, and but one part for those such as $\mu < 100$ or $\mu > 100$. In the latter two cases the critical region is in the lower tail or upper tail of the distribution of $\bar{x}$ around 100 respectively.

8. An <u>error of the second kind</u> is committed when we accept the null hypothesis when in fact it is not true. The probability of an error of the second kind is called β. But note carefully that β depends upon how far off from the null-hypothesis value the parameter actually is. For the example, if the alternative is $\mu > 100$, and $\sigma = 3$ and $n = 4$, then if μ actually is 101, β will be large, but if $\mu = 110$, β will be very small. Thus here β is a function of μ.
9. In general it is well to draw an <u>operating characteristic</u> (OC) <u>curve</u> giving the probability of acceptance of the null hypothesis for each value of the parameter. This tells how much discriminating power the test has. If insufficient then a larger sample size is needed. In statistical quality control, much use is made of OC curves.

To make significance tests we simply use the sampling distributions of the various sample statistics as given in the four subsections of Section 2.5. We can set up a critical region for z or t for testing $\bar{x}$ vs. μ, or for χ^2 for testing s^2 vs. σ^2. Then calculate an observed z, t or χ^2 and see if it lies in the critical region. Or, if we have attribute data, we could find a critical region for d or c. But usually it is easier to find the probability of as great or a greater discrepancy from the expected than was observed. If this is $\leq \alpha$ (or $\alpha/2$ if a two-tail test) we reject the hypothesis.

For example if $n = 100$, $p' = .05$ and the alternative hypothesis is $p' > .05$, and $\alpha = .01$, then what decision is made if $d = 12$? Tables show $P(12 \text{ or more} | n = 100, p' = .05) = .00427$. So the hypothesis $p' = .05$ is rejected in favor of $p' > .05$.

Elementary books give tests for two samples as well. We shall be more interested in single sample testing, double sampling (a second sample contingent on the first) and upon a series of samples. Also we shall discuss sequential sampling.

References

1. American Society for Quality Control, Standard A1-1971, Milwaukee, Wisc.
2. D. Mainland, L. Herrera and M. I. Sutcliffe, "Statistical Tables for Use with Binomial Samples-Contingency Tests, Confidence Limits, and Sample Size Estimates." New York University College of Medicine, NY, 1956.

CHAPTER 3

CONTROL CHARTS IN GENERAL

3.1. Introduction. In this chapter we shall introduce Shewhart control charts in general terms. Then in Chapters 4 through 8, various types of control charts and their applications are discussed. Control charts are of great use in the analysis and control of manufacturing processes, so as to produce quality that is satisfactory, adequate, dependable and economic.

3.2. Running Records of Performance. One often sees in industrial plants, running record graphs of performance, plotted up each day or whenever new data become available. The points may represent various kinds of quality performance such as dimensions, strengths, per cent rejections, or defects found at the final inspection. Or, the data may not be directly quality associated, such as, production or sales figures, absenteeism, number of accidents per week, power consumption, stock market prices, etc.

In Figure 3.1 we show four running records of quality performances. These are the plottings of data for three products, as given in Table 3.1. The top two graphs are measurement data for samples of $n = 5$ rollers. The top chart for $\bar{x}$'s is of course concerned with the process level, while the second for ranges, R, is concerned with process variability. The latter two charts are for attribute or counted data. The third shows the fraction defective of a certain test on electrical equipment, the sample size being about 1340 per day. The last is for misalignments, which is just one of

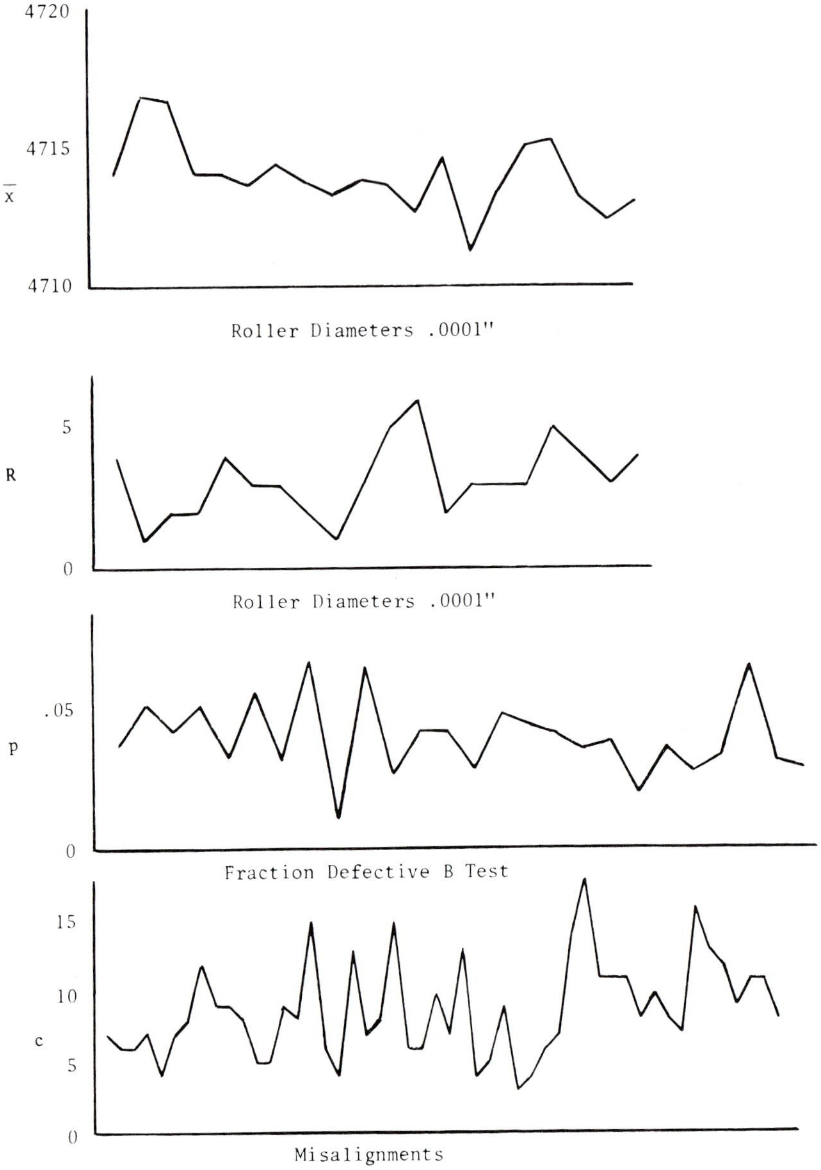

FIG. 3.1. Running record charts for data of Table 3.

many different categories of defects, for which a record was kept. Which of the four running record charts seems to show the greatest stability or control, that is, seems to have just a random pattern of points? Which

Table 3.1. Four Running Records of Quality Performance Over Time

Mean and range in .0001" of diameters of rollers for samples of n=5, specifications .4710" ± .0010"

Sample	$\bar{x}$	R
1	4714.0	4
2	4716.8	1
3	4716.6	2
4	4714.0	2
5	4714.0	4
6	4713.6	3
7	4714.4	3
8	4713.8	2
9	4713.2	1
10	4713.8	3
11	4713.6	5
12	4712.6	6
13	4714.6	2
14	4711.2	3
15	4713.4	3
16	4715.0	3
17	4715.2	5
18	4713.2	4
19	4712.4	3
20	4713.0	4

Fraction defective on "B" test, two factories, August, Sample size about 1340

Date	Frac. Def.
2	.0374
3	.0518
4	.0427
5	.0512
6	.0323
7	.0558
9	.0325
10	.0673
11	.0106
12	.0649
13	.0265
14	.0425
16	.0426
17	.0293
18	.0487
19	.0448
20	.0413
21	.0357
23	.0390
24	.0204
25	.0369
26	.0284
27	.0342
28	.0660
30	.0314
31	.0296

Number of misalignments c, in 50 aircraft sub-assemblies

Assembly	c	Assembly	c
1	7	26	7
2	6	27	13
3	6	28	4
4	7	29	5
5	4	30	9
6	7	31	3
7	8	32	4
8	12	33	6
9	9	34	7
10	9	35	14
11	8	36	18
12	5	37	11
13	5	38	11
14	9	39	11
15	8	40	8
16	15	41	10
17	6	42	8
18	4	43	7
19	13	44	16
20	7	45	13
21	8	46	12
22	15	47	9
23	6	48	11
24	6	49	11
25	10	50	8

seems to show the poorest control or most unstable performance? Try to guess.

Performance Varies. Certainly noticeable in all four graphs is variation. Results do vary from time to time, material to material, etc.

Variation is universal, it is part of life, therefore it is the part of wisdom to learn to live with it. In the first graph both high and low points are undesirable, while in the other three, the lower the point is the better we like it. A misconception we must guard against is that of concluding that since the points vary, the underlying production conditions are also varying. Such conditions may in fact be just as uniform as it is possible for human beings to make them, and still the points will vary, simply by chance.

Unusual Performance Calls for Action. In all four graphs, the purpose in plotting them is to keep track of the respective processes so as to take appropriate action whenever it is indicated. Thus in the first graph of averages, $\bar{x}$'s, resetting of the process is called for if an $\bar{x}$ strays too far away from the desired process average, in either direction. On the other three graphs too high a point may well be regarded as a danger signal and some action on the process may be needed to improve performance. Thus a high range R, for example means that the process is not consistently repeating itself.

What Performance is Unusual? This is the fundamental question in all such analyses of varying performance, and where does performance not vary? Common sense simply is not a reliable guide for avoiding two types of errors: (a) taking action when none is desirable, and (b) missing opportunities for taking appropriate action or learning what makes the process behave as it does. In order to minimize the chances of making these two types of errors, we need to use the control chart and other statistical tools. These risks are those called α and β in Chapter 2.

3.3. Random and Assignable Causes. In the preceding section we have begun to consider two more or less distinct types of causes. By random or chance causes we mean the whole host of small influences lying behind the particular measurement or result we happen to obtain. For example, in machining an outside diameter of some part, even when we try to hold all conditions as alike as possible, there are bound to be slight differences in diameter from one angular direction to another, from one position along the axis to another, and from one piece to another. The hardness of the stock will vary slightly even within a single piece. The rotational speed, the depth of cut and pressure of the tool will tend to vary, as will the lubrication, coolant, floor vibration, and temperature. Then after the piece is removed, the actually recorded measurement is obtained by an inspector with a gage. Different places along the piece, varying temperatures of gage and piece, presence of slight dust or varying thickness of oil film, varying pressure of gage upon the piece, the positioning of the piece in the gage, and finally the inspector's reading of the gage itself, can each cause results to vary. All of these and many others not mentioned may each have a relatively minor effect, but all taken together give variability to the resulting measurements. Note that the foregoing causes are operative even if we assume that all conditions are held as constant as is humanly possible. Thus there is bound to be some variation natural to every process, if only we try to take fine enough measurements. Such variation we call natural to the process, and say that it is due to chance causes. Ideally the results behave as though we were drawing numbered chips at random from a bowl containing a fixed population (replacing each chip before drawing the next).

Although no absolutely hard and fast distinction can be made between chance and assignable causes, they differ typically in the three characteristic respects shown in the accompanying tabulation.

	Number present	Effect of each	Worth seeking?
Chance causes	Very large number	Slight	Not worth looking for
Assignable causes	Very few, perhaps only one or none	Marked	Well worth seeking out, and possible to do so

In the study of a running record chart, our problem therefore becomes one of trying to tell when a high or low point on the graph (1) is an indication of the presence of an assignable cause of a change in the process, which should be sought out, or (2) can so readily occur by chance causes alone that there is no reliable evidence of the presence of an assignable cause. This is then the fundamental problem: When shall we assume that an assignable cause has been responsible for some apparently unusual performance, and when shall we attribute such performance to chance?

Some people think that the problem of deciding when to take action is an easy one. It is true that in some situations the assignable cause is so obvious that anyone can pick it up just by a glance at the chart. Unfortunately not all cases are so clear-cut, and also many samples that seem to be perfectly reliable indicators of assignable causes actually have arisen from chance causes and should therefore have been attributed to chance. Before going further the reader is challenged to try to pick out from among the points in Figure 3.1, which ones are reliable indicators of the presence of assignable causes. Without probability to guide you, it is a difficult and hazardous job.

To further emphasize the difficulty, the author has often shown audiences two running records of the same identical data plotted to two different scales. Nearly all will pick the one with a small vertical scale as showing the "best control." Then it is told them that the two are of the same data, and pointed out that if a decision depends <u>at all</u> upon the scale used in plotting then it cannot be a very objective decision.

3.4. The Control Chart for Interpreting Variations in Samples. The epoch-making discovery and development of the control chart were made in 1924 and the following years by a young physicist of the Bell Telephone Laboratories, Walter A. Shewhart. Wrestling with a problem made complicated by the presence of random variation, he came to realize that the problem was statistical in nature. Some of the observed variation in performance was natural to the process and unavoidable. But from time to time there would be variations which could not be so explained.

He reached the brilliant conclusion that it would be desirable and possible to set limits upon the natural variation of any process, so that fluctuations within these limits could be readily explained by chance causes, but any variation outside this band would indicate a change in the underlying process. There followed the development of charts for measurements and both kinds of attributes, the concept of the rational sub-group and its use in most effectively letting the process set the limits of natural variability, and a large amount of practical experimentation in trying out the new methods in actual plant problems. Then in 1931 came Shewhart's great "Economic Control of Quality of Manufactured Product," * in which the whole field was laid out including its theory, philosophy, applications, and, most pertinently, its economic aspects. Few fields of knowledge have ever been so completely explored and charted in the first exposition. The industrial field of quality control gets its name from this book. Control charts are often called "Shewhart control charts," Since all control charts stem from Shewhart's work, we shall use the shorter term, recognizing that "Shewhart" is the implied prefix for all of them.

---

* D. Van Nostrand Company, New York.

The basis of all control charts is the following: Any varying quantity (whether a measurement x, an average $\bar{x}$, a range R, a standard deviation s, a fraction defective p, or a count on the number of defects c) forms a distribution if chance causes alone are at work. Any such distribution has a mean and a standard deviation. Quite regardless of the shape of the distribution (unless extremely badly behaved), there will be, by chance causes only, very few points outside of the band between the mean minus three standard deviations and the mean plus three standard deviations. Hence, having set such limits, we have a band of normal variability for the statistical measure in question. Now suppose a point does go outside this band of normal variability. It is conceivable that such a point is just due to a rare "ganging up" of chance causes, and that no assignable cause was at work. But since industrial processes are so prone to contain assignable causes, it is a much better bet that the point outside the band is due to some assignable cause. Hence when such a point comes along, we assume there was some assignable cause at work and try to see what process conditions might have changed while this sample was produced and tested. Conversely, when a point lies inside the control band, we do not say that no assignable cause was present while the sample was produced and tested, but only that we have no reliable evidence for supposing that there was an assignable cause at work. Hence no action is taken. We attribute such points within the band as being due to chance causes only. Thus with control charts we save time looking for assignable causes when none are present and use this saved time for more careful search in those cases in which we do have reliable evidence of some nonchance factor at work. As we shall see, the control chart is a most powerful tool for making more fruitful our efforts in stabilizing and controlling our processes at desired performances.

Now let us consider again the data given in Table 3.1. Figure 3.2 shows the same running records as did Figure 3.1, but now we have added central lines

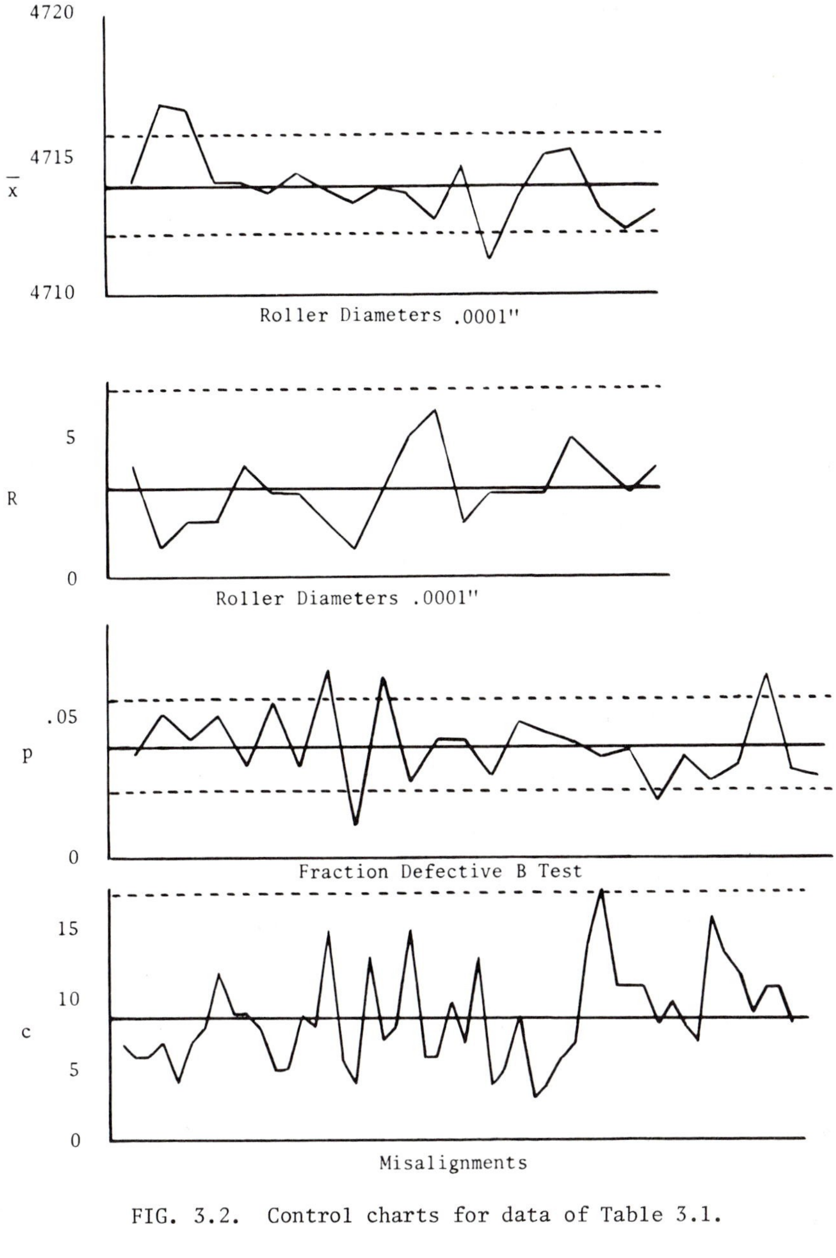

FIG. 3.2. Control charts for data of Table 3.1.

(solid) and control limit lines (dotted). These latter bound the respective bands of normal variability. As we see the chart for the ranges R is in perfect control, that is, the within-sample variability is apparently homogeneous The chart for misalignments shows only one point of the 50 marginally outside.

It is worth investigating to see why so many misalignments occurred on this subassembly. Although there is perhaps some reason for each misalignment, there are so very many opportunities for misalignments in each subassembly that the total number per sub-assembly behaves in quite a chance fashion (Poisson distribution).

The chart for $\bar{x}$'s reveals evidence of an assignable cause in the second and third points and also that for number 14, a low point. The B test results show five points indicating causes to be sought. The high points are taken as indications of "bad" assignable cause, while the two low points suggest "good" assignable causes (unless due to lax inspection). How did you, the reader, make out in your guessing?

3.5. Two Purposes of Control Charts. Up to this point in the chapter we have been primarily concerned with the _first purpose of control charts_, namely to _analyze past data_ for control. The idea is to try to see whether the results are homogeneous. Could they readily have come from a single population, that is, by chance variation from a stable process? Statistically, we are testing the hypothesis that we have, say, k samples from a single population with constant but _unknown_ parameter(s). Such an approach is commonly used until the process begins to show really good control. Note that the control limits are set by the process itself in this case.

A _second purpose of control charts_ is to compare observed results against a population whose parameter or parameters are known. The hypothesis being tested is that this population, completely specified, is the one from which the samples are drawn. Such an approach is used after exploratory work is done on the process, and reasonable standards can be set for it. In one sense, comparing each single sample point with the control band is a little significance test. Is the sample compatible or not? The α risk for each point is very small, perhaps .005 or less. But having _all_ of 20 sample points lying inside the control band has a much larger α risk, perhaps around .05 or .10. This too is a test of the hypothesis that the specified population with the given parameters is the one from which the samples were drawn, as well as being a test of homogeneity. This case of control charts is often called "standards given." Note that here the standards set the limits, not the current data, as was the case of "analysis of past data."

3.6. Economic Balance between Two Errors. In the preceding sections we have been concerned with two types of errors which the industrial worker can make. The first is to conclude that a sample differs significantly from the other samples or the standards when in reality the discrepancy is entirely due to chance causes. This type of error leads us to hunt for an assignable cause when none is present. Many undesirable process changes and unjustified blaming of personnel, tools or material stem from this error.

The other general type of error is to miss noting the presence of some assignable cause of either poorer or better performance. Such errors may well miss evidence of impending or actual trouble.

Much experimentation in the Bell System from 1924 onward seemed to indicate that the use of limits at plus and minus three standard deviations around the central line gives a good balance in industrial situations between the risks of the two errors. Subsequent industry-wide experience has born out the choice. If $\pm\ 2\sigma$ limits were used we would tend to be looking too often for non-existent or relatively weak assignable causes. Or if we were to use $\pm\ 4\sigma$ limits we would too often miss indications of assignable causes and opportunities for process improvement.

3.7. Meaning of a Process in Control and Potential Advantages. Basically we say a process is "in statistical control" when results behave like samplings from a single population. This can be thought of as drawings with replacement from a bowl of numbered chips. Or if there are enough chips, replacement becomes unimportant.

Some of the advantages which may accrue when a process is brought into good control are:

1. The process appears to be free from sources of variation worth identifying, and hence it is doing about all that we can expect of it. If this is not good enough, then we know that we shall have to make a more or less fundamental change in the process. Just "tampering" with it cannot be expected to help.

2. The act of getting a process into good statistical control ordinarily involves the identification and removal of undesirable assignable causes and possibly the inclusion of some good ones such as new materials or methods. Hence quality performance has commonly been much improved. Many a process which had been unable to hold specifications proves easily able to meet them when brought into control.

3. If a control chart for laboratory tests on uniform material shows lack of control, results from the test are suspect, because the lack of control shows that the laboratory conditions have not been satisfactorily standardized. On the other hand, if a control chart of reasonable length shows good control, we are justified in assuming homogeneity of laboratory conditions.

4. A process in good statistical control is predictable. It can be counted upon. Therefore if the chart for incoming material from a supplier

shows good control, we can justifiably decrease inspection, plotting each additional point on the chart, and merely watching for any evidence of lack of control. Thus we do not need to regard each submitted lot as a law unto itself. In the presence of good statistical control by the supplier all the previous lots supply evidence on the present lot, which is not safely the case if there has been lack of control.

5. If our own processes are in good statistical control, we can more safely guarantee our product. We know what it will do.

6. When testing is destructive, a chart in control gives us confidence in the quality of the untested product, but if the process is out of control, we are in a far weaker position for predictability on the remainder.

7. A chart in control in experimentation enables us to determine soundly the experimental error, which is not the case if the chart shows lack of control of variability.

8. The soundest way to cut inspection is through getting the process into control.

PROBLEMS

3.1. Make a list of chance causes and a list of possible assignable causes for some specific problem you are familiar with in your field of interest.

The following tables list quality performance data. Plot the $\bar{x}$ and R in the first two, p in the third or $\bar{c}$ in the fourth, as assigned, and comment on the running records.

Table P3.2. Pounds Per Square Inch To Operate Hydraulic Stop-Light Switch, Four Switches Per Sample

Average $\bar{x}$	Range R
73	10
83	20
75	15
71	5
76	5
71	5
77	20
77	5
66	10
59	20
76	20
80	20
95	10
79	25
84	15
77	20
75	0
76	5
76	10
80	15
72	20
71	5
96	25
86	15
86	10
76	5
82	5
81	10

Table P3.3. Final Mooneys (Measure Of Internal Viscosity) of GR-S Rubber Samples of Four Successive Measurements

Average $\bar{x}$	Range R
49.750	6.0
49.375	2.0
50.250	2.0
49.875	2.5
47.250	1.5
45.000	6.0
48.375	1.5
48.500	1.0
48.500	1.0
46.250	3.5
49.000	2.0
48.125	1.5
47.875	2.5
48.250	2.0
47.625	1.5
47.375	1.5
50.250	3.5
47.000	2.0
47.000	1.0
49.625	2.5
49.875	1.0
47.625	2.5
49.750	1.5
48.625	2.0

Table P3.4. Fraction Defective in Samples of 100 Auto Carburetors

Sample	Fraction p
1	.10
2	.04
3	.06
4	.12
5	.06
6	.08
7	.10
8	.12
9	.08
10	.07
11	.03
12	.04
13	.03
14	.04
15	.04
16	.10
17	.07
18	.05
19	.03
20	.06
21	.08
22	.10
23	.04

Table P3.5. Average Number of Defects Per Truck, Daily Production 225 Trucks

Date	Average No. Defects, $\bar{c}$
Sept. 6	(Labor Day)
7	.81
8	.80
9	.80
10	.62
13	.67
14	.67
15	.74
16	.92
17	.74
20	.86
21	.79
22	.71
23	.72
24	.86
27	.87
28	1.21
29	.87
30	1.40
Oct. 1	.78

References

1 W. A. Shewhart, "Economic Control of Quality of Manufactured Product." Van Nostrand-Reinhold, Princeton, New Jersey, 1931.

2 American Society for Quality Control, Standard B3-1970. Also American National Standards Institute, Z1.3-1970.

CHAPTER 4

CONTROL CHARTS FOR MEASUREMENTS

4.1. Introduction. In the preceding chapter we have discussed the general characteristics of control charts, and their usefulness, and emphasized two general purposes: (a) process control versus given standards, and (b) analysis of past data for control.

In the present chapter, we are concerned with the most frequently used control charts for measurements: $\bar{x}$ charts for the process level, and R charts and s charts for the process variability. In general one will use an $\bar{x}$ chart for averages, and only one of the R and s charts. However, we shall carry along both of the latter variability charts for comparison. There are a number of other control charts for measurements, some of which are discussed in Chapter 7.

4.2. Control Charts for Averages, Ranges and Standard Deviations, from Given Standards. In this section we consider the case in which we have given the two parameters μ and σ. These may result from substantial past history of good control at satisfactory performance, or possibly be in the nature of goals. In the latter case one might possibly use the mid-point of the specified tolerance band for the standard mean μ. But the standard σ should virtually always be the result of experience, for otherwise all probability interpretations of points outside the control limits and inside too are lost.

In order to illustrate statistical principles in this book we often resort to sampling experiments, or at least suggest some. In order to illustrate $\bar{x}$, R and s charts with μ and σ given we shall use Population A of Table 4.1.

Table 4.1. Approximately Normal Populations for Sampling Experiments

Number x	A	B	C	D	E	F	G
+11				1			
+10			1	1			
+ 9			1	1	1		
+ 8			1	3	3		
+ 7		1	3	5	10		
+ 6		3	5	8	23		
+ 5	1	10	8	12	39		
+ 4	3	23	12	16	48		
+ 3	10	39	16	20	39	1	
+ 2	23	48	20	22	23	3	
+ 1	39	39	22	23	10	10	1
0	48	23	23	22	3	23	3
- 1	39	10	22	20	1	39	10
- 2	23	3	20	16		48	23
- 3	10	1	16	12		39	39
- 4	3		12	8		23	48
- 5	1		8	5		10	39
- 6			5	3		3	23
- 7			3	1		1	10
- 8			1	1			3
- 9			1	1			1
-10			1				
N	200	200	201	201	200	200	200
μ	0	+2	0	+1	+4	-2	-4
σ_x	1.715	1.715	3.47	3.47	1.715	1.715	1.715

In this table are listed several convenient populations with which to experiment. These are the same, approximately normal distributions as were extensively used in the famous War Production Board courses in quality control by statistical methods, as developed by Working and Olds [1]. One of their most valuable contributions to the presentation of statistical quality control in particular and of statistical methods in general was the widespread and ingenious use of various sampling experiments.

Beads or chips may be marked with the numbers listed in the first column in appropriate frequency.* Any one population, or a combination of several may placed in a bowl, thoroughly mixed, and samples drawn.

Experimentation with beads or chips may possibly seem at first rather childish to some readers. But it has been the author's experience that a great many people find that such experiments illuminate the principles of random variation and sampling better than the use of equations and formulas. In fact it seems to prove valuable and interesting even for those with considerable mathematical sophistication. Even if an experiment is not performed, a description of how one might be designed to illustrate a technique, can be valuable, after one has seen a few experiments.

It is helpful for the reader to learn to associate the sample drawings, from populations like those in Table 4.1, with coded results from data from problems in his own field. This merely requires a little imagination. For example, we may consider Rockwell hardness on the C scale. If the specifications are 50 to 60, we may record our readings relative to the nominal 55. Thus 57 is listed as a +2 and 51 as a -4. Other examples are the following:

Carbon in steel, .20 to .30%

Moisture content in stack of lumber, 6.5 ± .5%

---

*It is well to have different colors for the different populations, so that they may be readily sorted out after mixing, should this happen intentionally or by mistake. Chips may be dyed by Easter egg dye, or differing colored India inks used. Beads 1 cm in diameter are convenient. Circular chips, from thick fiber, are sometimes waste output from stampings and can be obtained free.

Compression of springs to specified length, 12.0 ± .5 lb

Outside diameter of a roller, 2.214 ± .005 in.

Weight of an ingot, 5,300 ± 50 lb

Root-mean-square finish measurement, nominal 25, maximum 30

Edge widths of piston rings ground to .2120 ± .0005 in.

Resistance of coils, 800 ± 5 ohms

Timing of relays, 30 ± 5 sec

Weights of castings, 4.15 ± .05 oz

On temperature at which a thermostatic switch operates, 55 ± 5°

Purity of oxygen to nearest .05 per cent, 99.50 to 100.00 per cent

Thermometers ± .5° from standard.

4.2.1. A Sampling Experiment. Table 4.2 shows the results for drawings of 50 samples, each of $n = 5$ observations, x. Population A of Table 4.1 was used for all 250 drawings which were made with replacement after each individual x, so as to have always the same population*. Then for each sample $\bar{x}$, R and s were found, as listed in the last three columns of the table.

The three series of sample statistics are plotted in Figure 4.1. Note that the units on the vertical scales differ. They were chosen as shown, to give roughly the same variability on each graph. On examining the charts for $\bar{x}$'s and R's it is seen that the two are not at all correlated. This is as it should be because $\bar{x}$ is concerned with the process level, whereas, R is a measure of the sample variability. On the other hand, the R and s charts <u>are</u> correlated, the high and low points correspond perfectly, and the entire graphs would fairly well coincide if one were placed over the other. They

*Actually this is not necessary, since a sample of five does not deplete the population of 200 appreciably. But after a sample of five, replacement should be made.

Table 4.2. Random Samples of Five from Distribution A of Table 4.1, Drawings of One Number at a Time and Replacing Before the Next Number was Drawn

Sample No.	x_1	x_2	x_3	x_4	x_5	Average $\bar{x}$	Range R	Standard Deviation s
1	0	+1	0	0	-2	- .2	3	1.10
2	-1	+2	+1	+2	-1	+ .6	3	1.52
3	-1	+3	0	+1	-2	+ .2	5	1.92
4	-5	+1	0	0	-1	-1.0	6	2.35
5	0	+2	+1	+1	-2	+ .4	4	1.52
6	+3	-2	+2	+1	0	+ .8	5	1.92
7	-1	+3	+1	+1	-1	+ .6	4	1.67
8	+1	-1	-2	0	-3	-1.0	4	1.58
9	-1	0	+1	-1	+2	+ .2	3	1.30
10	-2	-5	-2	+2	+1	-1.2	7	2.77
11	0	-2	+1	+1	+1	+ .2	3	1.30
12	+1	+2	0	+1	-1	+ .6	3	1.14
13	0	-2	-5	0	-1	-1.6	5	2.07
14	-2	0	+1	0	-4	-1.0	5	2.00
15	+3	-4	0	0	0	- .2	7	2.49
16	+1	+1	+1	-1	+2	+ .8	3	1.10
17	0	+1	-1	0	+1	+ .2	2	.84
18	-2	-2	+3	-2	-1	- .8	5	2.17
19	-2	+2	+2	-2	+2	+ .4	4	2.19
20	0	0	+1	+1	-1	+ .2	2	.84
21	+2	+1	0	-1	+1	+ .6	3	1.14
22	-2	-2	-1	-3	0	-1.6	3	1.14
23	-4	+4	-2	0	0	- .4	8	2.97
24	-2	-1	-1	0	0	- .8	2	.84
25	0	0	+1	-2	0	- .2	3	1.10
26	-1	0	+1	+3	-1	+ .4	4	1.67
27	-4	+1	0	+1	+3	+ .2	7	2.59
28	+1	0	+4	0	-3	+ .4	7	2.51
29	-1	+1	-2	-1	+1	- .4	3	1.34
30	+2	+2	-2	-4	0	- .4	6	2.61
31	+2	-1	0	-3	0	- .4	5	1.82
32	0	-2	+4	+1	0	+ .6	6	2.19
33	+1	+1	0	+2	0	+ .8	2	.84
34	-1	-3	0	+1	+1	- .4	4	1.67
35	0	+2	-1	-2	+2	+ .2	4	1.79
36	0	-3	-1	+2	-3	-1.0	5	2.12
37	-1	+1	+1	-1	-1	- .2	2	1.10
38	0	-2	+1	+1	-2	- .4	3	1.52
39	-2	-3	+2	0	0	- .6	5	1.95
40	0	-3	0	0	0	- .6	3	1.34
41	+1	+1	-3	-1	+3	+ .2	6	2.28
42	-1	-1	+2	+1	-2	- .2	4	1.64
43	0	+3	+1	-4	-2	- .4	7	2.70
44	+2	+1	-1	-1	-1	.0	3	1.41
45	+1	0	-1	+1	-1	.0	2	1.00
46	+1	-1	+3	0	0	+ .6	4	1.52
47	+2	0	+1	-3	-1	- .2	5	1.92
48	-1	-2	0	+1	+3	+ .2	5	1.92
49	-1	-1	0	+1	0	- .2	2	.84
50	0	-2	0	-4	+3	- .6	7	2.60
Totals	..	..	..	..	..	-6.6	213	85.87

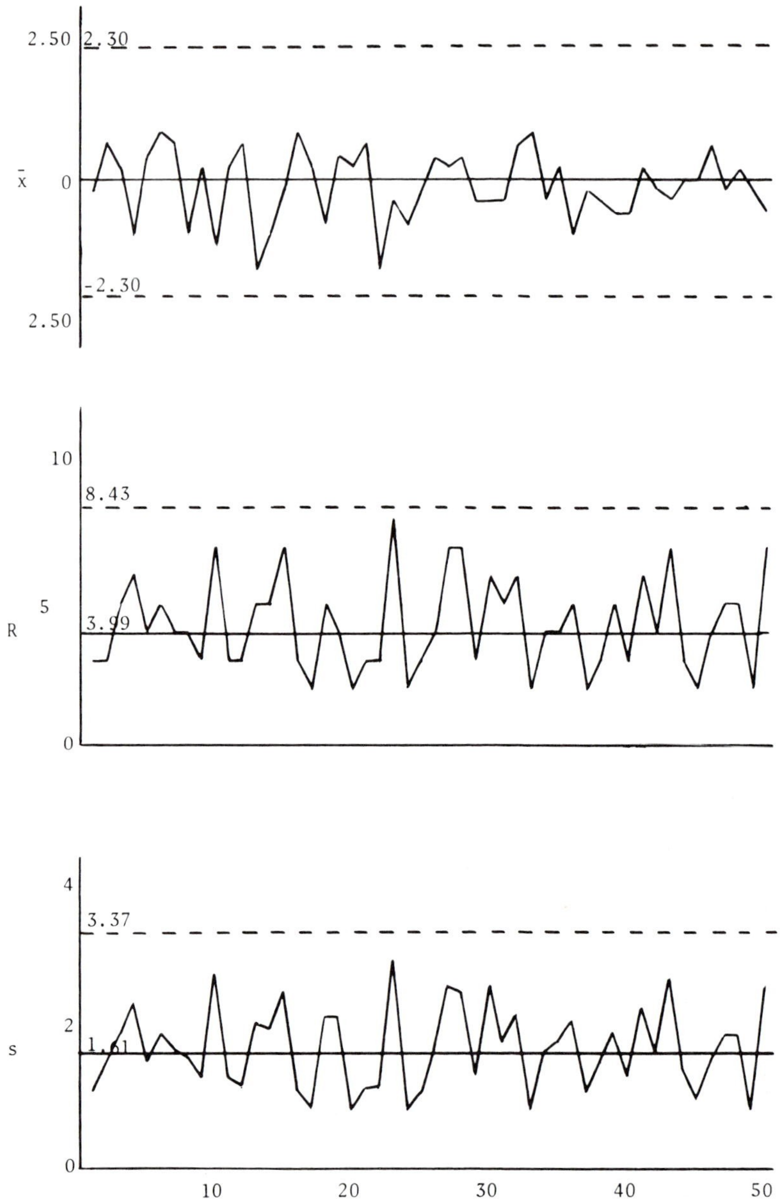

FIG. 4.1. Control charts for $\bar{x}$, R, and s for 50 samples of 5 each from Distribution A of Table 4.1. Central lines and control limits set from $\mu = 0$, $\sigma_x = 1.715$ (Population A).

both tell basically the same story (in different units) about process variability.

In this experiment we are illustrating the control chart case of "standards given." The standard values (parameters) for Population A of Table 4.1 are $\mu = 0$, $\sigma_x = 1.715$. Now with these, what central line and control limits shall we set for the $\bar{x}$ chart? By (2.6) we have the central line

$$ȼ_{\bar{x}} = \mu_{\bar{x}} = \mu \quad \text{standards given.} \tag{4.1}$$

As pointed out in Chapter 3, Section 3.6, limits for control charts are set at ± three standard deviations for the characteristic in question, around the mean. Thus we need $\mu \pm 3\sigma_{\bar{x}}$. But by (2.27), $\sigma_{\bar{x}}$ is smaller than σ_x by a factor of $\sqrt{n}$:

$$\sigma_{\bar{x}} = \sigma_x/\sqrt{n}\,. \tag{4.2}$$

Hence we have for $\mu \pm 3\sigma_{\bar{x}}$:

$$\text{Limits}_{\bar{x}} = \mu \pm (3/\sqrt{n})\sigma_x \quad \text{standards given} \tag{4.3}$$

If we define

$$A = 3/\sqrt{n} \tag{4.4}$$

then (4.3) becomes

$$\text{Limits}_{\bar{x}} = \mu \pm A\sigma_x. \tag{4.5}$$

The quantity A in (4.4) and (4.5) is a control chart constant. Values of A for n = 2 to 25 are given in Table V, in the back of the book, along with other control chart constants.

In our experiment, since $A = 3/\sqrt{5} = 1.342$, we have from $\mu = 0$, $\sigma_x = 1.715$

$$ȼ_{\bar{x}} = 0, \quad \text{Limits}_{\bar{x}} = 0 \pm 1.342\cdot(1.715) = \pm\, 2.30.$$

All $\bar{x}$ points lie well inside these limits, as was to be expected, since all data were from the same bowl (population of known μ and σ_x).

Next consider the plotting of the ranges, R. The appropriate expressions involving control chart constants are given at the bottom of Table V, being

$$ⳁ_R = d_2\sigma_x \quad \text{standard given} \tag{4.6}$$

$$\text{Limits}_R: \ LCL_R = D_1\sigma_x,\ UCL_R = D_2\sigma_x \ \text{standard given.} \tag{4.7}$$

See Chapter 5 for further explanation of these expressions. Suffice it to say that they give $E(R) \pm 3\sigma_R$. For our experiment we find

$$ⳁ_R = 2.326\cdot(1.715) = 3.99$$

$$LCL_R = 0(1.715) = 0 \qquad UCL_R = 4.918\cdot(1.715) = 8.43.$$

The one largest range among the 50 was 8, and hence again we have perfect control relative to the given standard $\sigma_x = 1.715$. (μ is not involved in an R chart). When $LCL_R = 0$, we do not bother to plot it on the chart.

Finally for the sample standard deviations, s, we have

$$ⳁ_s = c_4\sigma_x \qquad \text{standard given} \tag{4.8}$$

$$\text{Limits}_s: \ LCL_s = B_5\sigma_x,\ UCL_s = B_6\sigma_x \ \text{standard given.} \tag{4.9}$$

Since $\sigma = 1.715$ these give

$$ⳁ_s = .9400\cdot(1.715) = 1.61$$

$$LCL_s = 0\cdot(1.715) = 0 \qquad UCL_s = 1.964\cdot(1.715) = 3.37.$$

Again all sample standard deviations lie inside the control band indicating compatibility with $\sigma = 1.715$.

On all three charts, the probability of the next point lying outside of the control band is very small, and meanwhile the probability of all of 25 or 50 lying inside is still quite high, if the standards used in setting the limits are correct.

Now what happens if we have an assignable cause which changes the population, and our data are taken from the new population? For example, try Population B of Table 4.1. We show in Figure 4.2 the central line and limits

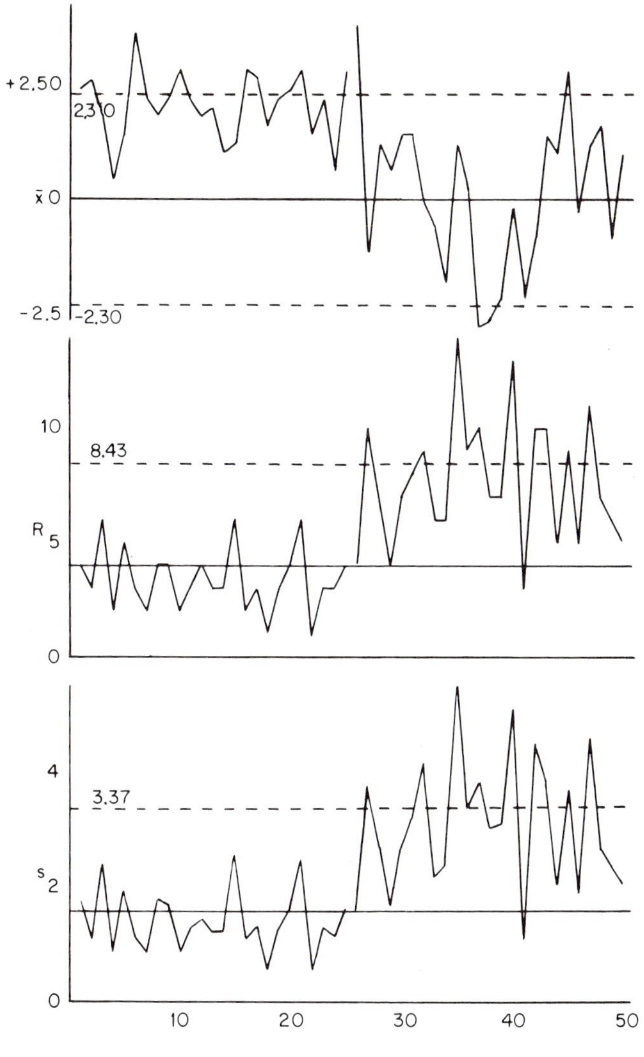

FIG. 4.2. Control limits for Population A, of Table 4.1 extended from Fig. 4.1, and compared to $\bar{x}$, R, and s for Populations B and C, respectively samples 1-25, 26-50. Data from Table 4.3.

set from $\mu = 0$, $\sigma_x = 1.715$, extended to the new data, and 25 samples of n = 5 run off. The data are given in Table 4.3. The very first two $\bar{x}$ points are above the upper control limit for $\bar{x}$'s, and in all 25 there are nine above. However, the first one, <u>all by itself</u>, is sufficient cause for an investigation. It is evidence of a process change (population change) from $\mu = 0$,

Table 4.3. Random Samples of Five, Numbers 1-25 from Population B (μ=+2, σ_x=1.715) and 26-50 from Population C (μ=0,σ_x=3.47) of Table 4.1.

Sample No.	x_1	x_2	x_3	x_4	x_5	Average $\bar{x}$	Range R	Standard Deviation s
1	+1	+3	0	+4	+4	+2.4	4	1.82
2	+3	+2	+4	+3	+1	+2.6	3	1.14
3	+3	+4	+1	-2	+3	+1.8	6	2.39
4	+1	+1	0	-1	+1	+ .4	2	.89
5	0	+2	-1	+4	+2	+1.4	5	1.95
6	+4	+4	+5	+3	+2	+3.6	3	1.14
7	+2	+2	+3	+1	+3	+2.2	2	.84
8	+1	+3	+3	+3	-1	+1.8	4	1.79
9	+5	+2	+2	+1	+1	+2.2	4	1.64
10	+2	+3	+2	+3	+4	+2.8	2	.84
11	+1	+2	+4	+1	+3	+2.2	3	1.30
12	+2	0	+1	+4	+2	+1.8	4	1.48
13	+3	+2	+3	0	+2	+2.0	3	1.22
14	+2	+1	-1	+2	+1	+1.0	3	1.22
15	-1	+4	+2	-2	+3	+1.2	6	2.59
16	+2	+4	+2	+4	+2	+2.8	2	1.10
17	+4	+4	+2	+1	+2	+2.6	3	1.34
18	+2	+2	+1	+1	+2	+1.6	1	.55
19	+3	+1	+2	+4	+1	+2.2	3	1.30
20	+5	+1	+2	+1	+3	+2.4	4	1.67
21	+2	+3	+7	+1	+1	+2.8	6	2.49
22	+1	+1	+1	+2	+2	+1.4	1	.55
23	+3	+3	+3	+2	0	+2.2	3	1.30
24	0	+1	-1	+1	+2	+ .6	3	1.14
25	+4	+3	+4	0	+3	+2.8	4	1.64
26	+2	+3	+6	+3	+5	+3.8	4	1.64
27	-1	-2	-5	-3	+5	-1.2	10	3.77
28	+3	0	0	+5	-2	+1.2	7	2.77
29	-2	0	+2	+2	+1	+ .6	4	1.67
30	+1	+6	+1	0	-1	+1.4	7	2.70
31	+1	+6	-1	-2	+3	+1.4	8	3.21
32	-2	+4	-4	-3	+5	.0	9	4.18
33	-1	-3	-1	-1	+3	- .6	6	2.19
34	-5	-3	-2	+1	0	-1.8	6	2.39
35	+5	+6	+1	+2	-8	+1.2	14	5.54
36	-1	+6	-3	-1	0	+ .2	9	3.42
37	-1	-5	-2	+2	-8	-2.8	10	3.83
38	+1	-6	-3	-5	0	-2.6	7	3.05
39	-1	+1	-6	-5	0	-2.2	7	3.11
40	0	+8	-4	-5	0	- .2	13	5.12
41	-2	-2	-4	-1	-2	-2.2	3	1.10
42	+5	-5	-3	+3	-4	- .8	10	4.49
43	+1	0	0	+8	-2	+1.4	10	3.85
44	-2	+3	0	+1	+3	+1.0	5	2.12
45	+6	+7	+1	+2	-2	+2.8	9	3.70
46	+2	-3	0	+1	-1	- .2	5	1.92
47	+1	-3	-3	+3	+8	+1.2	11	4.60
48	+2	+6	-1	+1	0	+1.6	7	2.70
49	-3	-2	-2	+3	0	- .8	6	2.39
50	+4	-1	-1	+2	+1	+1.0	5	2.12

$\sigma_x = 1.715$. Since the $\bar{x}$ chart has the point out, we presume that a shift in the true process level μ has occurred. But as we shall shortly see, an <u>increase</u> in <u>variability</u> σ_x can also cause a point out on the $\bar{x}$ chart, because it would give more spread to the $\bar{x}$'s.

Now how are the range points, 1 to 25? They all lie inside the control limits. This is to be expected, because σ_x is still at 1.715. The same is also true of the standard deviations s.

Samples 26-50 in Table 4.3 were drawn from Population C of Table 4.1. It has μ at 0, but twice the variability with $\sigma_x = 3.47$. Looking first at the R chart, we find that 10 of the 25 ranges lie above UCL_R figured from $\sigma_x = 1.715$. Also the point around which the ranges center has moved up to about 8. The first signal for action occurred on sample 27.

Next look at the $\bar{x}$'s for the same samples. We see that on the very first sample, number 26, a point went above the $UCL_{\bar{x}}$. One might erroneously conclude that this was due to the process level μ shifting up. Actually all we can confidently say is that such a point out on the $\bar{x}$ chart indicates a process change (population change). It can be from either a shift in μ or an <u>increase</u> in σ_x. Here we know it to be due to an increase in σ_x, because for Population C, $\mu = 0$, $\sigma_x = 3.47$. However, <u>usually</u> a point outside the $\bar{x}$ band does signal an assignable cause affecting the level μ, rather than increasing σ_x.

The chart for s's follows similarly to that for R's. Again with the scale shown the R and s charts practically duplicate each other and certainly give the same conclusions.

If control limits were to be set from $\mu = +2$, $\sigma = 1.715$, which are the parameters for Population B, the corresponding points would all be in control. The same is true for the data of Population C.

4.3. Control Charts for Averages, Ranges and Standard Deviations, Analyzing Past Data. Let us now turn to the case in which we do not have reasonable standard values to use, and wish to analyze a run of samples. This is the typical approach in the early analysis of a process. We may already have data available, or decide to take a series of samples as rapidly as convenient, and use them to test for process control. The hypothesis we are testing is that the samples come from a single population. Or in more applied terms, that the process variation is all due to chance causes, and that no assignable causes are at work. Take note that a process may be in perfect control and not meeting specifications at all well. All we are trying to do is to determine whether the process is running consistently. Then we compare what it is doing with what we should like it to do. "We first let the process speak for itself."

Let us again consider the data of Table 4.2. In order to get started in analyzing a process, a preliminary run of about 25 samples is taken. So let us use the first 25 samples of Table 4.2 and set central lines and control limits for $\bar{x}$, R and s charts. We use the following formulas for the case "analysis of past data," supposing we have k samples.

$$\text{¢}_{\bar{x}} = \Sigma_{i=1}^{k} \bar{x}_i/k = \bar{\bar{x}} \tag{4.10}$$

$$\text{¢}_R = \Sigma_{i=1}^{k} R_i/k = \bar{R} \tag{4.11}$$

$$\text{¢}_s = \Sigma_{i=1}^{k} s_i/k = \bar{s} \tag{4.12}$$

$$\text{Limits}_{\bar{x}} = \bar{\bar{x}} \pm A_2 \cdot \bar{R} \quad \text{or} \quad \bar{\bar{x}} \pm A_3 \bar{s} \tag{4.13}$$

$$LCL_R = D_3\bar{R} \qquad UCL_R = D_4\bar{R} \tag{4.14}$$

$$LCL_s = B_3\bar{s} \qquad UCL_s = B_4\bar{s}. \tag{4.15}$$

wherein we meet some more control chart constants. Note that these all require data from the observed samples, and make no use of the unknown parameters or standard values of the population. But (4.10) to (4.15) supply estimates of (4.1), (4.5)-(4.9), assuming control, for which we are testing.

For the first 25 samples of Table 4.2 we find

$$\bar{\bar{x}} = -4.2/25 = -.17$$

$$\bar{R} = 102/25 = 4.08$$

$$\bar{s} = 40.98/25 = 1.64.$$

All three of these are relatively quite close to the central line values set from μ and σ in the previous section. For the respective control limits, using Table V:

$$\text{Limits}_{\bar{x}} = -.17 \pm .577(4.08) = -.17 \pm 2.35 = -2.52, +2.18 \text{ from } \bar{R}$$

$$\text{Limits}_{\bar{x}} = -.17 \pm 1.427(1.64) = -.17 \pm 2.34 = -2.51, +2.17 \text{ from } \bar{s}$$

$$LCL_R = 0 \quad UCL_R = 2.115(4.08) = 8.63$$

$$LCL_s = 0 \quad UCL_s = 2.089(1.64) = 3.43.$$

If we were to use these limits on the plottings of Fig. 4.1 we would find all points in control just as before. We would now have a firm base (homogeneity or in-controlness) to justify estimating the parameters. This is done as follows:

$$\text{Est. of } \mu = \hat{\mu} = \bar{\bar{x}} \tag{4.16}$$

$$\text{Est. of } \sigma_x = \hat{\sigma}_x = \bar{R}/d_2, \text{ or} \tag{4.17}$$

$$\text{Est. of } \sigma_x = \hat{\sigma}_x = \bar{s}/c_4. \tag{4.18}$$

We then have for the data

$$\hat{\mu} = -.17 \quad (\text{vs. } \mu = 0)$$

$$\hat{\sigma}_x = 4.08/2.326 = 1.75 \quad (\text{vs. } \sigma_x = 1.715)$$

$$\hat{\sigma}_x = 1.64/.9400 = 1.74 \quad (\text{vs. } \sigma_x = 1.715).$$

Finally <u>since we have control</u>, we can say that nearly all of the individual measurements x will lie between

$$\hat{\mu} \pm 3\hat{\sigma}_x$$

which are for our data

$$- .17 \pm 3(1.75) = - 5.42, + 5.08.$$

These limits actually contain 100% of Population A. See Table 4.1.

Control lines set from the preliminary run of data may be extended and new points plotted immediately after each sample is taken. All of the next 25 samples of Table 4.2 show perfect control. It is good practice to revise control lines periodically. Suppose we now revise them on the basis of the first 50 samples

$$\bar{\bar{x}} = - 6.6/50 = - .13 \quad \bar{R} = 213/50 = 4.26 \quad \bar{s} = 85.87/50 = 1.72$$

$$\text{Limits}_{\bar{x}} = - .13 \pm .577(4.26) = - .13 \pm 2.46 \quad (\text{from } \bar{R})$$

$$\text{Limits}_{\bar{x}} = - .13 \pm 1.427(1.72) = - .13 \pm 2.45 \quad (\text{from } \bar{s})$$

$$LCL_R = 0 \quad UCL_R = 2.115(4.26) = 9.01$$

$$LCL_s = 0 \quad UCL_s = 2.089(1.72) = 3.59.$$

All 50 samples showed good control relative to these limits hence we can extend them without revision to watch as new data come in. Also we could again estimate σ_x:

$$\hat{\sigma}_x = \bar{R}/d_2 = 4.26/2.326 = 1.83$$

$$\hat{\sigma}_x = \bar{s}/c_4 = 1.72/.9400 = 1.83.$$

(Instead of being closer to σ_x than the earlier estimates, as we might have expected, these prove to be a bit further away, but still not bad.)

As a further example let us show the calculations for the $\bar{x}$, R data of Table 3.1. For $\bar{x}$'s we code from 4710 (.0001") and find

$$\bar{\bar{x}} = 4710 + (78.4/20) = 4710 + 3.92 = 4713.92 \quad (\text{in } .0001")$$

$$\bar{R} = 63/20 = 3.15 \quad (\text{in } .0001").$$

Then for these samples of n = 5, the above are the center lines and

$$\text{Limits}_{\bar{x}} = \bar{\bar{x}} \pm A_2\bar{R} = 4713.92 \pm .577(3.15) = 4712.10,\ 4715.74$$

$$LCL_R = D_3\bar{R} = 0 \qquad UCL_R = D_4\bar{R} = 2.115(3.15) = 6.66.$$

Since the $\bar{x}$ chart is not in control we do not have but just a single population, and hence we cannot set meaningful limits on the individual x's. Since the R chart was in control, however, we can estimate σ_x by

$$\hat{\sigma}_x = \bar{R}/d_2 = 3.15/2.326 = 1.35 \quad \text{(in .0001'')}.$$

Thus _if_ we can maintain control at the nominal (middle of the specification range) .4710", we can meet

$$.4710'' \pm 3(.000135'') = .47059'',\ .47141''.$$

Since these are so comfortably inside of the specification limits for x's, that is, .4700" to .4720", we can afford to let μ move around somewhat, perhaps permitting some wear in the cutting tool without resetting the process.

4.4. Comparison of the Two Charts for Variability. Throughout most of the two previous sections we carried both variability charts: R and s. In actual practice this is never done; we always settle on one or the other. For small samples up to, say, n = 8 or 10, the two charts are almost exactly equivalent. For samples of three or more the chart for s's is always just a bit more reliable, but above 10 the range begins to lose out to the standard deviation rather rapidly*. There are three uses of R's or s's we may study:

---

*One way to slow this deterioration is to use $\bar{R}$ for sub-samples instead of R for all. For example, if n = 40, use five ranges for sub-samples of n = 8, and use $\bar{R}$ of the five.

1. Control charts for within-sample variability.
2. Use of $\bar{R}$ or $\bar{s}$ to help in setting $\bar{x}$ limits.
3. If the variability chart shows good control, to estimate σ_x for the process by $\hat{\sigma}_x = \bar{R}/d_2$ or $\hat{\sigma}_x = \bar{s}/c_4$.

As we have been seeing, R's and s's for the three purposes are almost exactly equivalent. The author has seen dozens of such experiments, and results are about this close every time.

Now which shall we use? If work is to be done by hand or by a desk calculator ranges are much easier, so we use them. But if a digital computer or automatic line calculator is to be used then we prefer to use s, because it is a bit more accurate, and s is more easily programmed than R.

A final comment is that it is possible to use a control chart for s^2 instead of s. There are two objections to this: (a) When analyzing past data we would need to use $\overline{s^2}$, to test for homogeneity. But a single large s^2 will have a more marked effect on $\overline{s^2}$ than will the same sample s on $\bar{s}$. Thus the center line is more distorted, if there is non-homogeneity and with the inflated $\overline{s^2}$ we may miss the warning. (b) The distribution of s^2 is far more unsymmetrical than that for s.

4.5. Comparison of a Process with Specifications. In general measurable characteristics of all types such as dimensions, weights, surface finish, strength, chemical composition, impurities, physical properties or electrical characteristics, such as we might make $\bar{x}$ and R charts for, are subject to specifications. These specifications are most often in the form of maximum and minimum limits (U and L) for individuals x. Or more recently they might be set by a maximum σ_x and limits on μ. See also [2] for another modern approach. But for the purposes of this section let us be concerned with specifications L and U set for individual measurements x. All or at least some acceptably high percentage are to lie between L and U.

There are a number of complications and pitfalls in this important comparison. All it takes is clear thinking, but that is asking quite a bit! The first is that in general specifications are written for individual x's, not for averages $\bar{x}$'s. (Exceptions do occur such as the content weight for packages where the control may be on average content weights.) Then next we have cases where the process is in control or not in control. Thus there is the question of what process adjustments are feasible to make.

As a start let us say that there are four cases which occur and which must be treated quite differently. Failure to distinguish between these cases costs U.S. industry tens of millions of dollars annually at the very least, probably much more. The cases are:

1. Process in control, and safely meeting specifications.
2. Process not in control, but "safely" meeting specifications.
3. Process in control, but not safely meeting specifications.
4. Process not in control, and not safely meeting specifications.

What approaches shall we use in the respective cases? Case 1 is the happiest of the four? In this case we are doing well. We may, however, give some consideration to (a) decreasing the frequency of taking samples to save time and money, and/or (b) running at the most economical level for μ to save material or time. Case 2 is still satisfactory. Note the quotation marks on "safely." The reason for them is that the lack of control is a danger sign. One of the assignable causes present may suddenly come in much more strongly and cause pieces out of specifications.

Case 3 is one in which the process is stable but its capabilities are in question. Since we have control, we are justified in estimating the limits for individual x's by

$$\hat{\mu} \pm 3\hat{\sigma}_x = \bar{\bar{x}} \pm 3(\bar{R}/d_2) \text{ or } \bar{\bar{x}} \pm 3(\bar{s}/c_4) \quad \text{x limits} \tag{4.19}$$

Since we are not meeting specifications, one or the other or both of these natural limits are outside of the range L to U. Perhaps the trouble is entirely from the process level being off-center, and we can adjust μ. Otherwise we are in trouble. Just "tampering" with the process will not help. It will take some fundamental process change.

Finally in Case 4, the first step is to try to get the process into control, seeking the assignable causes, meanwhile trying to get μ to run closer to the "nominal" (L + U)/2. Large amounts of dollars, yen, ruples, pounds and francs are thrown away through not treating cases 3 and 4 differently. The common error in Case 3 is to try to reset frequently on insufficient evidence when the process spread $6\hat{\sigma}_x$ exceeds the specified tolerance U - L. The error in Case 4 is to fail to seek out the assignable causes so as to gain the best possible performance out of the process.

A very common error, especially for the beginner, is to compare <u>control limits for $\bar{x}$'s</u> with <u>specification limits for</u> x's. If the process is in control and the control limits for $\bar{x}$, such as, $\bar{x} \pm A_2\bar{R}$ lie between L and U, this does <u>not</u> say we are meeting specifications for x's. All it <u>does say</u> is that the $\bar{x}$'s are inside the specifications for x's. But where are the x's in such a case? They may well be outside L to U. This is because x's spread more widely than do $\bar{x}$'s, in fact $\sqrt{n}$ times as much. Why? See (2.27).

Let us take an example from industry. In caps for sealing food within glass jars, "gasket space" is the distance from top of rubber to top of cap. For samples of n = 5, $\bar{\bar{x}}$ = .1204", $\bar{R}$ = .0154". Among 24 samples only one $\bar{x}$ point was out of control (which should be investigated) and no R points. So control was quite good. For $\bar{x}$'s we have

$$\text{Limits}_{\bar{x}} = \bar{\bar{x}} \pm A_2\bar{R} = .1204'' \pm .577(.0154'') = .1115'', .1293''.$$

Now suppose specifications for x's are .1100" to .1300". The unwary might conclude that the specifications are being met. But assuming control, all we can say is that the great majority of <u>averages</u> $\bar{x}$ are between L = .1100" and U = .1300". What about the x's? For these we must estimate σ_x. This is done by

$$\hat{\sigma}_x = \bar{R}/d_2 = .0154''/2.326 = .0066''.$$

Then limits for x's set by the process are

$$\bar{\bar{x}} \pm 3\hat{\sigma}_x = .1204'' \pm 3(.0066'') = .1204'' \pm .0198'' = .1006'', .1402''$$

These are well <u>outside</u> of the specifications .1100" to .1300". See Figure 4.3 which pictures the relation. To estimate the percentage of x's outside specifications, we may use (2.24)

$$z = \frac{U - \bar{\bar{x}}}{\hat{\sigma}_x} = \frac{.1300'' - .1204''}{.0066''} = 1.45 \text{ Probab. above } .073$$

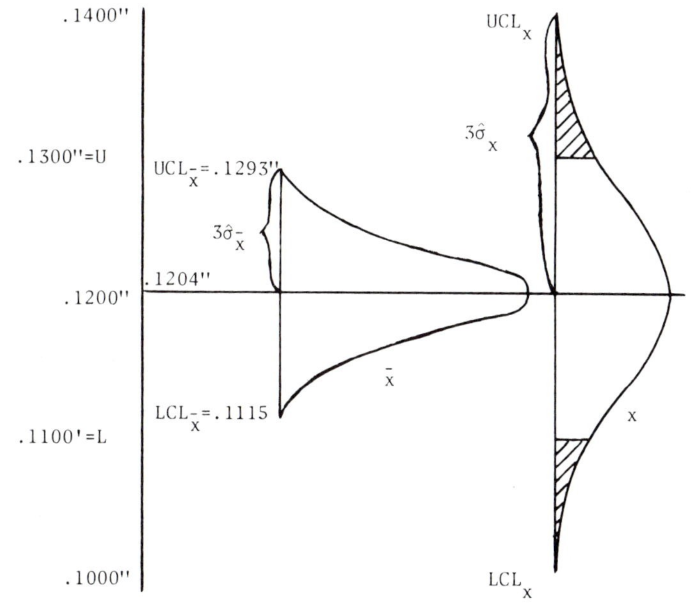

FIG. 4.3. Comparison of an in-control process with specifications. $\bar{\bar{x}}$ = .1204", $\bar{R}$ = .0154". Distributions and three sigma limits for $\bar{x}$'s and x's are shown. Specifications for x's are .1100" to .1300". Although practically all $\bar{x}$'s will be within these limits, the shaded areas on the x distributions show the proportion out.

$$z = \frac{L - \bar{\bar{x}}}{\hat{\sigma}_x} = \frac{.1100'' - .1204''}{.0066''} = -1.58 \quad \text{Probab. below } \underline{.057}$$

$$\text{Proportion out } .130$$

Thus about 13% will be outside of these specifications.

In further discussion of process vs. specifications, consider Figure 4.4, in which we assume the process in good statistical control with distributions of x's as shown, estimated from $\bar{\bar{x}}$ and $\bar{R}$.

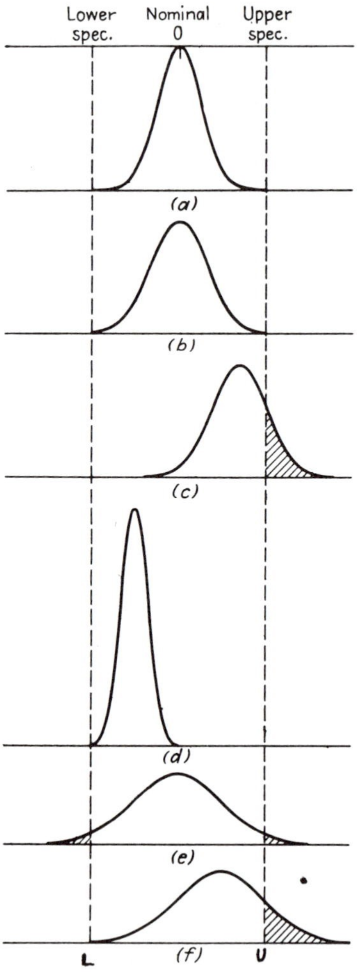

FIG. 4.4. Six cases of process distributions in relation to specification. Normal distribution of x's. Reproduced with permission from I. W. Burr, "Engineering Statistics and Quality Control", McGraw-Hill, New York, 1953, p. 119.

In case (a) the process is well able to meet the specifications, because the distribution is perfectly centered and the natural spread of the process, namely $6\hat{\sigma}_X$ is a fair amount less than the "tolerance" U-L. In fact it is possible to let the process mean drift somewhat and we shall still be meeting the specifications. There is some latitude. In case (b), however, the natural process limits $\hat{\mu} \pm 3\hat{\sigma}_X$ are just barely within the range U to L, hence close control over μ needs to be maintained. In case (c), the shaded area shows the proportion of pieces above U. The width of the distribution would be all right if μ were moved down to the nominal. Another more difficult possibility would be to reduce greatly the process variability while maintaining the same average μ.

In case (d) the process spread $6\hat{\sigma}_X$ is much less than the specified tolerance, thus permitting us to run close to the lower specification with adequate protection against pieces being below the lower specification L. We might want to use this approach if we were particularly concerned with a minimum specification for strength, or for content weight in a container. We also might wish to set a lathe working on an outside diameter at such a level initially, so that, as the tool wears and diameters tend to increase, we can let μ increase to a maximum extent. In this way we can get a maximum length of time before retooling or resetting.

In case (e), although the average is right on the nominal, there is still a substantial proportion of pieces outside of specifications in <u>both</u> directions. The only solution is a cut in process variability (unless the specifications can be relaxed). Since we have assumed good control, we shall need to make some rather drastic change in the process, such as going to another type of machine, if we are to cut the variability sufficiently.

In case (f) we are not making any appreciable number of pieces below L but have many above U. Better centering would help somewhat in reducing the total percentage out of specifications. But it is possible, for example on an outside diameter, that those pieces below L are scrap, whereas those above U can be reworked. Thus in (f) when we balance costs of inspection, reworking and scrapping we might find that the μ value shown minimizes the total cost. This case is the old expensive one of sort 100 per cent, rework, sort again 100 per cent, and rework, etc. The only real solution is to make a radical change in the process, or else to see whether the tolerance can be increased.

A final warning is that unless you have these concepts clear, one can hardly trust you to practice statistical quality control in a plant. So reread this section as often as necessary.

4.6. Continuing the Charts. In the initial stages of a control-chart application, we are primarily concerned with letting the process do the talking, that is, we collect the data, plot them, make the calculations, draw the lines, and interpret the results. As a consequence of indicated lack of control, we seek out the assignable causes responsible and may possibly find one or more from the preliminary run of data. In any case we face the problem of continuing the chart. Usually we merely extend the center line and control limits on both the $\bar{x}$ and R (or s) charts. There arises the question, however, as to whether to include the data corresponding to points which lay outside the control band. Data produced while assignable causes were operative should be eliminated and $\bar{\bar{x}}$ and $\bar{R}$ revised if <u>both</u> of the following conditions are met: (a) the assignable cause for such performance was found, and (b) it was eliminated. Now if the assignable cause behind the unusual performance was not found or, having been found, nothing has been done to remove it, then such data are still as typical of the process as any other and should be retained. In fact whenever any significant change is made in the process, all data produced before the change are no longer typical of the revised process and should be discarded.

Periodically the center lines and control limits should be revised, perhaps after every 25 samples, and in any case after a process change, as soon as 20 or 25 samples under the new conditions have accumulated. In this way up-to-date, meaningful control lines are maintained. Periodically, also, comparison with specifications should be made so that when the process becomes stabilized with $\bar{\bar{x}} \pm 3\bar{R}/d_2$ lying within specifications we can substitute the control chart for 100 per cent sorting.

After a good period of production in control at satisfactory quality levels, it may well be possible to decrease inspection by lengthening the time between samples. Then any out-of-control points or undesirable trends are cause for taking more frequent samples and for investigation of the reason for the off performance.

Many of the points of this and previous sections will be illustrated in the following section.

4.7. Illustrative Examples. Example 1. Weaver [3] gives an interesting example of the use of control charts in the steelmaking industry. The problem concerns the weights of ingots and the per cent of yield in rolling. Molten metal from open-hearth furnaces is poured into ingot molds, which are removed after a sufficient period of cooling. After being reheated, the ingots are then rolled into bars, perhaps four, each 25 ft long and 5 in. square. Each such bar might then be rolled into six billets, 2 in. square and 2.4 ft long. There is some unavoidable loss, but unless the weights of the ingots are well controlled, there may be a considerable loss in the process.

With a view to improving the yield, Weaver used control charts to study the weights of the ingots for a certain type of steel. In order to avoid as much of the multiple cut loss as possible, the ingot had been designed to weigh 5,300 lb. Figure 4.5 shows some typical data for April

1945, each point representing the weights of four ingots chosen at random from a single heat from an open-hearth furnace. Only a portion of the month's data is shown, but the center lines and control limits are for the whole month.

It is immediately apparent that the average weight, 5,620 lb, is 320 lb above the desired 5,300 lb. Moreover, the average range was 190 lb, and one range was over 440 lb, showing a large variation among the weights of ingots within a single heat of steel. The charts showed lack of control and therefore gave promise that the situation could be improved if the assignable causes could be found. Weaver continues (page 8):

"It was obvious that ingot weights were much too erratic and all too heavy for the intended purpose. Naturally, the controlling factor in the ingot weight must be the volume of the mold into which it is poured. These molds were being obtained from two suppliers, and a check of their dimensions showed that while one supplier was doing much better than the other, both were making molds oversize in dimensions. Mold suppliers were contacted, specifications and tolerances discussed, and three months later the situation had improved as shown in [the middle of Figure 4.5].

"The average overweight had decreased from 320 pounds to 190 pounds, the variation from lot to lot was reduced, and the variation within a given lot was now only a 134 pound average [that is, $\bar{R}$] compared to the original 190 pounds. We set our sights for a 5300 pound ingot with not over 100 pound average variation from one to another, made further corrections in the process, talked further with our mold suppliers, and by April, 1946, the results were remarkable. [See the last part of Figure 4.5.]

"Our ingots now averaged 5296 pounds and had an average variation one to another of only 82 pounds. All this consumed just one year's time, but it was well worth the effort. This brings us to an extremely important part of any quality control program: the evaluation of results, and, of course, the expres-

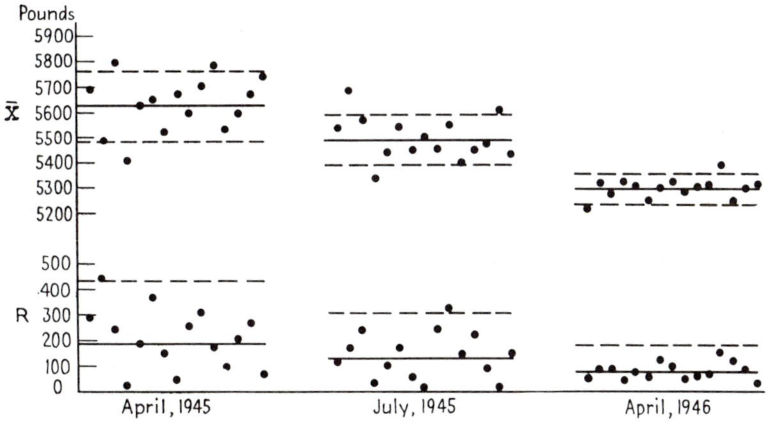

FIG. 4.5. Control charts for $\bar{x}$ and R for weights of steel ingots, each sample point containing four from a single heat. Progress is shown toward the desired weight of 5300 lb and toward better control. Limits set from data from entire month. Reproduced with permission from I. W. Burr, "Engineering Statistics and Quality Control", McGraw-Hill, New York, 1953, p. 122.

sion which registers most readily with anyone, be he workman, foreman, superintendent, or president, is dollars. It was a simple matter to calculate the savings in pounds per ingot and, from the number of ingots poured during the year, the total saving in tons. The value of a ton of steel at this point being well known, the savings for one year could be evaluated. This particular application turned up a saving of slightly over $175,000.

"The savings from this improvement meant improved yield due to decreased loss, and actually put more money in the pockets of the foremen participating in the incentive bonus plan, for yield is one of the major items affecting this plan. How much trouble do you suppose we had on the next project with these foremen? Did we receive cooperation? Were they enthusiastic? No answer required!"

The control charts helped in this work by forcefully picturing the high average weight and the large average variability within ingots of a single lot.

Also, the erratic performance as shown by the lack of control was a clear indication that improvement was possible. The charts helped check improvement during the progress of the work and gave an excellent idea of the results obtained, as seen in Fig. 4.5.

Example 2. Our second example is concerned with the deflector for a rear bearing. Figure 4.6 pictures the piece in question. The characteristic causing the trouble was the concentricity (distance between centers) of the 9.254-in. outside diameter in relation to the 6.085-in. inside diameter. The upper specification for this concentricity was .0020 in. In December almost 80 per cent of the pieces exceeded this upper specification. Accordingly, average and range charts were run, revealing the condition shown in the first part of Figure 4.7. Data are given in Table 4.4. The one point out of control on the $\bar{x}$ chart was some indication of an assignable cause of trouble, but since such a high percentage of defective pieces lay above the specifica-

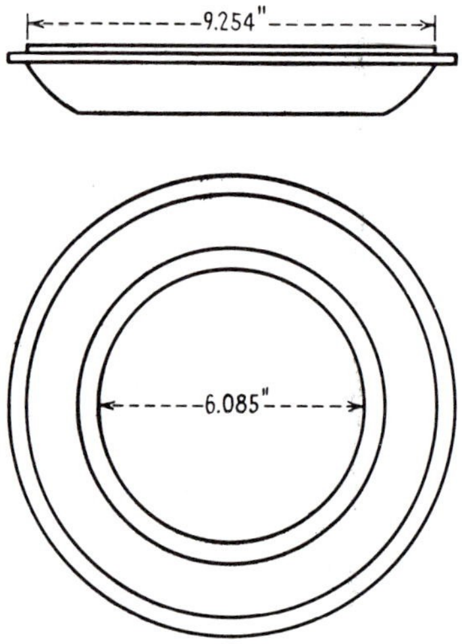

FIG. 4.6. Deflector for a rear bearing. The dimension in question is the concentricity of the 9.254" outside diameter in relation to the 6.085" inside diameter. The upper specification is .0020". Reproduced with permission from I. W. Burr, "Engineering Statistics and Quality Control", McGraw-Hill, New York, 1953, p. 123.

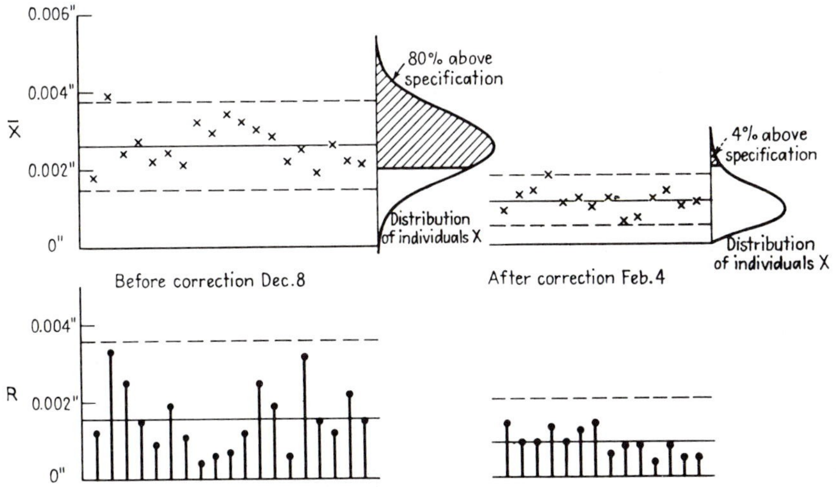

FIG. 4.7. Control charts for $\bar{x}$ and R for the concentricity of the deflector for a rear bearing, shown in Figure 4.6. The effect of process changes is clearly shown in the two sets of charts and the two distributions of individual measurements x. Data are given in Table 4.4. Reproduced with permission from I. W. Burr, "Engineering Statistics and Quality Control", McGraw-Hill, New York, 1953, p. 123.

tion, rather drastic steps were obviously necessary. Additional $\bar{x}$ and R chart studies were made on the three operations affecting the concentricity, and it was found that each of the operations contributed to the final condition.

Investigation determined the assignable causes, and corrective action was instituted. The following steps were taken:

1. The machine repair department reground the pot fixture used for locating on the 9.254-in. pilot OD (outside diameter).
2. A new type of holding fixture was designed and built for locating on the rough-bore ID (inside diameter).
3. A compressed-air method was devised properly to remove chips during the machining of the serrations in the 6.085-in. ID. It also served as a coolant.
4. A new operator was assigned to the machine which cut the serrations and maintained the 6.085-in. ID.

Table 4.4 Deflector - Rear Bearing, Concentricity of Diameters Shown in Figure 4.6. Specification: .0020 In. (Maximum). Sample Size Four.

Before correction, Dec. 8,		After correction, Feb. 4,	
Average $\bar{x}$, in.	Range R, in.	Average $\bar{x}$, in.	Range R, in.
.0018	.0012	.0009	.0014
.0039	.0033	.0013	.0009
.0024	.0025	.0014	.0009
.0027	.0015	.0018	.0013
.0022	.0009	.0011	.0009
.0024	.0019	.0012	.0012
.0021	.0011	.0010	.0014
.0032	.0004	.0012	.0006
.0029	.0006	.0006	.0008
.0034	.0007	.0007	.0008
.0032	.0012	.0012	.0004
.0030	.0025	.0014	.0008
.0028	.0019	.0010	.0005
.0022	.0006	.0011	.0005
.0025	.0032	.0159	.0124
.0019	.0015		
.0026	.0012		
.0022	.0022		
.0021	.0015		
.0495	.0299		
$\bar{\bar{x}}$ = .00261 in.	$\bar{R}$ = .00157 in.	$\bar{\bar{x}}$ = .00114 in.	$\bar{R}$ = .00089 in.
UCL = .00375 in.	UCL = .00358 in.	UCL = .00179 in.	UCL = .00203 in.
LCL = .00147 in.	LCL = 0	LCL = .00049 in.	LCL = 0

After the corrective action was accomplished, the process was restudied in February, and the results are shown in the second part of Figure 4.7. Subsequent investigation and study indicated that the process was in an even better state of control and that only occasionally was there a piece beyond the specification.

The two frequency graphs shown in Figure 4.7 indicate the distribution of individual measurements in the respective cases. The first one is drawn as a normal curve. Determining the percentage of pieces above the upper specification is rather hazardous because of the lack of control. Treating the case as though it were in control and from a normal population, however, we first find $\hat{\sigma}_x = \bar{R}/d_2 = .00157/2.059 = .000763$, then

$$z = \frac{\text{specification} - \bar{\bar{x}}}{\hat{\sigma}_x} = \frac{.0020 \text{ in.} - .00261 \text{ in.}}{.000763 \text{ in.}} = -.80$$

From Table I we then find that about 79 per cent of the pieces can be expected to lie above .0020 in.

For the data after correction we cannot very well assume normality because the average is too close to zero, below which a concentricity cannot go. Hence a skewed curve has been drawn and the percentage above .0020 in. estimated as 4 per cent.

It may also be mentioned that we are showing in the graphs of these illustrative examples a number of different ways of plotting the control-chart points. In Figure 4.7 two different kinds of point designation for the two charts were used. The crosses for the average points and the vertical bars with a large dot for the ranges contrast well so that, if used consistently in a company, one can always tell at the first glance whether he is looking at an $\bar{x}$ or an R chart. In cases where we have points so far from the control band that the two charts tend to overlap, it is convenient to have two different kinds of points plotted. The vertical bars are rather effective for ranges because the length of the bar shows the distance between the smallest and largest measurements in the sample.

Example 3. The third example is on the weight of charge of an insecticide dispenser. Inventory checks showed that about $14,000 worth of fluid was being given away free each month because of overfill of the dispensers. Such average overfill may, however, have been necessary in order safely to meet the minimum specification requirement. Upon his return from a short course in statistical quality control, an engineer was given the problem to work on. His first 50 samples of data are shown in Table 4.5. The data for Dec. 13 and Dec. 14 constituted the preliminary run. From them

$$\bar{\bar{x}} = 463.5 \quad \text{and} \quad \bar{R} = 25.8 \text{ g}$$

Table 4.5. Charge Weight of Insecticide Disperser, in Grams
Specifications: 454 ± 27g

Sample No.	Date	Observed charge weights, g.				Average $\bar{x}$	Range R	Remarks
1	Dec. 13	476	478	473	459	471.5	19	
2		485	454	456	454	462.2	31	
3		451	452	458	473	458.5	22	
4		465	492	482	467	476.5	27	
5		469	461	452	465	461.8	17	
6		459	485	447	460	462.8	38	
7		450	463	488	455	464.0	38	
8		Lost	478	464	441	461.0	37	Sample of 3
9		456	458	439	448	450.2	19	
10		459	462	495	500	479.0	41	
11	Dec. 14	443	453	457	458	452.8	15	
12		470	450	478	471	467.2	28	
13		457	456	460	457	457.5	4	
14		434	424	428	438	431.0	14	
15		460	444	450	463	454.2	19	
16		467	476	485	474	475.5	18	
17		471	469	487	476	475.8	18	
18		473	452	449	449	455.8	24	
19		477	511	495	508	497.8	34	
20		458	437	452	447	448.5	21	
21		427	443	457	485	453.0	58	
22		491	463	466	459	469.8	32	
23		471	472	472	481	474.0	10	
24		443	460	462	479	461.0	36	
25		461	476	478	454	467.2	24	End of preliminary data

The specifications were an average of 454 g (1 lb) and a minimum of 427 g. A shop bogey of 481 g was also imposed as a maximum. Even though the average charge weight was considerably above 454 g, there were some not meeting the lower specification. Among those tested, there was a 424 in sample 14, and doubtless more among those not tested. The control chart (Fig. 4.8) shows two points out of control on the $\bar{x}$ chart and one point practically on the upper control limit of the R chart. These signs of instability in the process are indications that there are assignable causes present which if found and removed will enable closer control of the weights. Investigations

Table 4.5 (continued)

26	Dec. 17	450	441	444	443	444.5	9	
27		454	451	455	460	455.0	9	
28		456	463	Lost	445	454.7	18	Sample of 3
29		447	446	431	433	439.2	16	
30		447	443	438	453	445.2	15	
31		440	454	459	470	455.8	30	
32		480	472	475	472	474.8	8	
33		449	451	463	453	454.0	14	
34		454	455	452	447	452.0	8	
35		474	467	477	451	467.2	26	
36		459	457	465	444	456.2	21	
37		465	475	456	468	466.0	19	
38		458	450	451	451	452.5	8	
39		447	417	449	445	439.5	32	
40		453	442	456	453	451.0	14	
41	Dec. 18	471	467	461	455	463.5	16	
42		462	454	462	468	461.5	14	
43		474	471	471	463	469.8	11	
44		461	454	468	452	458.8	16	
45		473	453	465	475	466.5	22	
46		474	455	486	490	476.2	35	
47		466	471	482	474	473.2	16	
48		447	454	476	486	465.8	39	
49		473	488	482	475	479.5	15	
50		460	450	461	445	454.0	16	

were started and the center lines and control limits extended. The next 25 samples are shown. There are three points below the lower control limit for $\bar{x}$'s, giving some concern as to whether the lower specification is being met. The ranges appear to have decreased somewhat, which is an encouraging sign.

Some of the assignable causes which the engineer found were obstructions in the tubes, defective cutoff mechanism, and lack of control in timing and pressure. Within a month the rate of loss from overfill per month was down to about $12,000, and inside of three months it was down to about $2,000 per month, with much safer meeting of the minimum specification.

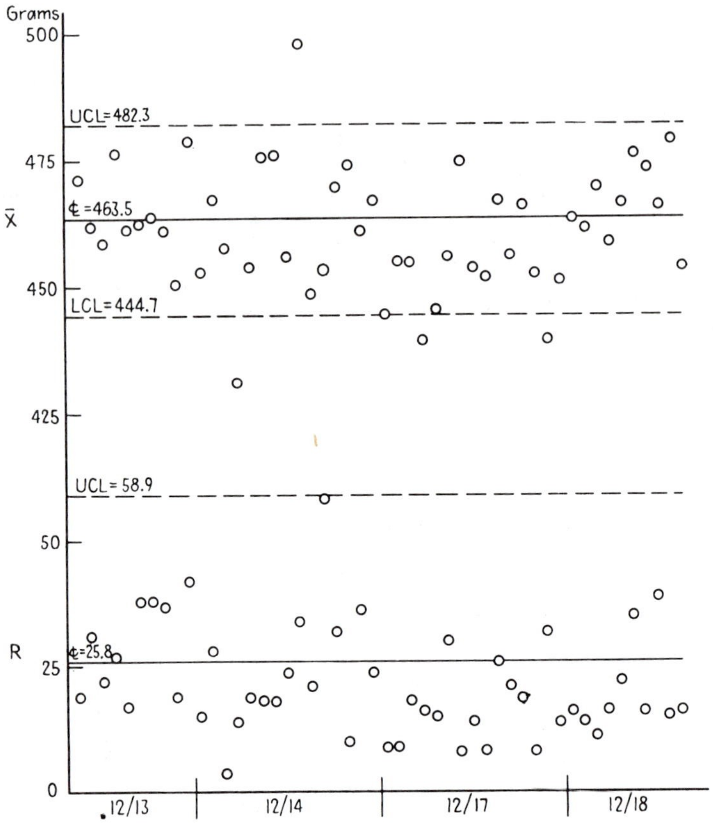

FIG. 4.8. Control charts for $\bar{x}$ and R for charge weights in grams of insecticide dispensers. Specifications 427 to 481 g. Data from Table 4.5. Sample size 4. The preliminary period covered the data of Dec. 13 and 14. Reproduced with permission from I. W. Burr, "Engineering Statistics and Quality Control", McGraw-Hill, New York, 1953, p. 126.

A few comments on the statistical aspects of the problem are in order. What average level would be safe for $\bar{\bar{x}}$ so as to avoid having any charge weights below the lower specification? As long as there is lack of control, no level is really safe! But if we could get the process in control and $\bar{R}$ were still as large as 25.8, then we would estimate the standard deviation for individuals,

$$\hat{\sigma}_x = \frac{\bar{R}}{d_2} = \frac{25.8}{2.059} = 12.5$$

Hence, safely to meet the lower specification we should have $\bar{\bar{x}}$ at $427 + 3\hat{\sigma}_x =$ 464.5. Figure 4.9 shows that if the process level is maintained at 464.5 g and control is achieved, there will be very few individual weights x below 427 g. Here "very few" means the proportion of cases below a z of - 3 for a normal curve, which we have assumed here. Thus the proportion is .0013 by Table I. Such a proportion is, of course, arbitrary but is a commonly used figure in industry. If, for example, we could afford to have about .5 per cent of the charge weights below 427 g, then we could let $\bar{\bar{x}}$ be somewhat lower than 464.5 g. From Table I, we find by interpolation that when $z = -2.575$, the area below is .0050. Hence we would have 427 + 2.575(12.5), or 459.2 g, as the desired process level. It is obvious that control must be obtained and the process variability decreased in order safely to meet the lower specification and yet average 454 g, which is as low as one is allowed to average.

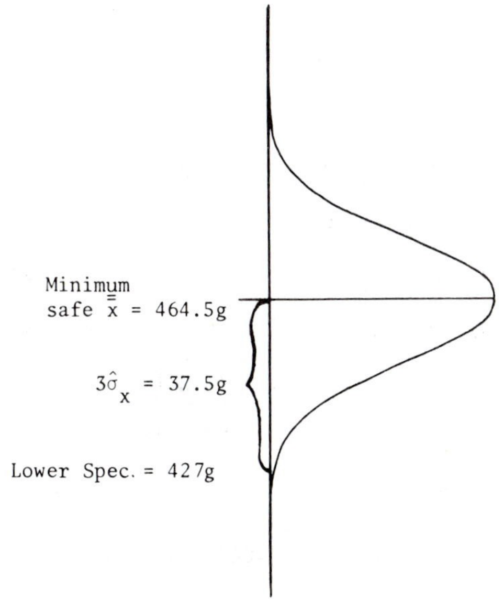

FIG. 4.9. Normal curve distribution of individual charge weights, placed at such a level as just to meet the lower specification.

There is also the question of the missing values. The two samples of three were treated as though the $\bar{x}$ and R were for samples of four. This is not technically correct, as such points should have different control limits and even the center line for the range should be different. Since neither sample was exceptional, the graphs were not complicated by changing the lines. The correct way to handle such samples is given in Chapter 7.

Example 4. The final example to be given in this section is the famous rheostat knob problem, which was extensively used in the wartime quality control courses under Working and Olds [1], and which is still being used because it is such an instructive example.

The knob consisted of a metal insert on which a plastic molding was pressed. The fit of the knob in assembly was largely dependent upon the dimension from the plane of the back of the plastic disk to the far side of a pinhole. Both the metal insert, which was bought from an outside supplier, and the molding operation influenced the dimension. The engineering specifications for the dimension were .140 $\pm$ .003 in. They were being inspected 100 per cent against this specification with a go and not-go gage.

Since the rate of rejection was very high, it was decided that a study would be made with $\bar{x}$ and R charts. In order to obtain numerical measurements, a right-triangular wedge-shaped gage was designed to slip between the pin and plastic disk, it being so calibrated that the dimension was read directly from a scale, according to how far in the wedge could be pushed. Samples of five knobs every hour were drawn, and the gage used to obtain the data as recorded in Table 4.6. Proceeding as usual

$$\bar{\bar{x}} = \frac{3.7974 \text{ in.}}{27} = .14064 \text{ in.} \qquad \bar{R} = \frac{.233 \text{ in.}}{27} = .00863 \text{ in.}$$

from which the control limits are

$\bar{x}$: $\bar{\bar{x}} \pm A_2\bar{R}$ = .14064 in. $\pm$ .577(.00863 in.) = .1357 in., .1456 in.

R: $D_3\bar{R} = 0$ $D_4\bar{R}$ = 2.115(.00863 in.) = .0183 in.

All points for Mar. 18 and 19 in Fig. 4.10 are inside the control bands; in fact, control is unusually good even if chance variation alone were at work. The engineer was therefore in a good position to estimate the process capabilities and compare them with the specifications. From the foregoing

$$\bar{\bar{x}} = .14064 \text{ in.} \quad \hat{\sigma}_x = \frac{\bar{R}}{d_2} = .00371 \text{ in.}$$

so that the "natural specifications" for individual knob dimensions are

$$.14064 \text{ in.} \pm 3(.00371 \text{ in.}) = .1295 \text{ in. and } .1518 \text{ in.}$$

Since these lay way outside of the engineering specifications, there was bound to be a large percentage of out-of-specification knobs. (Actual count gave 14 below and 28 above among the 135 measured, or a total of over 30 per cent.)

But this was already known; so what had the control chart done on the measurements which could not already be told by the go and not-go gage? There were two things. In the first place, the chart said that the process was doing about all that could be expected of it without a rather radical change in the process. Pressure from supervision could not be expected to help. Above all, it indicated that process adjustments downward when a knob above .143 in. was encountered and upward when one below .137 in. was found would not help, but would serve only to give even more knobs out of specifications. The second thing the control chart did was to tell what sort of specifications the process could be expected to meet, that is, .130 to .152 in.

In a circumstance such as this there are only three things to do, (1) make a more or less fundamental change, such as a redesign or use of a different machine, (2) ask for a revision of the specifications, or

Table 4.6. Rheostat Knobs, Data on Dimension from Back of Knob to Far Side of Pinhole in Thousandths of an Inch. Specification: .140 ± .003 in.

Date	Hour produced	Measurements on each of five items in series					Average for sample	Range for sample
		x_1	x_2	x_3	x_4	x_5		
Mar. 18, 1943.......	12-1	140	143	137	134	135	137.8	9
	1-2	138	143	143	145	146	143.0	8
	2-3	139	133	147	148	139	141.2	15
	3-4	143	141	137	138	140	139.8	6
	5-6	142	142	145	135	136	140.0	10
	6-7	136	144	143	136	137	139.2	8
	7-8	142	147	137	142	138	141.2	10
	8-9	143	137	145	137	138	140.0	8
	9-10	141	142	147	140	140	142.0	7
	10-11	142	137	145	140	132	139.2	13
Mar. 19, 1943.......	7-8	137	147	142	137	135	139.6	12
	8-9	137	146	142	142	140	141.4	9
	9-10	142	142	139	141	142	141.2	3
	10-11	137	145	144	137	140	140.6	8
	11-12	144	142	143	135	144	141.6	9
	12-1	140	132	144	145	141	140.4	13
	1-2	137	137	142	143	141	140.0	6
	2-3	137	142	142	145	143	141.8	8
	3-4	142	142	143	140	135	140.4	8
	4-5	136	142	140	139	137	138.8	6
	5-6	142	144	140	138	143	141.4	6
	6-7	139	146	143	140	139	141.4	7
	7-8	140	145	142	139	137	140.6	8
	8-9	134	147	143	141	142	141.4	13
	9-10	138	145	141	137	141	140.4	8
	10-11	140	145	143	144	138	142.0	7
	11-12	145	145	137	138	140	141.0	8
Mar. 20, 1943.......	7-8	137	145	139	142	142	141.0	8
	8-9	140	142	145	145	144	143.2	5
	9-10	141	144	146	147	146	144.8	6
	10-11	138	144	145	146	141	142.8	8
	11-12	137	137	146	141	136	139.4	10
	12-1	142	144	147	137	141	142.2	10
	1-2	137	146	139	144	138	140.8	9
	2-3	142	140	140	145	141	141.6	5
	3-4	141	136	146	141	145	141.8	10

Table 4.6 (continued)

Date	Hour produced	Measurements on each of five items in series x_1	x_2	x_3	x_4	x_5	Average for Sample	Range for Sample
Mar. 24, 1943.......	7-8	143	141	139	137	141	140.2	6
	8-9	143	143	139	141	145	142.2	6
	9-10	142	135	138	146	142	140.6	11
	10-11	141	135	145	141	142	140.8	10
	11-12	140	145	141	146	143	143.0	6
	12-1	135	142	141	140	141	139.8	7
	1-2	140	140	139	141	147	141.4	8
	2-3	142	139	139	143	141	140.8	4
Apr. 28, 1943.......	5-6	146	148	148	144	147	146.6	4
	6-7	146	146	148	149	149	147.6	3
	7-8	145	144	146	148	149	146.4	5
	8-9	146	144	144	145	149	145.6	5
	9-10	146	145	148	147	149	147.0	4
	10-11	146	148	147	146	149	147.2	3
Apr. 29, 1943.......	5-6	145	149	147	148	149	147.6	4
	6-7	146	148	147	147	147	147.0	2
May 11, 1943.......	5-6	143	144	144	143	146	144.0	3
	6-7	145	143	142	144	148	144.4	6
	7-8	144	143	144	149	147	145.4	6
	8-9	146	144	143	145	148	145.2	5
	9-10	144	143	144	147	146	144.8	4
	10-11	145	146	145	143	147	145.2	4
	5-6	141	142	141	144	146	142.8	5
	6-7	145	147	145	145	147	145.8	2
	7-8	145	145	146	146	147	145.8	2
	8-9	145	146	145	144	146	145.2	2

(3) sort the pieces 100 per cent. Since (2) would be least expensive, a number of pieces outside of the specifications were tried in assembly and seemed to work perfectly satisfactorily. Some were even found further from .140 in. than any among the data of Mar. 18 and 19, and no trouble was encountered. Accordingly the engineering department was asked to review

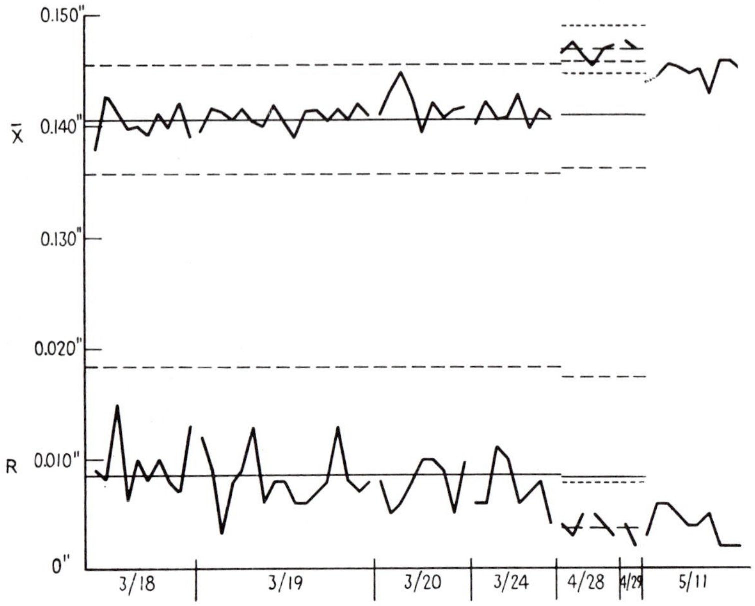

FIG. 4.10. Charts on the rheostat knob data given in Table 4.6. The preliminary run of data was on March 18 and 19. Control limits were extended from these through the rest of March (center lines solid, control limits dashed). At the end of March these lines were revised and extended ready for new data. In April both charts show out of control, the R chart by all the R's being way below the center line: new process conditions. The new data are, however, in control relative to their own limits (center lines by long dashes, limits by dots). Samples of five taken each hour during production. Reproduced with permission from I. W. Burr, "Engineering Statistics and Quality Control", McGraw-Hill, New York, 1953, p. 136.

the specifications in the light of these findings. After checking the fit with various knobs, it was decided to change the specifications to .125 to .150 in. (Such a large relaxing of tolerances shows courage in the engineering department.)

Although the natural specifications of the process were .1295 to .1518 in., which do not lie wholly within the new specifications, the agreement was regarded as sufficiently good to justify removing the 100 per cent sorting and instead to continue the control chart as sufficient evidence for safety. [Assuming normality we could use $z = (.150 \text{ in.} - .14064 \text{ in.})/.00371 \text{ in.} = 2.52$ and estimate that only about .6 per cent would be too high.]

The chart is continued in Fig. 4.10 for Mar. 20 and 24, and good control continues. At the end of this time production was stopped. Because of the good control at satisfactory quality performance, it was possible at that point to set standard values for the process and use these to set control limits for future production whenever it occurred. Using all the data at hand, and adopting Greek letters for population or standard values rather than the circumflex for estimates, we have

$$\mu = \bar{\bar{x}} = \frac{6.2038 \text{ in.}}{44} = .14100 \text{ in.}$$

$$\bar{R} = \frac{.362 \text{ in.}}{44} = .00823 \text{ in.}$$

$$\sigma_x = \frac{\bar{R}}{d_2} = \frac{.00823 \text{ in.}}{2.326} = .00354 \text{ in.}$$

Using these together with formulas (4.5) to (4.7),

$$¢_{\bar{x}} = \mu = .14100 \text{ in.} = .14100 \text{ in.}$$

$$¢_R = d_2\sigma_x = .00823 \text{ in.}$$

$$\text{Limits}_{\bar{x}} = \mu \pm A\sigma_x = .1362 \text{ in.}, .1458 \text{ in.}$$

$$\text{Limits}_R\text{:} \quad D_1\sigma_x = 0 \quad D_2\sigma_x = .0174 \text{ in.}$$

These are shown in Fig. 4.10 as continuations after the date Mar. 24.

In the meantime work was done with the vendor to try to get him to reduce the variability of the insert, which apparently was more responsible for variation in the dimension than was the plastic molding operation. Results for Apr. 28 and 29 immediately showed out of control on both charts. All but one of the $\bar{x}$ points was above the upper control limit, which is an extremely clear-cut indication of a shift in process level. At the same time there was an almost equally clear indication of a decrease in process variability, which was most welcome. All points on the range chart were well below the old center line. It was therefore perfectly obvious that there was a new process condition. The big question was: Would it meet the specification of .125 to .150 in.? First, to check control, even though there are too few samples to place great reliance on the results,

$$\text{¢}_{\bar{x}} = \bar{\bar{x}} = \frac{1.1750 \text{ in.}}{8} = .14688 \text{ in.}$$

$$\text{¢}_{R} = \bar{R} = \frac{.030 \text{ in.}}{8} = .00375 \text{ in.}$$

$$\text{Limits}_{\bar{x}} = \bar{\bar{x}} \pm A_2\bar{R} = .1447 \text{ in.}, .1490 \text{ in.}$$

$$\text{Limits}_{R}\text{:} \quad D_3\bar{R} = 0 \quad D_4\bar{R} = .0079 \text{ in.}$$

There is good control within the new process; hence the natural specifications can be determined.

$$\bar{\bar{x}} \pm \frac{3\bar{R}}{d_2} = .14688 \text{ in.} \pm .00484 \text{ in.} = .1420 \text{ in.}, .1517 \text{ in.}$$

There is again some question about the upper specification, with some danger of a few knobs being beyond the upper specification. This danger had apparently been remedied by the time the next production, on May 11, took place. There is a shift downward toward a safer process level, and the data for that date are in control with respect to themselves. It was subsequently possible to decrease the amount of inspection to considerably less often than one sample of five per hour.

4.8. Kinds of Assignable Causes. There are several different types of assignable causes which may occur in a process or laboratory and which affect our control charts. It is therefore desirable to summarize them. Basically an assignable cause brings about a change in the population from which our measurements are being drawn. The following are some of the different effects which, in practice, assignable causes can have upon the population: (1) change in process average or level; (2) change in process variability or spread; (3) change in curve shape; (4) steady shift in one of the foregoing; (5) cyclic or erratic shift in (1), (2), or (3); (6) mistakes. We shall be concerned with the way in which these effects show on a control chart. It must be remembered that they do not commonly appear in a perfectly clear-cut manner and may be quite obscure.

4.8.1. Change in Process Average. We are here concerned with rather sudden changes in level. Such, for example, would occur from starting up a machine cold in the morning, a change in source of material such as iron ore, acid or brass stock, a change in method or machine, a chipped tool, sudden presence of an impurity, etc. Grant [4] gives an interesting example on steel castings, in which two $\bar{x}$ charts (on yield point and per cent elongation) showed sudden changes simultaneously. The cause of the undesirable shift in level was finally traced to the source of quench water. As soon as the original source was restored, the charts came back into control.

Changes in process average can show only on the $\bar{x}$ chart and not on the R chart.* (A change in mean μ, with constant σ, gives same size of R's.) Probably the commonest way it appears is by a point going out of control on either the high or the low side of the control band. The other way for such an effect to make its presence known is by a run of about 10 consecutive points either above or below the center line, even though no point goes out of the control band. Such a run is ample evidence of a shift in level.

4.8.2. Changes in Process Variability. Whenever the process variability increases significantly, we can expect sooner or later to have points above the upper control limit for ranges. We may also have a run above the center

*This is an invariable rule, unless one draws his sample of n pieces from a lot which contains pieces made both before and after a potential change such as a resetting or retooling. We must always guard against this by taking our sample from pieces produced under as homogeneous conditions as possible. If a sample contains pieces produced both before and after a change in process level, we are quite likely to have a high range, perhaps even out of control.

line for ranges. On the other hand, if the sample size is not over six, we have a lower control limit of zero, and hence there is no way in which a point can lie below the lower control limit for ranges. Hence the only way in which a decrease in process variability can show up is by a run below the center line for ranges.

However, a change in process variability affects not only the range chart but also affects the $\bar{x}$ chart. Thus, as we saw in Sec. 4.2, when the population variability _increases_ we may get points out of control on either one or both sides of the band. On the other hand, a _decrease_ in process variability may appear on the $\bar{x}$ chart as a series of points all exceptionally close to the center line. Hence, although evidences of lack of control on the R chart may be taken as indications of changes in process variability (or lack of homogeneity in sampling) a point out of control on the $\bar{x}$ chart may be from an increase in process variability _or_ from a shift in level (or a combination of the two). Such a point should therefore be interpreted in conjunction with the range chart.

4.8.3. Change in Curve Shape. The $\bar{x}$ and R control charts do not in general detect changes in curve shape when such changes are shifts in the shape of the underlying population. One would need large samples, perhaps of 200 or more measurements each, for such a study. But what is akin to a change in curve shape is an assignable cause which may give a considerably higher value to about every tenth piece. Then if our sample of 5 is taken over the last 100 produced, we are in effect sampling from the old population with a "bump" added onto it a long way from the center. A sample from this no longer homogeneous population may give a point out on either the $\bar{x}$ or the R chart, or both. If, for example, the mechanism which is designed to segregate the first and last gaskets from each tube being cut into gaskets becomes defective, such an effect can occur.

4.8.4. Steady Shift in One of the Foregoing. Steady progressive changes are among the commonest phenomena in industry, and occur quite often in the laboratory. Tool wear, deterioration of abrasives, chemical and physical changes such as viscosity, furnace condition, and seasonal changes are familiar examples. Although some may cause trends in variability, most of them cause trends in process level. Hence we usually see evidence of such influences by slanting trends, either upward or downward, on the $\bar{x}$ chart. This subject is taken up in more detail in Chapter 7.

4.8.5 Cyclic and Erratic Shifts. Many times the process level is affected by cycles. These may arise from daily, weekly, or seasonal effects, or for psychological, chemical, or mechanical reasons. Many process industries have cycles due to raw-materials changes. Industrial accidents and hospital calls often follow cycles. Industries in which there are extensive tooling changes each week-end may show cycles. The general appearance on the $\bar{x}$ chart, where almost all such cyclic phenomena show up, will have the aspect of an ordinary $\bar{x}$ chart superimposed on a cyclic curve rather than along a straight line for a center. It will in general show many points out of control, at both the tops and bottoms of the cycles. A stimulating example in the machine shop is given by McCoun [5]. One $\bar{x}$ chart showed beautiful control at apparently very satisfactory quality, but it was found that there were many pieces out of specification and far away from where the in-control chart predicted. The measurements on each of 100 consecutive flywheels (outside diameters) showed a perfectly clear-cut cycle. It was then found that all the previous samples for the in-control $\bar{x}$ chart had been taken near the bottom of each cycle. The period between consecutive samples had happened to be precisely in step with this cycle. Hence the $\bar{x}$ chart was completely misleading. This example points up the value which may accrue by taking a series of consecutive individual measurements and the importance of not taking one's samples of n pieces at too rigid and uniform a time interval.

If the cycle is long enough in time so that we get about 10 or more $\bar{x}$'s per cycle, the cyclic character will show up within two or three cycles, but if we get only two or three samples per cycle, or only one as in the case just discussed, then the cyclic appearance may not show up. If a cyclic effect is to be expected for any reason, the solution is to take more frequent samples or to measure every individual.

Assignable causes which act in an erratic manner are common in industrial processes and research. They may be operative for several consecutive samples or on only one and may affect either the $\bar{x}$ chart or the R chart, or both. In any case they cause a temporary change in the population, and of course the greater the change, i.e., the stronger the assignable cause, the greater the probability of an indication on the charts.

One of the commonest types of erratic assignable causes is that of too frequent resetting of a machine or process. Often, changes of setting are done on insufficient evidence, perhaps on only one piece or two. When this is the case, the adjustment is frequently in the wrong direction and seldom by the right amount. All that is accomplished by such resetting is that the product output is more variable than it otherwise would be. For example, in one large company, as an experiment, a process whose level had been adjusted 68 times in one 8-hr shift was let run for 8 hr without any adjustment at all. (The engineering department had agreed to pay for the losses in out-of-specification pieces.) The result was less scrap produced than previously, when all the frequent adjustments had been made!

4.8.6. Mistakes. These are just plain blunders, in using the equipment or measuring devices, misrecordings, using wrong formulas, misplotting, and errors in calculation. They are unpredictable, they should never occur, but unfortunately they do. They must be guarded against with every possible precaution and check. It can be very embarrassing for a point out of the control band to turn out to be a mistake in calculation by the quality-control man!

There is a valuable article by Olmstead [6], in which he discusses the various statistical tools (control charts and others) which may be used to investigate the different types of assignable causes.

4.9. When to Set Standard Values — Process Capability. In Example 4 of Section 4.6, standard values for the process were set, for the purpose of watching future production. It was possible to set standard values only because two conditions were met, that is, (1) there was a sizable history of in-control production, which implies, first, that we are justified in assuming that the data come from only one population and, second, that we have a reasonably adequate amount of data from which to estimate the population characteristics; and (2) the performance summarized by those population characteristics is satisfactory with respect to the specifications.

The standard or population values μ and σ_x are set from all the data at hand which were used in $\bar{\bar{x}}$ and $\bar{R}$, from which

$$\bar{\bar{x}} = \mu$$

$$\frac{\bar{R}}{d_2} = \sigma_x \quad \text{or} \quad \frac{\bar{s}}{c_4} = \sigma_x$$

The "natural process limits", $\mu \pm 3\sigma_x$, set in this manner <u>by the process</u>, must lie within the engineering specification limits L and U, or at least not significantly outside. However, when such natural process limits $\mu \pm 3\sigma_x$ lie about on or barely inside the range L to U, then the process average μ must be tightly controlled to the nominal (L + U)/2. If such is the actual situation, then we can use μ and σ_x in (4.1), (4.5) and (4.6), (4.7) or (4.8), (4.9) to set control lines for future production. These are exactly the same as would be obtained from the case "analysis of past data" using (4.10) to (4.15).

Now there is still some controversey as to how we should define "process capability." This is because each range R or standard deviation s is calculated from a sample of material produced over a short period of time during which the process average is likely to be virtually constant. Thus if there is control on the R or s chart, as we are assuming,

$$\text{Short term } \sigma_x = \bar{R}/d_2 \quad \text{or} \quad \bar{s}/c_4 \tag{4.20}$$

On the other hand for the entire production of an eight-hour shift, say, the standard deviation is likely to be larger than that in (4.20). This is because of variation in level μ, from imperfect control on the $\bar{x}$ chart.

$$\text{Long-term } \sigma'_x = \text{standard deviation of entire output} \tag{4.21}$$

Depending upon how closely the process average μ is controlled by the $\bar{x}$ chart, the long-term σ'_x may be $1.2\sigma_x$ to $1.5\sigma_x$. Therefore, in order to meet specification limits L to U fairly comfortably, we should have, say,

$$U - L \geq 8\sigma_x$$

rather than $U - L \doteq 6\sigma_x$.

In the latter case we would use (4.1), (4.5) and (4.6), (4.7) or (4.8), (4.9) for control lines for future production. But if $U - L \geq 8\sigma_x$ and we are not interested in going further in discovering assignable causes on the $\bar{x}$ chart, we can use the methods of Section 7.2, employing "modified limits" or slanting $\bar{x}$ limits.

Therefore many industrial people, when considering the capability of a machine or other process to meet a tolerance of U - L, require of the short-term σ_x:

$$8\sigma_x \leq U - L \tag{4.22}$$

or of the anticipated long-term σ'_x:

$$6\sigma'_x \leq U - L \tag{4.23}$$

Thus many regard $8\sigma_x$ or $6\sigma_x'$ as the "practical process capability", which must not exceed the tolerance U - L.

4.10. Use of Runs for Out of Controlness. Points beyond the control limits are the common way to obtain an indication of an assignable cause. One other way, fairly often used, is that of a long run of points on just one side of the centerline of a chart. Obviously a run of but three in succession on one side of a center line will occur much too often to give a clear signal of an assignable cause. Equally obvious is that 20 in succession on one side would be an extremely clear indication of an assignable cause. Where do we stop between? Let us see. Consider first an $\bar{x}$ chart. Commonly we assume that the $\bar{x}$'s are normally distributed even if the x's are not. Then we would expect about half of the $\bar{x}$'s on one side and half on the other (although there might be a slight chance of an $\bar{x}$ right on the center line). What is the probability of the next k all being <u>below</u> the center line? The probability for each would be 1/2, assuming no assignable cause. Thus by (2.19) P(next k lying <u>below</u> $\bar{\bar{x}}$) = $(1/2)^k$. This is also P(next k lying <u>above</u> $\bar{\bar{x}}$). Each event is to be considered an indication of an assignable cause. Thus the P(next k all lying on the same side) = $2(1/2)^k = (1/2)^{k-1}$. Now suppose, as some writers suggest, that we use k = 7 for the number on one side to provide a clear signal. Then P(signal) = $(1/2)^6 = 1/64$. Now in 25 consecutive $\bar{x}$ points there would 19 potential $\bar{x}$'s to start such a run. This would seem to the author to give too great a chance for such a run to occur with no assignable cause at work. Thus it would rather greatly increase the risk α of thinking an assignable cause is at work when none is. Therefore the present author recommends that runs of 8 to 10 on one side be required for an indication on the $\bar{x}$ chart.

With R or s charts the situation is a bit different, because, owing to the lack of symmetry $P(R < \bar{R}) > P(R > \bar{R})$, and similarly for s's. Thus we might well use 7 in succession above $\bar{R}$ and 9 in succession below $\bar{R}$. Or it is possible, as some have done, to use the center line at the median range, rather than at the mean range $\bar{R}$, which would make the case of runs like that for $\bar{x}$'s.

One final word is that some authors suggest a whole host of ways of concluding that an assignable cause was at work. The more such warning signals one uses (a) the greater the chance we have of finding an assignable cause when one is at work, that is, the lower the β risk becomes, but (b) the greater the α risk becomes, and (c) the more complicated the chart study and interpretation becomes. If enough warning criteria are used practically all data will be "out of control."

The present author recommends just the use of 8 to 10 (take your choice) on one side of a center line in addition to use of three-sigma control limits.

4.11. Summary. The formulas for $\bar{x}$, R and s charts are here included in one single outline to show the unity of the formulas, in the hope that the student by seeing them all together may be able to learn them by heart, for in so doing he will also understand them better.

I. Purpose: analysis of past data
   A. Using averages and ranges we have $\bar{\bar{x}}$, $\bar{R}$
      1. Averages: $\bar{x}$
         a. Center line: $\bar{\bar{x}}$
         b. Limits: $\bar{\bar{x}} \pm A_2\bar{R}$
      2. Ranges: R
         a. Center line: $\bar{R}$
         b. Limits: $D_3\bar{R}$, $D_4\bar{R}$
      3. Individuals: x—for comparison with specifications if process is in control
         a. Center line: $\bar{\bar{x}}$
         b. Natural limits: $\bar{\bar{x}} \pm (3\bar{R}/d_2)$

B. Using averages and standard deviations, we have $\bar{\bar{x}}$, $\bar{s}$

1. Averages: $\bar{x}$

a. Center line: $\bar{\bar{x}}$

b. Limits: $\bar{\bar{x}} \pm A_3\bar{s}$

2. Standard deviations: s

a. Center line: $\bar{s}$

b. Limits: $B_3\bar{s}$, $B_4\bar{s}$

3. Individuals: x—for comparison with specifications if process is in control

a. Center line: $\bar{\bar{x}}$

b. Limits: $\bar{\bar{x}} \pm (3\bar{s}/c_4)$

II. Purpose: control from standards, μ, σ_x

A. Using averages and ranges

1. Averages: $\bar{x}$

a. Center line: μ

b. Limits: $\mu \pm A\sigma_x$

2. Ranges: R

a. Center line: $d_2\sigma_x$

b. Limits: $D_1\sigma_x$, $D_2\sigma_x$

B. Using averages and standard deviations

1. Averages: $\bar{x}$

a. Center line: μ

b. Limits: $\mu \pm A\sigma_x$

2. Standard deviations: s

a. Center line: $c_4\sigma_x$

b. Limits: $B_5\sigma_x$, $B_6\sigma_x$

C. Individuals: x

1. Center line: μ

2. Limits: $\mu \pm 3\sigma_x$

As we have already pointed out in Chapter 3, each of the varying quantities $\bar{x}$, R, s or x forms a distribution <u>if</u> the samples are drawn from the same population. The formulas in the foregoing outline are all designed to give, as well as possible, the average of the distribution of the statistic in question for a center line, and for the control limits points which are three standard deviations of the statistic in question above and below its average. Thus if the process is in control, which means that the results behave as though the samples were randomly drawn from the same (normal) population, than we shall not more than one time in a hundred get a value of the sample statistic outside of these control limits. On the other hand if the samples are not all from the same population then there is a greater chance of a sample point being outside of the control band. And if the assignable cause is strong we can expect a point out in even a short run of samples while this cause is at work.

Thus we have from sampling theory formulas for the following limits or else estimates of them.

$$E(\bar{x}) \pm 3\sigma_{\bar{x}}$$

$$E(R) \pm 3\sigma_R$$

$$E(s) \pm 3\sigma_s$$

$$E(x) \pm 3\sigma_x$$

In the next chapter more will be said about these distributions and the control chart constants.

Chapter 8 will be devoted to applications of control charts. Meanwhile the illustrative examples and the problems will serve to acquaint the reader with different aspects of the job of applying the methods. It must always be emphasized that success in the use of control charts depends directly upon the care used in drawing samples. In so far as possible all measurements included in a single sample should have arisen under as nearly identical conditions as possible. In this way the variation within the sample reflects random causes only. Then subsequent samples can be taken under different conditions, careful records being kept of what those conditions were: time, personnel, material, process resetting, gage sets, etc. Thus one can hope to pin down the cause of an out-of-control point through noting what conditions for this sample were different from those for others which were in control. The subject of the effect of different kinds of sampling upon control charts will be taken up in Section 7.5.

PROBLEMS

4.1. Briefly describe a potential application in some field of interest to you. What might you hope to learn?

4.2.(a). For the data of Table 3.2, make $\bar{x}$ and R charts. (b). Draw what conclusions you can. (c). If justified, estimate μ and/or σ_x. State justification. (Pressure to operate hydraulic switch.)

4.3.(a). For the data of Table 3.3, make $\bar{x}$ and R charts. (b). Draw what conclusions you can. (c). If justified estimate μ and/or σ_x. State justification. (Final Mooneys for rubber).

4.4. The following data give the results for the presample Mooneys [measure of internal viscosity of Government Reserve styrene (GR-S) for rubber making]. These samples of n=4 correspond respectively to the 24 samples of Table 3.3 and Problem 4.3. (a). Make $\bar{x}$ and R charts for the data and draw what conclusions you think justified. (b). To what precision were the original measurements made? (c). <u>If assigned</u>, correlate these 24 $\bar{x}$'s with those of Problem 4.3.

Samples 1-8 $\bar{x}$	R	Samples 9-16 $\bar{x}$	R	Samples 17-24 $\bar{x}$	R
33.250	6.5	31.750	2.5	30.250	2.5
33.375	.5	35.125	1.5	29.000	.5
32.500	2.5	34.875	3.0	30.375	2.0
30.250	8.5	33.875	2.0	30.375	1.5
34.125	5.5	34.375	3.0	29.375	2.0
29.500	1.5	32.125	2.0	31.125	1.5
35.000	5.0	31.625	1.5	30.875	4.0
32.000	5.5	31.125	4.0	30.250	4.5
				766.500	73.5

4.5. In [7] is given a set of thicknesses of polystyrene films calculated from wave lengths and spacing of interference fringes between 3.6 and 5.0 microns. Thicknesses in mm x 10^4.

Film Sample	Repeated Measurement x_1	x_2	x_3	$\bar{x}$	R	s
1	884	888	882	884.7	6	3.06
2	773	764	770	769.0	9	4.58
3	807	797	821	808.3	24	12.06
4	886	894	897	892.3	11	5.69
5	817	823	813	817.7	10	5.03
6	787	808	788	794.3	21	11.85
7	882	898	894	891.3	16	8.33
				5857.6	97	50.60

(a). Make control charts for $\bar{x}$ and R. (b). How is control on the R chart? (c). Would you be justified in estimating σ? Why? If so, make the estimate. (d). How is control on the $\bar{x}$ chart? Is this surprising?

4.6. Same as Problem 4.5, but $\bar{x}$ and s charts.

4.7. The following data represent six readings for each of 10 groups of cement-mortar briquettes. [8]

Draw the control charts for s, R and $\bar{x}$, computing the limits for the last in two ways. Can the within-group variabilities be regarded as homogeneous, i.e., are the R and s charts in control? Can one readily assume that all groups are from one homogeneous lot?

Group No.	x_1	x_2	x_3	x_4	x_5	x_6	$\bar{x}$	R	s
1	390	380	445	360	375	350	383.3	95	33.4
2	578	500	470	520	530	450	508.0	128	45.6
3	530	540	470	560	460	470	505.0	100	43.3
4	623	532	547	600	600	594	582.7	91	35.2
5	596	528	540	562	590	530	557.7	68	30.0
6	345	322	312	358	375	310	337.0	65	26.5
7	488	550	530	568	420	530	514.3	148	53.3
8	625	625	610	615	610	600	614.2	25	9.7
9	722	727	690	700	705	700	707.3	37	14.2
10	800	798	750	724	720	726	753.0	80	37.1

4.8. This problem is taken from an article by Ott [9]:

"During the manufacture of phonograph pick-up assemblies in a plant of Sonotone Corporation, an electrical characteristic of the completed assemblies was more variable than desired.

"In an effort to manufacture the units with greater uniformity, attention was directed toward a ceramic component of the assembly. These component pieces were cut in the plant from ceramic strips purchased from an outside vendor; some 25 individual pieces were obtained from each ceramic strip. The production foreman and engineer agreed that it was worthwhile to determine how the variability of assemblies containing pieces cut from different strips would compare with the variability corresponding to different pieces from a single strip.

"It was decided arbitrarily to cut 7 pieces from each of 11 different strips. The eleven sets were used to complete eleven assemblies by putting them through the regular production process..... The 11 sets of ceramic pieces were made into completed assemblies by one group of operators in about an hour's production time. This was done deliberately since the purpose of the production experiment was to determine possible effects resulting from different strips of ceramics. If different sets had been done by different shifts and on different days, it was entirely possible that unintentional differences in processing might cause significant variations which would then be confused with the effects being considered in this simple but effective experiment.... The following readings of the electrical characteristics of the final assemblies were then read and recorded.

Set	x_1	x_2	x_3	x_4	x_5	x_6	x_7	$\bar{x}$	R
1	16.5	17.2	16.6	15.0	14.4	16.5	15.5	16.0	2.8
2	15.7	17.6	16.3	14.6	14.9	15.2	16.1	15.8	3.0
3	17.3	15.8	16.8	17.2	16.2	16.9	14.9	16.4	2.4
4	16.9	15.8	16.9	16.8	16.6	16.0	16.6	16.5	1.1
5	15.5	16.6	15.9	16.5	16.1	16.2	15.7	16.1	1.1
6	13.5	14.5	16.0	15.9	13.7	15.2	15.9	15.0	2.5
7	16.5	14.3	16.9	14.6	17.5	15.5	16.1	15.9	3.2
8	16.5	16.9	16.8	16.1	16.9	15.0	16.1	16.3	1.9
9	14.5	14.9	15.6	16.8	12.9	16.6	10.9	14.6	5.9
10	16.9	16.5	17.1	15.8	15.7	13.0	15.0	15.7	4.1
11	16.5	16.7	16.3	14.0	14.9	15.6	16.8	15.8	2.8

"The range of Sample No. 9 was large and the average was low, essentially from the influence of the 10.9 reading. However, the production engineer recognized this low reading as resulting from an electrical short in the assembly, and advised that this one reading be disregarded."

Show the control chart analysis you would use for these experimental data.

4.9. This problem is concerned with the bearing diameter for a blower rotor shaft, for which the specification is .9841 in. + .0000 in., -.0002 in., .9839 to .9841 in.

"Preliminary study revealed that 35.1 per cent were below the low specification, also that 2.6 per cent were above the upper limit. [See first part of table.] Investigation showed that the operator was not exercising enough caution to maintain this close tolerance. As a result of the wide spread of individual pieces inspection was performing 100 per cent sorting of the part. In addition to salvage and rework, some 6 per cent were being scrapped.

"It was decided that the operator should be relieved of the responsibility of backing the wheel away from the work, and an Arnold gage, which automaticall backs the wheel off when the predetermined diameter is reached, was installed on the job. The results of using this type of gage are shown in the second part of the table.

"Inspection was reduced to 5 pieces out of each tote pan of 60. A point is plotted on the $\bar{x}$ and R charts. If the points are in control the pan is accepted; if out of control the pan is sorted 100 per cent."

Draw the charts from the data, and comment on the preceding analysis, with n = 5. Can you set a standard value for σ for the later data? What standard μ would you suggest for continuation? What precision of measurement x was used?

Early Data in .0001" from .9839":

Sample	$\bar{x}$	R	Sample	$\bar{x}$	R	Sample	$\bar{x}$	R
1	.64	1.7	8	.36	2.8	15	.56	1.7
2	.10	1.3	9	.50	1.6	16	- .02	2.2
3	-.04	1.9	10	.58	1.0	17	.60	2.0
4	-.36	3.2	11	.78	2.6	18	- .04	2.2
5	.68	1.4	12	.28	1.4	19	- .04	3.8
6	1.06	1.0	13	.86	1.2	20	- .54	2.9
7	.38	1.8	14	.32	2.7		+6.66	40.4

Later Data in .0001" from .9839"

Sample	$\bar{x}$	R	Sample	$\bar{x}$	R	Sample	$\bar{x}$	R
1	1.42	.3	8	1.00	.2	15	1.52	.3
2	1.22	.5	9	1.54	.3	16	1.40	.5
3	1.24	.3	10	1.28	.3	17	1.52	.4
4	1.16	.5	11	1.36	.2	18	1.38	.4
5	1.36	.6	12	1.54	.1	19	1.20	.6
6	1.06	.7	13	1.32	.5	20	.98	.8
7	1.58	.2	14	1.48	.2		26.56	7.9

4.10. The data of Table P4.7 give results for eccentricity (distance between two center points) for a needle valve. (a) Plot the first 25 $\bar{x}$'s and R's and calculate control lines. (b) Comment on results as to control and meeting of maximum specification. (c) What control lines would you use for the next 25 samples? (d) Plot them and comment. (e) Revise and

Table P4.7. Eccentricity Between Conical Point and Pitch Diameter of a Needle Valve, in .0001 in. (Maximum Specification: .0100 in.) Bore-matic. Spindle 1.

Operator	Time	Date	x_1	x_2	x_3	$\bar{x}$	R
M.D.	1 P.M.	May 3	32	30	30	31	2
M.D.	7 A.M.	May 5	37	18	37	31	19
	9		50	35	36	40	15
	11		57	24	75	52	51
	1 P.M.		49	6	24	26	43
M.D.	6 A.M.	May 6	67	25	25	39	42
	8		52	56	53	54	4
	10		18	39	47	35	29
	1 P.M.		40	51	51	47	11
M.D.	7 A.M.	May 7	31	61	28	40	33
	9		15	10	35	20	25
	11		27	49	19	32	30
	1 P.M.		51	34	40	42	17
H.D.	3		19	32	10	20	22
M.D.	7 A.M.	May 8	30	16	50	35	34
	9		15	30	50	32	35
	11		32	46	29	36	17
	1 P.M.		39	19	34	31	20
H.D.	6		42	40	30	37	12
	9		70	16	57	48	54
M.D.	7 A.M.	May 9	12	19	23	18	11
	9		34	14	40	29	26
	11		58	36	41	45	22
	1 P.M.		7	11	8	9	4
H.D.	3 P.M.	May 13	40	12	12	21	28
	6		66	58	19	48	47
M.D.	8 A.M.	May 14	29	37	74	47	45
	10		20	9	15	15	11
	1 P.M.		30	38	88	52	58
H.D.	3		83	57	90	77	33
	6		40	100	82	74	60
	9		24	63	8	32	55
M.D.	7 A.M.	May 15	9	41	5	18	36
	9		44	80	66	63	36
	11		102	10	41	51	92
	1 P.M.		43	51	53	49	10
H.D.	6		21	27	49	32	28
M.D.	7 A.M.	May 16	18	19	31	23	13
	9		7	22	13	14	15
	11		43	48	19	37	29
	1 P.M.		67	35	5	36	62
A.W.G.	3		35	51	33	40	18
M.D.	7 A.M.	May 19	39	35	49	41	14
	9		66	18	56	47	48
	11		39	13	43	32	30
	1 P.M.		14	27	10	17	17
M.D.	7 A.M.	May 20	23	71	27	40	48
	9		17	24	60	34	43
	11		55	31	32	39	24
M.D.	7 A.M.	May 26	6	34	39	26	33

suggest μ and σ if reasonable. (f) The distribution of x's is unsymmetrical. Would this affect the control chart constants used?

4.11. The following data are for the effects of minor alloy additions on corrosion resistance of aluminum bronze [10]. Ten of 22 elements were randomly chosen for this problem. Ten observations per sample.

Element	Symbol	$\bar{x}$	s
Cobalt	Co	31.77	2.25
Nickel	Ni	30.96	1.69
Tellurium	Te	31.99	.76
Magnesium	Mg	31.88	1.22
Iron	Fe	31.62	1.48
Manganese	Mn	30.89	1.57
Titanium	Ti	31.10	1.76
Silver	Ag	30.13	1.44
Bismuth	Bi	28.44	.97
Selenium	Se	29.44	1.23

Make $\bar{x}$ and s charts, analyze results and comment.

4.12. Show how one could tell what the sample size is from the marked control limits and central lines on Figure 4.8.

4.13. For the data in Table 4.3, samples 1-25, without plotting points find center lines and control limits for $\bar{x}$'s and R's analyzing as past data. Is process in control? Do the same using standards given: $\mu = 2.00, \sigma = 1.715$, for Populations B.

4.14. Same as Problem 4.13 but for samples 26-50. (Standards given for Population C: $\mu = 0$, $\sigma = 3.47$.)

4.15. How would a decrease in σ_x with no change in μ show up on $\bar{x}$ and R (or s) charts?

4.16. If $\bar{x}$ limits lie outside one or both specification limits for x's, do you need to test for control to tell whether you are meeting specifications? If they lie inside do you need to check control to tell?

References

1. H. Working and E. G. Olds. "Manual for an Introduction to Statistical Methods of Quality Control in Industry, Outline of a Course of Lectures and Exercises." Office of Production Research and Development, War Production Board, Washington, D. C., 1944.
2. I. W. Burr, Specifying the desired distribution rather than maximum and minimum Limits. Indust. Quality Control 24 (No. 2), 94-101 (1967).
3. W. R. Weaver, The foreman's view of quality control. Indust. Quality Control 5 (No. 2), 6-14 (1948).
4. E. L. Grant. "Statistical Quality Control." Mc-Graw Hill, N. Y., 1952. Example 8, p. 117.
5. V. E. McCoun, The case of the perjured control chart. Indust. Quality Control 5 (No. 6), 20-23 (1949).
6. P. S. Olmstead, How to detect the type of an assignable cause, Part I, Clues for particular types of trouble, Indust. Quality Control 9 (No. 3), 32-38 (1952). Also Part II, Procedure when probable cause is unknown, Indust. Quality Control 9 (No. 4), 22-32 (1953).
7. E. C. Copelin, "Attempts to Establish Performance Tests for Infrared Spectrophotometers." Ph.D. Thesis, Purdue University, p. 45 (1958).
8. E. S. Pearson, The percentage limits for the distribution of range in samples from a normal population. Biometrika 24, 404-417 (1932).
9. E. R. Ott, Variables control chart in production research, Practical Aids Department of Indust. Quality Control 6 (No. 3), 30-31 (1949).
10. R. H. Hoefs. "The Effect of Minor Alloy Additions on the Corrosion Resistance of Alpha Aluminum Bronze." Ph.D. Thesis, Purdue University (1953).

CHAPTER 5

BACKGROUND OF CONTROL CHARTS FOR MEASUREMENTS

5.1. Introduction. This chapter is designed to explain further the control chart constants we have met in Chapter 4; to give some formulas and some interrelations between them. Such study will strengthen the student's grasp of the control chart constants and their role.

There are two methods by which control chart constants may be found for a given population. We may proceed empirically, through a large collection of random samples, tabulated and analyzed. Such results are only approximate, but this method may be the only one available. See Section 5.2.

The second method is to derive the constants by theoretical considerations. This approach is used in Section 5.3. It may yield exact or approximate results.

5.2. The Empirical Approach to Sampling Distributions. A method which is always available is to draw a large number of samples of n x's each with replacement if a finite population, or by use of tables of random numbers if an infinite population. Then one tabulates the distributions of the obtained sample statistics $\bar{x}$, R and s, and finds the respective means and standard deviations of the three. These would be estimates of the true values $E(\bar{x})$, $E(R)$, $E(s)$; $\sigma_{\bar{x}}$, σ_R, σ_s. Using the obtained estimates as in Section 5.3.1. we would obtain approximations to the control chart constants for the population of x's at hand. This approach is necessary if theory is not available.

The approach was used by Lois Niemann [1]. She drew 4000 samples, each of n = 4, from a strongly unsymmetrical population (constructed by 1000 numbered beads, skewness $\alpha_3 = 1.15$). Using available known theory on $\bar{x}$'s but finding

$\bar{R} = \hat{\mu}_R$, $s_R = \hat{\sigma}_R$, $\bar{s} = \hat{\mu}_s$, $s_s = \hat{\sigma}_s$ she estimated the control chart constants appropriate for samples from this population. They were found to be remarkably close to those for a normal population of $\bar{x}$'s, differing at most by 10%. The calculations were done on desk calculators, this being before the flowering of the electronic computer.

The foregoing approach is quite a bit more feasible with a modern electronic computer. But still one must fight the fact that errors in such empirical statistical results vary inversely, only as the square root of the number of values of the statistic which were obtained. The method is useful where theory is as yet unavailable.

5.3. Control Charts for Samples from Normal Populations of x's. We next use available normal curve sampling theory to derive the control chart constants we need.

5.3.1. Formulas for Control Chart Constants. Suppose we have a population of x's with given μ and σ_x. Then whether normal or not we have

$$E(\bar{x}) = \mu_{\bar{x}} = \mu_x \quad \text{and} \quad \sigma_{\bar{x}} = \sigma_x/\sqrt{n}. \tag{5.1}$$

Then just as in Section 4.2

$$\text{₵}_{\bar{x}} = \mu_x \tag{5.2}$$

$$\text{Limits}_{\bar{x}} = \mu_{\bar{x}} \pm 3\sigma_{\bar{x}} = \mu_x \pm 3\sigma_x/\sqrt{n} = \mu_x \pm A\sigma_x, \; A = 3/\sqrt{n}. \tag{5.3}$$

This formula for A is correct for any population having a finite σ_x. Check a couple of cases in Table V.

In order to derive the other control chart constants we assume normality of x's. If this is assumed then we can evaluate c_4 and c_5 as in Section 5.3.2 to give

$$E(s) = \mu_s = c_4\sigma_x \tag{5.4}$$

$$\sigma_s = c_5\sigma_x. \tag{5.5}$$

Note that c_4 is not 1 in Table V, but approaches 1 as n increases. Using these we have

$$\text{¢}_s = c_4\sigma_x \tag{5.6}$$

$$\text{Limits}_s = \mu_s \pm 3\sigma_s = c_4\sigma_x \pm 3c_5\sigma_x = (c_4 \pm 3c_5)\sigma_x. \tag{5.7}$$

We define the limits by multipliers of σ_x:

$$\text{Limits}_s = B_5\sigma_x,\ B_6\sigma_x, \tag{5.8}$$

from which

$$B_5 = c_4 - 3c_5,\quad B_6 = c_4 + 3c_5. \tag{5.9}$$

Next suppose that μ_x, σ_x are unknown. We could not then use (5.3) and (5.8). But if we have a collection of samples yielding $\bar{\bar{x}}$ and $\bar{s}$, we assume control tentatively and use them.

Since s's and therefore $\bar{s}$'s average $c_4\sigma_x$, but not σ_x, $\bar{s}$ is a biased estimate of σ_x. For example, if n = 5, $\bar{s}$ averages $.940\sigma_x$, that is, it underestimates σ_x <u>on the average</u>. But if we divide $\bar{s}$ by .940, then $\bar{s}/.940$ will average σ_x, and thus provides an unbiased estimate. Hence we take

$$\hat{\sigma}_x = \bar{s}/c_4 \tag{5.10}$$

to replace σ_x in (5.3), (5.6) and (5.7). Thus

$$\text{Limits}_{\bar{x}} = \hat{\mu}_x \pm 3\hat{\sigma}_x/\sqrt{n} = \bar{\bar{x}} \pm 3\bar{s}/(c_4\sqrt{n}) = \bar{\bar{x}} \pm A_3\bar{s}. \tag{5.11}$$

Therefore

$$A_3 = 3/(c_4\sqrt{n}). \tag{5.12}$$

Verify this relation for n = 4, say, in Table V. Further

$$\text{¢}_s = c_4\hat{\sigma}_x = c_4\bar{s}/c_4 = \bar{s}. \tag{5.13}$$

What could be more natural than $\bar{s}$ for the center line? Continuing

$$\text{Limits}_s = B_5\hat{\sigma}_x,\ B_6\hat{\sigma}_x = (B_5/c_4)\bar{s},\ (B_6/c_4)\bar{s} = B_3\bar{s},\ B_4\bar{s} \tag{5.14}$$

the B_3, B_4 constants being defined as appropriate multipliers for $\bar{s}$. Therefore

$$B_3 = B_5/c_4 \qquad B_4 = B_6/c_4. \tag{5.15}$$

Or, using (5.9)

$$B_3 = (c_4 - 3c_5)/c_4 = 1 - (3c_5/c_4), \; B_4 = 1 + (3c_5/c_4) \tag{5.16}$$

Now for the constants associated with the range and $\bar{R}$ in particular we proceed the same way exactly. Assume

$$E(R) = \mu_R = d_2\sigma_x \tag{5.17}$$

$$\sigma_R = d_3\sigma_x. \tag{5.18}$$

Then an unbiased estimate of σ_x from $\bar{R}$ is

$$\hat{\sigma}_x = \bar{R}/d_2.$$

Using this to replace σ_x in (5.3) along with $\hat{\mu}_x = \bar{\bar{x}}$ we have

$$\text{Limits}_{\bar{x}} = \bar{\bar{x}} \pm 3\bar{R}/(d_2\sqrt{n}) = \bar{\bar{x}} \pm A_2\bar{R} \tag{5.19}$$

$$A_2 = 3/(d_2\sqrt{n}). \tag{5.20}$$

Now also we have for the case standard σ_x given

$$℄_R = d_2\sigma_x \tag{5.21}$$

$$\text{Limits}_R = \mu_R \pm 3\sigma_R = d_2\sigma_x \pm 3d_3\sigma_x = (d_2 \pm 3d_3)\sigma_x = D_1\sigma_x, D_2\sigma_x, \tag{5.22}$$

defining D_1, D_2 as multipliers of σ_x for the limits. Thus

$$D_1 = d_2 - 3d_3, \; D_2 = d_2 + 3d_3. \tag{5.23}$$

Now using $\hat{\sigma}_x = \bar{R}/d_2$ to replace σ_x in (5.21)-(5.23) yields

$$℄_R = \bar{R}$$

$$\text{Limits}_s = D_1\hat{\sigma}_x, D_2\hat{\sigma}_x = (D_1/d_2)\bar{R}, \; (D_2/d_2)\bar{R} = D_3\bar{R}, D_4\bar{R}. \tag{5.24}$$

So

$$D_3 = D_1/d_2 \qquad D_4 = D_2/d_2. \tag{5.25}$$

Or, using (5.23)

$$D_3 = (d_2 - 3d_3)/d_2 = 1 - (3d_3/d_2) \quad D_4 = 1 + (3d_3/d_2). \tag{5.26}$$

It is easily seen from the foregoing that the basic constants needed are just the four c_4, c_5, d_2, d_3.

5.3.2.* Derivation for Standard Deviations. If samples are drawn randomly and independently from a normal population with standard deviation $\sigma_x = \sigma$, then it can readily be proved that

$$\frac{(n - 1)s^2}{\sigma^2} = \chi^2 \tag{5.27}$$

follows the χ^2 distribution with n - 1 degrees of freedom. A proof may be found in any mathematical statistics book.

The density function for this χ^2 variable is

$$f(\chi^2) = \frac{(\chi^2)^{[(n-1)/2]-1} e^{-\chi^2/2}}{2^{(n-1)/2}\Gamma[(n-1)/2]} \qquad 0 < \chi^2 < \infty$$

The reader may not be familiar with the gamma function in the denominator. Its purpose here is to make the integral of $f(\chi^2)$ from 0 to ∞ to be 1. By definition

$$\Gamma(k) = \int_0^\infty w^{k-1} e^{-w} dw \qquad k > 0 \tag{5.28}$$

which depends upon the exponent k for its value. In particular

$$\Gamma(1) = \int_0^\infty e^{-w} dw = -e^{-w}\Big]_0^\infty = 1. \tag{5.29}$$

A convenient recursion relation

$$\Gamma(k+1) = k\Gamma(k) \qquad k > 0 \tag{5.30}$$

is obtained by integration by parts:

*Sections designated by an asterisk may be omitted without destroying the continuity.

$$\Gamma(k+1) = \int_0^\infty w^k e^{-w}\, dw \quad u = w^k,\ dv = e^{-w}\, dw$$
$$du = kw^{k-1}\, dw,\ v = -e^{-w}$$

$$\Gamma(k+1) = -w^k e^{-w}\Big]_0^\infty + \int_0^\infty kw^{k-1} e^{-w}\, dw$$
$$= 0 + k\Gamma(k)$$

From these two relations we have, taking k as any positive integer n

$$\Gamma(n) = (n-1)! \tag{5.31}$$

Thus the gamma function takes factorial values for positive integers and may be regarded as an interpolation formula between them.

In the χ^2 distribution we need $\Gamma(1/2)$ which proves to be $\sqrt{\pi}$. A sketch of the proof follows:

$$\Gamma(1/2) = \int_0^\infty w^{-1/2} e^{-w}\, dw > 0$$

Let $w = x^2$, $dw = 2xdx$

$$\Gamma(1/2) = \int_0^\infty 2e^{-x^2}\, dx$$

This cannot be evaluated directly, so we use a trick and find its square

$$[\Gamma(1/2)]^2 = \int_0^\infty 2e^{-x^2}\, dx \int_0^\infty 2e^{-y^2}\, dy$$
$$= 4\int_0^\infty \int_0^\infty e^{-(x^2+y^2)}\, dx\, dy$$

Then transforming to polar coordinates by

$$x = r \sin \theta,\ y = r \cos \theta,\ dxdy = r\, dr\, d\theta,\ x^2 + y^2 = r^2$$

the integral over the first quadrant becomes

$$[\Gamma(1/2)]^2 = 4\int_0^\infty \int_0^{\pi/2} re^{-r^2}\, dr\, d\theta$$
$$= 2\int_0^\infty e^{-r^2}\, 2r\, dr \int_0^{\pi/2} d\theta$$
$$= 2[-e^{-r^2}]_0^\infty \cdot (\pi/2) = \pi$$

Since $\Gamma(1/2) > 0$

$$\Gamma(1/2) = \sqrt{\pi}. \tag{5.32}$$

Now for the moments of s, we use the density function of χ^2 as follows, using (5.27),

$$f(\chi^2)d\chi^2 = \frac{\left[\frac{(n-1)s^2}{\sigma^2}\right]^{(n-1)/2-1} e^{-\frac{(n-1)s^2}{2\sigma^2}} d\,\frac{(n-1)s^2}{\sigma^2}}{2^{(n-1)/2}\,\Gamma[(n-1)/2]}.$$

Then, distributing the $2^{(n-1)/2}$ as needed, the expectation of s^i is:

$$E(s^i) = \int_0^\infty \frac{s^i\left[\frac{(n-1)s^2}{2\sigma^2}\right]^{(n-1)/2-1} e^{-\frac{(n-1)s^2}{2\sigma^2}} d\,\frac{(n-1)s^2}{2\sigma^2}}{\Gamma[(n-1)/2]}.$$

Now let $w = \frac{(n-1)s^2}{2\sigma^2}$, $s = \sigma\cdot\sqrt{\frac{2w}{n-1}}$

$$E(s^i) = \int_0^\infty \frac{\sigma^i\, w^{(n-1+i)/2-1}\, 2^{i/2}}{\Gamma[(n-1)/2](n-1)^{i/2}}\, dw,$$

and using (5.28)

$$E(s^i) = \frac{\sigma^i\, 2^{i/2}\,\Gamma[(n-1+i)/2]}{(n-1)^{i/2}\,\Gamma[(n-1)/2]}. \tag{5.33}$$

Taking i = 1, yields

$$E(s) = \mu_s = \sigma\sqrt{\frac{2}{n-1}}\,\frac{\Gamma(n/2)}{\Gamma[(n-1)/2]} = c_4\sigma.$$

Hence

$$c_4 = \sqrt{\frac{2}{n-1}}\,\frac{\Gamma(n/2)}{\Gamma[(n-1)/2]}. \tag{5.34}$$

For example if n = 7, using (5.30)-(5.32)

$$c_4 = \sqrt{\frac{2}{7-1}}\frac{\Gamma(7/2)}{\Gamma(3)} = \frac{(5/2)(3/2)(1/2)\sqrt{\pi}}{2!\sqrt{3}}.$$

If, however, i = 2, (5.33) yields

$$E(s^2) = \frac{\sigma^2 2}{n-1}\frac{\Gamma[(n+1)/2]}{\Gamma[(n-1)/2]} = \frac{2\sigma^2}{n-1}[(n-1)/2]\frac{\Gamma[(n-1)/2]}{\Gamma[(n-1)/2]}$$

$$E(s^2) = \sigma^2. \tag{5.35}$$

(This is true also for non-normal populations.)

Making use of the theorem $\sigma_x^2 = E(x^2) - [E(x)]^2$ we have

$$\sigma_s^2 = E(s^2)-[E(s)]^2 = \sigma^2 - c_4^2\sigma^2 = (1-c_4^2)\sigma^2.$$

Hence

$$\sigma_s = \sigma_x\sqrt{1-c_4^2} = \sigma_x c_5.$$

Thus

$$c_5 = \sqrt{1-c_4^2}. \tag{5.36}$$

We can therefore readily evaluate c_4 and c_5.

5.3.3.* The Distribution of Range. This general problem requires approximate integration for all but the very simplest cases, in fact double or triple integration. Letting the density function for the x's be f(x), the cumulative or distribution function F(x) is defined by

$$F(x) = P(X \le x) = \int_0^x f(v)dv.$$

Now take a sample of n x's randomly and independently from the population. Instead of calling the <u>first</u> one x_1, call the <u>smallest</u> x_1, the second smallest x_2, up to the largest $x_n = x_1 + R$, since the range is $x_n - x_1$. Now the overall probability, to within higher order infinitesimals, for the random variable X_1 to lie in x_1, $x_1 + dx_1$, random variable R' to lie in R, R+dR and the other n-2 between is given by the following:

$$n\ f(x_1)dx_1\cdot(n-1)f(x_1+R)dR\cdot[F(x_1+R)-F(x_1)]^{n-2} \tag{5.37}$$

The first part says that there are n choices among the n x's for the lowest x_1 (it to be in the given interval). Then the second part has n-1 choices for the x to be the largest one and in the interval x_1+R to x_1+R+dR. The third part uses

$$P(x_1 < X < x_1 + R) = F(x_1 + R) - F(x_1),$$

and this is the probability of the all other x's lying between x_1 and $x_1 + R$. Now to find the density function of R, g(R), we need to integrate (5.37) from $x_1 = -\infty$ to $x_1 = \infty$ which eliminates x_1. Then finally we need

$$E(R),\ E(R^2)\ \text{by}\ \int_0^\infty Rg(R)dR,\ \int_0^\infty R^2g(R)dR.$$

Then we use $\sigma_R^2 = E(R^2) - [E(R)]^2$. In the case of the normal curve for f(x) this requires approximate triple integration. However, despite the difficulties E. S. Pearson [2] had quite full results available for a normal distribution by 1932, building on earlier work by Tippett [3]. This work, facilitating the use of ranges instead of standard deviations for small samples, was a great boon to the users of control charts for measurements and helped a great deal in their spread to applications.

From such work we have the numerical values of d_2 and d_3 and can find all the other control chart constants.

Fairly recent work by the author [4], showed that for population distributions departing considerably from the normal curve, the control chart constants are not greatly affected, for small sample sizes (up to n = 10). Hence we can use the control chart constants of Table V with much confidence.

PROBLEMS

5.1. Given n = 8, evaluate c_4, c_5 from (5.34), (5.36). Then take d_2 and d_3 from Table V. Using these as needed in appropriate formulas verify the other control chart constants in Table V, for n = 8.

5.2. Same as Problem 5.1. but use n = 9.

5.3. Use $\sigma_s^2 = E(s^2) - [E(s)]^2$ to show that $E(s) < \sigma$, that is, that $c_4 < 1$. (Hint: We know $E(s^2) = \sigma^2$. What would happen if $E(s) = \sigma$?)

References

1. Lois J. H. Niemann, "An Experiment in Sampling from a Pearson Type III Distribution." M.S. Thesis, Purdue University (1949).

2. E. S. Pearson, The percentage limits for the distribution of range in samples from a normal population. Biometrika 24, 404-417 (1932).

3. L. H. C. Tippett, On the extreme individuals and the range of samples taken from a normal population. Biometrika 17, 364-387 (1925).

4. I. W. Burr. The effect of non-normality on constants for $\bar{x}$ and R charts, Indust. Quality Control 23 (No. 11), 563-569 (1967).

CHAPTER 6
CONTROL CHARTS FOR ATTRIBUTES

6.1. Introduction. This chapter is concerned with control charts for attribute or countable data. Such data are of two basic kinds, but both involve "defects", and this can easily be confusing. Let us make the distinction clear.

Definition [1]. A "defect" is an instance of a failure to meet a requirement imposed on a unit with respect to a single quality characteristic. (The term defect thus covers a wide range of possible severity; on the one hand it may be merely a flaw or a detectable deviation from some minimum or maximum limiting value, or on the other, a fault sufficiently severe to cause an untimely product failure.)

In inspection or testing, each piece or unit is checked to see if it does or does not contain any defects. Then we may describe the outcome of the inspection in two distinct ways.

In the first place we may examine each piece and record it as a "good" one if it contains <u>no</u> defects, or record it as a "defective" one if it contains <u>any</u> defects. In this scoring, every piece is either a good one or a defective one. After inspecting n pieces we will have found say d of them to be defectives and n-d of them to be good ones. We emphasize that a defective piece may have but one defect or several, but it still counts as just one defective.

On the other hand we may count and record the number of defects, c, we find on a single piece. This count may be 0,1,2,... Such an approach of counting defects on a piece becomes especially useful if most of the pieces contain one or more defects. Then the count tells not only that the piece is a defective, but how defective it is. Examples of defect counting might be (a) defects in a complicated electronic or mechanical assembly, (b) typographical errors on a page, (c) accidents in a month, (d) flaws on a sheet of paper or metal. We may count the defects on a single unit or on a sample of n units, recording the count as c defects. We thus count defects on a field of opportunity for defects to occur.

Again it is emphasized that a defective piece may have only quite a minor defect or a major or critical one; it all depends upon the piece in question and the class of defects being inspected for.

We may always convert a measurable characteristic of a piece to an attribute by setting limits, say, L and U for x. Then if x lies between, the piece is a good one, or if outside, it is a defective one. But the converse is not true. For example, an electric light bulb, when tested either lights or it does not, giving respectively a good or a defective. But there is no way to put a measurement x on this bulb. It simply works or it does not.

6.2. Control Charts for Defectives. Let us first consider the two kinds of charts available for defectives. We shall first take the simplest case, namely, when the sample size n is constant. Thus we have in each sample n pieces to inspect for a single type of defect, or for all of the defects in some class. There might be 20 or more different kinds of defects in the class. But still each piece is either a defective (contains one or more defects), or a good one (contains no defects in the class). Then

$$n = \text{number of pieces inspected.} \tag{6.1}$$

$$d = \text{number of defective pieces in the n} \tag{6.2}$$

$$p = d/n = \text{fraction defective in the sample} \tag{6.3}$$

$$np = \text{number of defective pieces in the } n^* \tag{6.4}$$

$$q = 1 - p = \text{fraction effective or good in the sample.} \tag{6.5}$$

Now also we have a true or population fraction defective for the production process, for which we use the primed notation (rather than π or some other Greek letter). This follows the notations in [1].

$$p' = \text{fraction defective in the population} \tag{6.6}$$

$$q' = 1 - p' = \text{fraction effective in the population.} \tag{6.7}$$

Then we may plot on a control chart successive values of p, giving a <u>fraction defective chart</u>, or of d = np, giving a chart for <u>defectives</u>. With constant n's it makes little difference which we use, but perhaps more people will understand a chart for p, especially using a percent scale.

Let us recall here the conditions for the binomial distribution to apply, as given in Section 2.5.3:

1. Two possibilities: good and defective pieces
2. Constant fraction defective p' for each piece
3. Independence of pieces, that is, the occurrence of a defective or a good one does not change p' or q' for the next one to be defective or good. Thus the defectives do not tend to "come in bunches."

These conditions are assumed whenever we make control charts for p or np. If any of them are violated, the control limits and interpretation thereof do not apply. Some such counter cases will be mentioned in Section 6.2.5, later on.

6.2.1. Control Charts for Fraction Defective, Standard Given. Let us first consider the simple case of a p chart, when p' is given and n is

---

*We have here two notations for the number of defectives: np and d. Charts for this number are most often called "np charts."

constant. Then using Section 2.5.3 on the binomial distribution, $E(p) = p'$, $\sigma_p = \sqrt{p'(1-p')/n}$. Therefore we have the center line and three-sigma limits:

$$\text{₵}_p = p' \tag{6.8}$$

$$\text{Limits}_p = p' \pm 3\sqrt{p'q'/n}. \tag{6.9}$$

Now how do we know p'? Just as we have explained for measurement charts, standards can be set when we have a recent period of good control at satisfactory performance. Also in the case of p charts we may set a standard p' as a goal to be desired. Or, at the end of a month we may review the past month with those involved with the process, eliminating some out of control results or no longer typical data, and set a standard p' for the next month.

As an example of a p chart with a standard given, consider Table 6.1. It lists the number of defective carburetors, that is, a defective had one or more defects, excluding leaks. There were perhaps 20 different possible defects. Each sample consisted of 100 consecutive carburetors inspected at the end of the production line. This is in a real sense a sampling of production conditions obtaining while these 100 were produced. The table also gives the fraction defective p.

For a long time a goal of 2 percent had been set, but this had seemed impossible of attainment. At the time these data were collected, all samples being on the same day, the fraction defective had actually begun to reach the 2 per cent level, through continuous follow-up of all assignable causes in assembly and production. Figure 6.1 shows the p chart for the data. Using the standard fraction defective $p' = .02$ and the sample size of $n = 100$ in (6.8) and (6.9) gives

$$\text{₵}_p = .02$$

$$UCL_p = .02 + 3\sqrt{\frac{.02(.98)}{100}} = .02 + .042 = .062$$

$$LCL_p = .02 - .042 = \text{—}$$

Table 6.1. Inspection Results on a Type of Carburetor at End of Assembly Line, All Types of Defects Except Leaks, Jan. 27, Two Shifts. Sample Size n = 100

Sample No.	No. defectives d	Fraction defective p	Sample No.	No. defectives d	Fraction defective p
1	4	.04	19	4	.04
2	5	.05	20	2	.02
3	1	.01	21	1	.01
4	0	.00	22	2	.02
5	3	.03	23	0	.00
6	2	.02	24	2	.02
7	1	.01	25	3	.03
8	6	.06	26	4	.04
9	0	.00	27	1	.01
10	6	.06	28	0	.00
11	2	.02	29	0	.00
12	0	.00	30	0	.00
13	2	.02	31	0	.00
14	3	.03	32	1	.01
15	4	.04	33	2	.02
16	1	.01	34	3	.03
17	3	.03	35	3	.03
18	2	.02			
				73	

The lower control limit would be −.022 if we used the formula, so we disregard it just as we disregarded zero lower control limits for ranges. The interpretation is that when $p' = .02$ and $n = 100$, the event of a sample with no defectives is not rare enough to warrant any action.

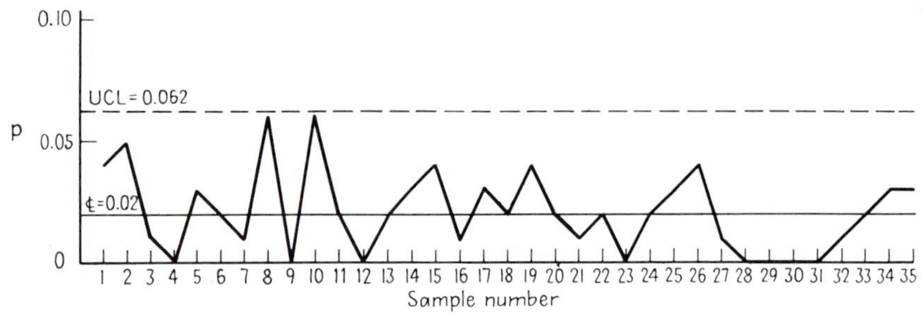

FIG. 6.1. Fraction defective chart on final inspection results at end of assembly line on a type of carburator. All types of defects except leaks. Work of two shifts on Jan. 27. Sample size 100. Reproduced with permission from I. W. Burr, "Engineering Statistics and Quality Control", McGraw-Hill, New York, 1953, p. 217.

Thus we could not regard such a point as an indication of the presence of some assignable cause which has temporarily lowered p'. The appropriate binomial, n = 100, p' = .02, gives by (2.30) P(d=0) = .133, and hence by chance causes alone about one sample in seven or eight will have no defectives. Such a point is then fairly common and is not a reliable indication of an assignable cause. The binomial distribution is rather unsymmetrical for such a case, having points considerably farther above the expected p' = .02 than below.

As judged by Figure 6.1, the process at this stage can justifiably be called stable, with a fraction defective of .02. The goal of 2 per cent has clearly been attained this day, and the good control gives confidence that it can continue to be met.

6.2.2. Control Chart for Number of Defectives, Standard Given. In this case we are again given a standard value p', and a sample size of n, assumed constant here for the present. But now we plot the number of defectives, np or d. Such a chart is commonly called an "np chart," although the author would prefer the one-letter designation "d chart."

Using the data of Table 6.1 as an example we plot the 35 values of d, the number of carburetors with at least one defect. The chart would look absolutely identical to Figure 6.1, except that the vertical scale would be for np or d, and since n=100, .05 is replaced by 5, .10 by 10, and so on. Also we have from the binomial distribution (2.30) that E(d) = np' and $\sigma_d = \sqrt{np'(1-p')}$ from (2.31) and (2.32). Therefore we have

$$℄_{np} = np' \tag{6.10}$$

$$\text{Limits}_{np} = np' \pm 3\sqrt{np'(1-p')}. \tag{6.11}$$

Now using p' = .02, n = 100 yields

$$℄_{np} = 2$$

$$\text{Limits}_{np} = 2 \pm 3\sqrt{100(.02)(.98)} = 2 \pm 3(1.4) = -, 6.2.$$

These control lines correspond exactly to those for p: .02, .062, if we multiply by n = 100, for (6.10) and (6.11) are exactly n times (6.8) and (6.9). The same conclusions of course apply as for the p chart.

Still a third vertical scale could be used namely a per cent scale, obtained by multiplying the p scale by 100%. Then p = .05 corresponds to 5%, etc. If it is desired to use a <u>per cent defective</u> scale, it is best to do all the calculations in fraction defective, and then convert to a per cent scale for those who will view it.

6.2.3. Charts for p and np, Analysis of Past Data. In Section 6.2.4 we shall take up the cases of unequal sample sizes, but for the present we continue with equal n's.

In the present case, for either type of chart, we do not know p'. Since we are testing the hypothesis that all samples came from the same binomial population, we assume p' has an unknown but constant value. The best estimate of p', from k samples, is $\bar{p}$, that is

$$\bar{p} = \frac{\sum_{i=1}^{k} d_i}{\sum_{i=1}^{k} n_i} = \frac{\text{total defectives}}{\text{total of all samples}}. \tag{6.12}$$

This formula is used whether or not the n's are equal. Assuming equal n's, however, this may also be written

$$\bar{p} = \frac{\Sigma_{i=1}^{k} p_i}{k} \qquad \text{n's equal} \qquad (6.13)$$

Now we substitute $\bar{p}$ for p' in (6.8) to (6.11) obtaining

$$₵_p = \bar{p} \qquad (6.14)$$

$$\text{Limits}_p = \bar{p} \pm 3\sqrt{\bar{p}(1-\bar{p})/n} \qquad (6.15)$$

$$₵_{np} = n\bar{p} \qquad (6.16)$$

$$\text{Limits}_{np} = n\bar{p} \pm 3\sqrt{n\bar{p}(1-\bar{p})}. \qquad (6.17)$$

For an example consider the data of Table 6.2. These are for k=30 samples, each of n=50 sheets of paper, inspected against a certain standard. The p values are plotted in Figure 6.2.

Now for $\bar{p}$ we use (6.12)

$$\bar{p} = \frac{72}{30(50)} = .0480.$$

We could also have used (6.13)

$$\bar{p} = \frac{1.44}{30} = .0480.$$

Thus the center line for the p chart is .0480. For the control limits we have by (6.15)

$$\text{Limits}_p = .0480 \pm 3\sqrt{.048(.952)/50} = .0480 \pm .0907 = -, .1387.$$

Drawing the two control lines we see that there are no significant runs relative to the center line, but that sample number 12 is out of the control band signaling an assignable cause even though it is but one out among 30 samples. We should look for the cause of such performance. What conditions were different then?

Now for an np chart, we would plot the whole-number counts, d, of defectives. Then using $\bar{p}$ = .0480, n = 50 in (6.16) and (6.17) we find

Table 6.2. Samples of 50 Sheets of Book Paper, 25 by 38 In., Tested for Sheet Formation Against a Carefully Selected Standard Sheet, so that if the Sheet Tested Has Poorer Formation than the Standard, It Is Called a Defective

Day No.	No. defectives d	Fraction defective p	Day No.	No. defectives d	Fraction defective p
1	1	.02	16	3	.06
2	0	.00	17	0	.00
3	1	.02	18	3	.06
4	4	.08	19	3	.06
5	1	.02	20	2	.04
6	5	.10	21	4	.08
7	2	.04	22	3	.06
8	1	.02	23	1	.02
9	3	.06	24	2	.04
10	3	.06	25	3	.06
11	3	.06	26	0	.00
12	7	.14	27	2	.04
13	3	.06	28	5	.10
14	2	.04	29	2	.04
15	2	.04	30	1	.02
				72	1.44

$$\text{Ȼ}_{np} = 50(.0480) = 2.4$$

$$\text{Limits}_{np} = 50(.0480) \pm 3\sqrt{50(.048)(.952)} = 2.4 \pm 4.53 = -, 6.93.$$

Thus d = np = 7 is the only sample value showing out of control. Hence we have identical results to those for the p chart, as indeed we must.

To emphasize that the np chart is just a scale change from the p chart, observe the right hand scale on Figure 6.2, for np's. The values are just n=50 times as great as for the p values on the left scale.

If on investigation we find the assignable cause and remove it from the process, then we should revise $\bar{p}$ to 65/1450, changing the control lines accordingly and extend them for new production. But if we have not found the cause, or having found it, do nothing to prevent its recurrence, then this sample is still as typical of the process as any, and should be left in. In this case then, our $\bar{p}$ will not be an estimate of but a single p', but instead an estimate of an average (weighted) of two values of p', one when the assignable cause is not at work and the other when it is.

Another point to be mentioned is that the definition of a defective as used may not mean at all that 4.8 per cent of the sheets of book paper are unusable. A rather high standard of sheet formation may have been chosen to increase artificially the percentages of "defectives." This kind of approach is often done so that smaller samples can be used to look for the presence of assignable causes. If defectives are extremely rare and the average number per sample is <u>much</u> less than 1, then the chart will have little power to give a warning of an assignable cause. We can

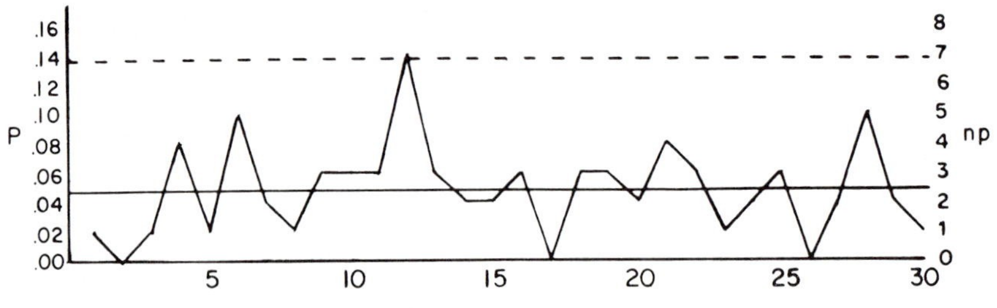

FIG. 6.2. Fraction defective chart for samples of 50 sheets of book paper tested for "sheet formation" against a minimum quality standard sheet. Data from Table 6.2. Also interpretable as an np chart by using scale on right side.

in such a case (a) greatly increase n, or instead (b) use an artificially severe criterion of a "defective" with the same sample size. In either case we gain more power to pick up assignable causes when present.

As a further example, let us consider some data from a random experiment, designed to illustrate some of the principles. The drawings were made from a box of 1 cm diameter beads, using a paddle with 50 holes. (Paddles with various numbers of holes can be used effectively in such experiments, and also to illustrate acceptance sampling by attributes, which we take up later on.) In the present experiment, the box contained 1600 yellow beads, and enough red ones to provide desired population fractions defective, such as, $p' = 102/1702 = .060$ for samples 1 to 75. Theoretically the law governing the probability of $d = 0,1,2 \cdots$ defectives in $n = 50$ is the hypergeometric law, since in 50 consecutive drawings, each of 50 beads, the probability p' varies slightly after each drawing of a single bead, depending on the previous draws. However, since 50 is such a small proportion of 1702, we can readily use the binomial law (2.30) as an excellent approximation. Hence a p chart approach is applicable. After each 50 the beads are returned, and remixed.

The results are shown in Table 6.3. Treating the first 25 samples of 50 as the preliminary data we find

$$\bar{p} = 81/1250 = .0648$$

$$\text{Limits}_p = .0648 \pm 3\sqrt{.0648(.9352)/50} = .0648 \pm .1044 = -, .169.$$

Using these lines we find the first 25 samples to be in perfect control. The fraction defective, may or may not be satisfactory from a practical viewpoint. The process is stable (as we certainly expected). The control lines are extended for the next 25 samples and good control continues. There is, however, a run of nine points in succession below the center line (samples 36 through 44). This might or might not be regarded as

Table 6.3. Defectives Data Drawn Randomly from Various Populations, Simulating a Production Process, Analysis of Past Data. True Fraction Defective Shown in Remarks Column

Sample No	Sample Size n	Number of Defectives d	Fraction Defective $p = d/n$	Remarks
43	50	1	.02	
44		2	.04	
45		4	.08	
46		2	.04	
47		1	.02	
48		6	.12	
49		3	.06	
50	50	4	.08	Revised $\bar{p}$ = 155/2500 = .062
	1250	74	1.48	Limits_p = -,.164 n = 50
51	200	9	.045	Limits_p = .011,.113 n = 200
52	200	6	.030	
53		14	.070	
54		15	.075	
55		12	.060	
56		11	.055	
57		17	.085	
58		6	.030	
59		11	.055	
60		8	.040	
61		14	.070	
62		14	.070	
63		15	.075	
64		14	.070	
65		13	.065	
66		14	.070	
67		5	.025	
68		14	.070	
69		6	.030	
70		6	.030	
71		14	.070	
72		10	.050	
73		16	.080	
74		12	.060	
75	200	15	.075	Revised $\bar{p}$ = 446/7500 = .0595
		291	1.455	
76	50	9	.18	Limits_p = .0595 $\pm$ .1004 = —,.160
77		5	.10	Evidence of assignable cause
78		6	.12	on sample 76
79		7	.14	p' = .12 on samples 76-100
80		7	.14	
81		7	.14	
82		8	.16	
83		4	.08	
84		9	.18	
85	50	4	.08	

Table 6.3 (continued)

Sample No.	Sample Size n	Number of Defectives d	Fraction Defective p = d/n	Remarks
1	50	3	.06	p' = .06
2	50	4	.08	
3		5	.10	
4		1	.02	
5		5	.10	
6		5	.10	
7		3	.06	
8		3	.06	
9		2	.04	
10		1	.02	
11		3	.06	
12		7	.14	
13		1	.02	
14		3	.06	
15		4	.08	
16		2	.04	
17		5	.10	
18		4	.08	
19		2	.04	
20		1	.02	
21		4	.08	
22		6	.12	
23		1	.02	
24		4	.08	
25	50	2	.04	End of preliminary data
	1250	81	1.62	$\bar{p}$ = .0648 Limits_p = —, .169
26	50	1	.02	
27		2	.04	
28		3	.06	
29		3	.06	
30		8	.16	
31		5	.10	
32		2	.04	
33		2	.04	
34		5	.10	
35		6	.12	
36		3	.06	
37		3	.06	
38		2	.04	
39		3	.06	
40		2	.04	
41		1	.02	
42	50	0	.00	

Table 6.3 (continued)

Sample No.	Sample Size n	Number of Defectives d	Fraction Defective p = d/n	Remarks
86	50	8	.16	
87		9	.18	
88		7	.14	
89		7	.14	
90		5	.10	
91		6	.12	
92		2	.04	
93		6	.12	
94		7	.14	
95		4	.08	
96		6	.12	
97		13	.26	
98		7	.14	
99		3	.06	For samples 76-100 is
100	50	4	.08	$\bar{p}$ = 160/1250 = .128
		160	3.20	
101	50	0	.00	Continue limits set from
102		2	.04	samples 1 to 75, that is,
103		1	.02	-,.160 for n = 50
104		1	.02	p′ = .03 for samples 101-125
105		4	.08	
106		1	.02	
107		1	.02	
108		2	.04	
109		1	.02	
110		1	.02	
111		2	.04	
112		2	.04	
113		0	.00	
114		2	.04	
115		3	.06	Run of 10 in succession on
116		1	.02	or below center line.
117		2	.04	
118		1	.02	
119		3	.06	
120		2	.04	
121		0	.00	
122		1	.02	
123		4	.08	
124		4	.08	
125	50	2	.04	
		43	.86	

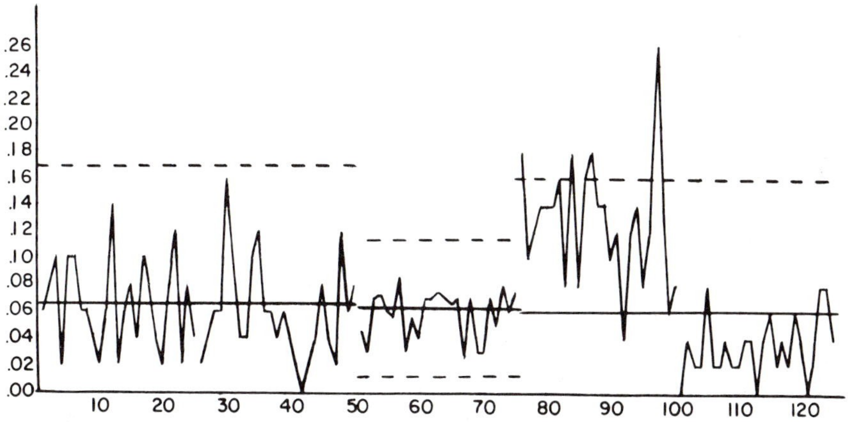

FIG. 6.3. Samples 1-25 preliminary data: p'=.06, n=50, $\bar{p}$=.0648. Limits extended. Sample 26-50: p'=.06, n=50, revised $\bar{p}$ = .0620. Samples 51-75, p'=.06, n=200, limits from $\bar{p}$=.062. Revised $\bar{p}$=.0595. Limits extended for samples 76-100: p'=.12, n=50. Out of control. Limits from $\bar{p}$=.0595 extended through samples 101-125: p'=.03, n=50. Data from Table 6.3.

significant by the person in charge.* If so regarded it would indicate a desirable assignable cause, unless it proves to be lax inspection.

Suppose that at this point it is decided to use n = 200. We revise the limits on the basis of the first 50 samples, obtaining

$$\bar{p} = 155/2500 = .0620$$

$$\text{Limits}_p = .0620 \pm 3\sqrt{.0620(.9380)/200} = .0620 \pm .0512 = .0108,.1132.$$

We see from the limits that a single point can now be outside the control band on <u>either</u> side which was not so before. But none were outside, thereby signaling the continuation of perfect control. The student should note how

*For a binomial distribution with p' = .06, n = 50 we may find P(3 or less in 50) = .6473, that is, considerably above .5, thus enhancing the probability of such a run as compared to that for a .5 probability.

much less variation there was in plottings for n = 200 than for n = 50. This of course follows from (2.34).

We now again revise the center line and limits and revert to n = 50.

$$\bar{p} = 446/7500 = .0595$$

$$\text{Limits}_p = .0595 \pm 3\sqrt{.0595(.9405)/50} = .0595 \pm .1004 = -, .160.$$

Immediately we have evidence of an assignable cause on sample number 76. The defectives total nine, giving p = .18. Action should be taken at once. However, we have continued the drawings from sample 76 to 100 with p' at .12. The next few samples after 76 are back in control, but 82 shows a p value on the limit and other such warnings follow.

We now assume that the assignable cause has been found after sample 100, and appropriate action taken. The data 76-100 are eliminated as not applicable, and the old limits from samples 1 to 75 are extended for sample 101 and onward. The points seem to be running low. At 115 there have been ten consecutive points on or below the center line, which indicates an assignable cause of low fractions defective. Actually samples 101-125 were made with p' = .03.

This experiment illustrates how a p chart might run, and shows how evidence occurs when p' is changed from .06 up or down, with samples of 50.

6.2.4. Varying Sample Sizes for p and np Charts. Quite often in practice, the sample size for data on defectives varies. Commonly when this is the case, it is because the sample represents results from sorting 100% the entire production of a shift or a day, or possibly one hour. Since the production is not exactly constant, n varies. How shall we handle this problem, and which chart, p or np, shall we use?

Let us consider the data in Table 6.4. These results are for September following those given in Figures 3.1 and 3.2 for results of applying the B test to an electrical assembly in August. The fraction defective, p = d/n, for each day is listed, and plotted in Figure 6.4.

Table 6.4. B-test Failures of Electrical Equipment for September. Sample Size Variable. Fractions Defective Plotted in Figure 6.4. The Data Follow those of Table 3.1.

Date	Sample Size	Defectives	Fraction Defective
1	1600	41	.0256
2	1390	34	.0245
3	1510	34	.0225
4	1640	17	.0104
6	1780	34	.0191
7	1670	7	.0042
8	1750	15	.0086
9	2000	39	.0195
10	1640	30	.0183
11	1710	14	.0082
13	1780	15	.0084
14	1510	21	.0139
15	1780	16	.0090
16	1600	57	.0356
17	1670	31	.0186
18	1540	34	.0221
20	1670	30	.0180
21	1450	16	.0110
	29,690	485	

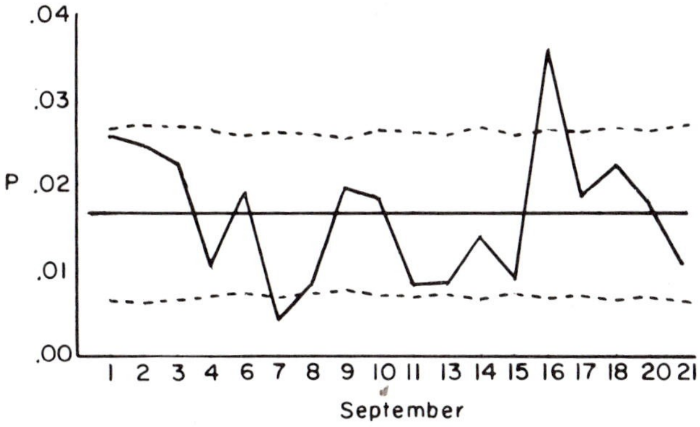

FIG. 6.4. B-test failures for electrical equipment for September. Data from Table 6.4. Variable sample size. $\bar{p}$ = .0163.

Using (6.12), (6.14) and (6.15), we find

$$ \text{Ȼ}_p = \bar{p} = 485/29{,}690 = .0163 $$

$$ \text{Limits}_p = \bar{p} \pm 3\sqrt{\bar{p}\bar{q}/n_i} \qquad i = 1,2,\ldots,18 $$

$$ = .0163 \pm 3\sqrt{.0163(.9837)/n_i} = .0163 \pm .380/\sqrt{n_i}\,. $$

The last form is quite convenient and permits rapid repeated calculation of the $\pm$ values to find appropriate limits. In Figure 6.4 we have plotted the individual control limits for each day's p. Control is much better than for August with a much better $\bar{p}$. But there is still evidence of an undesirable assignable cause on the 16th, and of a good one on the 7th, unless the latter proves to be lax testing. A desirable feature of this p chart is that, by (6.14), the center line is always at $\bar{p} = .0163$, regardless of n. Thus runs on one side of the center line may be checked for, and also the higher the point the worse the observed quality.

On the other hand, suppose that we make an np chart by plotting the d = np values. Now by (6.16) the center line is

$$ \text{Ȼ}_{np} = n_i\bar{p} = .0163\, n_i \qquad i = 1,2,\ldots,18 $$

and it will vary with each sample size. The limits are

$$ \text{Limits}_{np} = .0163 n_i \pm .380 \cdot \sqrt{n_i}\,. $$

We note that the width of the control band for a p chart becomes narrower as n_i increases, but for an np chart it becomes wider as n increases (varying as $\sqrt{n_i}$). But the chief source of change lies in the center line which varies as n_i itself. No longer can we easily check for runs, and also a point being below another does not necessarily mean a better p. For this reason we almost always use a p chart if the n's vary much at all.

Now let us reconsider Figure 6.4. The control limits are quite uniform. How might we simplify our calculations? Can we make do with "average" limits? The answer would seem to be yes. Now

$$ \bar{n} = \Sigma_{i=1}^{k} n_i/k \tag{6.18} $$

which gives us here 29,690/18 = 1650. This yields limits of

$$.0163 \pm .380/\sqrt{1650} = .0163 \pm .0094 = .0069, .0257.$$

This one set of average limits may be used for all the p points. The decisions are practically all obvious. But for points close to an average limit, either just inside or just outside we may wish to use the actual n. For example for Sept. 1 the point is inside, but close. Using the exact n = 1600 still leaves the point inside. One might wish to check also the exact limits for Sept. 8, 11 and 13. As shown in the figure all three are in control and would be for the average lower limit too.

Now on the other hand an np chart would have a large amount of variation among the limits, much of which is due to the center line varying around, as well as the width of the band. Using $\bar{n}$ for all would not be feasible here.

About how much variation from $\bar{n}$ can we tolerate? In our particular data here the smallest n = 1390 which is 16% below $\bar{n}$, while the largest n = 2000 is 21% above $\bar{n}$. It would seem that we might go even further.

A rule of thumb is that if our n's do not differ over 30% from $\bar{n}$ then for a p chart we may use $\bar{n}$ for all in this grouping of n's, checking any points very close to the limit. (For an np chart, the author suggests not over 5% variation of n from $\bar{n}$.)

It is perhaps worth emphasizing that when the n's vary, we do not find $\bar{p}$ by $\Sigma p/k$, as this would give equal weight to all sample p's whether n is large or small. Instead we use (6.12). But note that (6.12) may also be written

$$\bar{p} = \frac{\Sigma_{i=1}^{k} n_i p_i}{\Sigma_{i=1}^{k} n_i} \tag{6.19}$$

which clearly shows $\bar{p}$ to be a weighted average of the p's.

6.2.5. Not All Per cent Defective Data is Analyzable by a p Chart. Sometimes one meets per cent defective data which should not or cannot be analyzed by a p chart. Three types involve violations of the conditions for a binomial distribution to apply.

1. If we have extremely large sample sizes, such as the entire output of a plant, then the limits may be ridiculously narrow. For example, if the total output of glass jar caps per week is 4,000,000 and $\bar{p} = .020$ non-functional defectives, then the limits would be

$$.020 \pm 3\sqrt{.02(.98)/4{,}000{,}000} = .020 \pm .0002 = .0198,\ .0202$$

It is virtually impossible for a single weekly p to lie within such limits. The reason is that p' is not constant among such diversity of product. Such data can be analyzed by charts for individuals x and moving range, see Section 7.1.

2. When a number of parts are made at once, such as small metal castings, stampings, plastic pieces, rubber molded parts, etc., then the incidence of defectives is likely not to be independent. If one is bad, those near it are likely to be also, because of defective material or conditions. Sometimes such proportions defective can be analyzed to advantage by using $\bar{x}$ and R charts. That is, treat each sample proportion as a single x.

3. A moderately common case is that of per cent spoilage of bulk product. For example we may have 5000 gallons of paint of which 40 gallons are removed from the final product. The proportion defective is 40/5000 = .008. Can we treat such results as p values? The answer is yes, <u>but only if</u> we have a series of 5000 one gallon containers, each one of which is declared good or defective. But such is not likely to be the case. The paint may well be in a large vat and the undesirable product removed. The chief trouble is that there is no logical sample size. If gallons, n = 5000; if pints, n = 40,000, and if cubic feet, n = 668. What limits should one use? Thus the binomial distribution is inapplicable. Again one can call such a fraction defective an x and use $\bar{x}$ and R charts to analyze.

6.2.6. Further Comments on Charts for Defectives. As mentioned before, people from top management on down through foremen to the inspector and machine operator seem to understand percentages better than decimal fractions defective. So it is well to make the calculations in decimal fractions and plot them as such, but then to use the scale in per cents.

In calculations, especially those for p charts, it is easy to make a decimal point error! Hence great care should be exercised in calculating limits, or you are likely to have limits too wide by $\sqrt{10}$ or too narrow by $1/\sqrt{10}$. Either should "look" queer. One can also forget the multiplier 3 in front of the square root!

In choosing a sample size there are many practical considerations. One is that n should be small enough so that production and testing conditions can be presumed relatively uniform, in order that the sample can be called a "rational subgroup." Otherwise the chart loses much power to tell when production conditions changed. On the other hand the sample size needs to be large enough so that we will average at least .5 defective for $n\bar{p}$. If $\bar{p}$ is very small this will require a large n. The author knows of one case where $n\bar{p}$ was only 1 or 2 with n = 10,000, and the np chart was not helping much, because too many things could happen while 10,000 were made. But after developing a way to obtain a measurement, x; $\bar{x}$ and R charts for n = 4, enabled them to find the causes and eliminate them, giving much further improvement.

Usually the fraction defective or per cent defective chart is used, especially if the sample size varies appreciably. However, if n is constant, then use of an np = d chart enables one to plot the whole numbers d instead of bothering to figure p = d/n and plotting decimals. If the n is messy like 312 or 39 this is a saving. Such charts are especially useful in conjunction with acceptance sampling by attributes, where a record is kept of results on the first sample from each lot.

If it is desired to compare the current month's performance against a standard p' (or the old $\bar{p}$) we can do the following in addition to a control chart for the current month. Suppose there k samples of n each. Then we may treat $\bar{p}$ for the month as though it were a p for a big sample of nk pieces.

$$\text{Limits}_{\bar{p}} = p' \pm 3\sqrt{p'q'/kn} \qquad (6.19)$$

If $\bar{p}$ is outside either limit we have clear evidence of either an improvement or deterioration. Note, however, that this test is rather meaningless unless there was reasonable control around $\bar{p}$. (This is really a significance test on $\bar{p}$.)

Finally, we fairly often have quite low $\bar{p}$'s or p''s, in which case the $\bar{q}$ or q' term is close to 1 and its square root even closer to 1. Hence we may omit such a term from the calculations and have for example

$$\text{Limits}_{np} \doteq n\bar{p} \pm 3\sqrt{n\bar{p}} \qquad \bar{p} \text{ small} \qquad (6.20)$$

6.3. Control Charts for Defects. The second main type of chart for attributes is that for the <u>number of defects</u> on a sample or unit of product. In the charts for defects, the "c chart," we are concerned with a count of the defects found on, say, a complicated sub-assembly. The average number of defects of all types found in final inspection may be, say, 12 in such sub-assemblies. For them it can easily be true that almost never does one come along having no defects at all. Thus if we were to try to plot a fraction defective chart, p would always be 1.00 = n/n, since each sub-assembly contains at least one defect and is therefore a "defective." Such a string of 1.00 p values would tell us nothing! Instead when we <u>count</u> the number of defects c, we have whole numbers centering at 12, but running both above and below, and measuring quality performance in a meaningful way.

6.3.1. Control Charts for Defects, Standard Given. We take up first the case where we have a given standard number of defects per "area of opportunity" for defects.

$$c' = \text{standard number of defects.} \qquad (6.21)$$

Such a standard value is commonly the result of experience after a period of reasonably good control. Or it may be some requirement or goal, but such values must be reasonable, or we shall lose most probability interpretations of points out.

The Poisson distribution is the basic model used for a c chart, that is, given c', the probability for each c value to occur is given by (2.35). In order for the Poisson distribution to apply we must have the conditions given in Section 2.5.4., namely that (a) the fields of opportunity for defects to occur must be equal, (b) the defects must occur independently of each other, and (c) the maximum possible number of defects, (max c), must be far more than c'. Assuming these conditions satisfied, then we are given c' and we use (2.36) and (2.37) yielding

$$₵_c = c' \qquad (6.22)$$

$$\text{Limits}_c = c' \pm 3\sigma_c = c' \pm 3\sqrt{c'}. \qquad (6.23)$$

We may now plot these three lines, or two if $c' - 3\sqrt{c'} < 0$, and are prepared to take action on the very first sample, and of course subsequent ones. A point outside the limits is an indication of the c value being incompatible with c'. This indicates an assignable cause of the true average number of defects being either above or below the standard c', according to which limit is transgressed.

Note that unless c' is realistic, it would be entirely possible for the process to be homogeneous, that is, in control relative to some other average number of defects, and yet to show many points outside limits set from the standard c'. For example, suppose $c' = 2$, then the upper control limit is at $2 + 3\sqrt{2} = 6.2$. But if the actual process average is stable at 6 defects per unit, then many points will be above UCL_c. On the other hand all the c values might lie within the limits $c' \pm 3\sqrt{c'} = 6 \pm 3\sqrt{6} = -, 13.4$, and therefore be in control.

We could illustrate a c chart with given c' by drawing chips or beads from a bowl, marked according some Poisson population. Then draw from a different Poisson population, etc. Such populations are easily made up from tables of Poisson distributions giving $P(c)$, $c = 0,1,\ldots$ About 500 chips should be used so that at least a few rare c's are in the collection or bowl, which could give a point out of control.

6.3.2. Control Charts for Defects, Analysis of Past Data. We now come to the more common case of analysis of past data. Here we do not yet have a standard c'. We are, however, testing the hypothesis that there is an unknown but constant c', and under this assumption, the best estimate of c' from k sample c's is $\bar{c}$:

$$\text{Est. of } c' = \bar{c} = \frac{\Sigma_{i=1}^{k} c_i}{k}. \tag{6.24}$$

Then we use $\bar{c}$ and have the provisional or approximate lines from substituting $\bar{c}$ for c' in (6.22) and (6.23).

$$\text{℄}_c = \bar{c} \tag{6.25}$$

$$\text{Limits}_c = \bar{c} \pm 3\sqrt{\bar{c}}. \tag{6.26}$$

These are drawn through the points constituting the preliminary run of, say, 25 sample c values. Appropriate action is taken, $\bar{c}$ possibly revised, and it and limits found, against which we can plot each new c value as it is found.

As an example, let us take the data of Table 6.5. These are for counts of the total number of defects found per day on electrical switches. Several different types of defects were inspected for, c being the total number of all types of defects found by inspection of all of the 3000 switches produced daily. Hence when c = 90 defects were found, this could make 90 of the switches defectives, or, because some of the switches might contain more than one of the defects, there might be less than 90 defective switches. Here either a c chart, or an np or p chart, could be used. The quality control man used a c chart.

Figure 6.5 shows the data plotted. For analysis we have from (6.25), (6.26):

$$\text{¢}_c = \bar{c} = 1198/21 = 57.0$$

$$\text{Limits}_c = \bar{c} \pm 3\sqrt{\bar{c}} = 57.0 \pm 3\sqrt{57.0} = 34.4,\ 79.6.$$

Considerable lack of control is indicated by the control limit lines, both potentially good and bad assignable causes.

Careful records were kept on how many times each particular defect occurred and action taken to improve the status of the most prevalent or costly ones. It is possible to run a separate c chart on each of the more common types of defects. It is interesting that December of the previous year

Table 6.5. Total Defects of All Kinds on Day's Productions of 3000 Electrical Switches. December, 21 Working Days.

Day Number	1	2	3	4	5	6	7	8	9	10	11	
Total Defects c	30	56	47	86	44	23	16	64	80	54	73	
Day Number	12	13	14	15	16	17	18	19	20	21		
Total Defects c	65	76	69	53	58	30	91	90	36	57		Total 1198

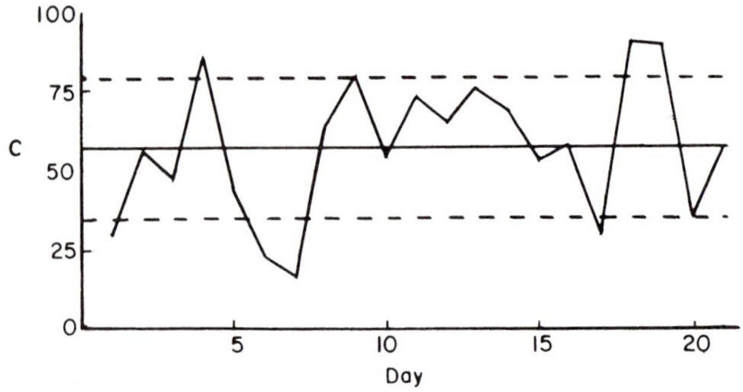

FIG. 6.5. Total defects of all kinds on day's production of 3000 electrical switches. December, 21 days. Data from Table 6.5.

had $\bar{c}$ of 406, or about seven times as many defects per day. But there is still work to be done as indicated by the points out of control.

It is perhaps worth noting that the control limits (6.26) for a c chart are similar to the approximate limits (6.20) for an np chart. This is because, when $\bar{q}$ is near 1, $\bar{p}$ is therefore small and we are reaching the conditions under which the Poisson distribution is an excellent approximation to the binomial.

6.3.3 Some Examples of c Chart Data. Many examples of possible applications of c charts occur in industry, not all of which are directly in quality control. Some are: errors or defects of any assembly or completed product, such as, automobiles, trucks, farm implements, radios, airplanes, electronic panels or bicycles; pinholes in a test area of painted surface; visual defects on enameled ware; surface defects in steel plate or tinplate; slivers on lengths of copper wire; breakdowns of insulation in 10,000' lengths of covered wire at (excess) test voltage; typographical errors per page of print; bacterial colonies on an agar plate; accidents or hospital calls per week; lost-time accidents per year; absenteeism; defects on yard goods, shirts, etc.; bubbles or hard spots in a dozen glass jars; counts of instances of any of 20 kinds of defects in jar caps in say 10,000; gas holes or other defects in castings; foreign matter in an assembly. In using a c chart, however, one should keep in mind the requirements justifying the Poisson distribution, given in Section 2.5.4. The next section describes how to avoid trouble in some cases with unequal "areas of opportunity" for defects.

6.3.4. Chart for Defects per Unit. We next discuss a type of chart which is exactly equivalent to the c chart, but is in a more convenient form sometimes, especially if the "area of opportunity" or sample size varies. Moreover by means of this approach the overall picture can be given without too much detail. Hence such a chart is often useful to management.

Let us illustrate by an example. A plant making trucks began to analyze the number of errors or defects, in the final inspection at the end of the assembly line. The average number of defects per truck from September 16 through November 1 was 1.95. Since daily production was 95 trucks and the period includes 35 working days, the total production was about 3325, hence there were about 6484 errors altogether, so that the average number per truck is 6484/3325 = 1.95.

Let us now introduce some notations.

n = number of units of product per sample (6.27)

c = total number of defects in sample of n units (6.28)

$u = c/n$ = average number of defects per unit over n units (6.29)

$\bar{u} = (\Sigma_{i=1}^{k} u_i)/k$ = average of u's for k samples (6.30)

u' = population or process average defects per unit. (6.31)

In the example being discussed, there would be 35 u values in the 35 days. September 16 might have had 190 errors in 95 trucks, or u = 190/95 = 2.00. The average of the 35 such u values was 1.95. After this past data, the plant began to plot the u points daily. The data are given in Table 6.6.

Table 6.6. Average Number of Errors Per Truck, Assembly Line 1, Station 2. Average Daily Production: Nov. 4 to 20, 95 Trucks; Nov. 21 to Jan. 10, 130 Trucks

Date	Average no. errors u	Date	Average no. errors u	Date	Average no. errors u
Nov. 4	1.20	Nov. 25	2.25	Dec. 17	1.85
5	1.50	26	2.50	18	1.82
6	1.54	27	2.05	19	2.07
7	2.70	29	1.46	20	2.32
8	1.95	Dec. 2	1.54	23	1.23
11	2.40	3	1.42	24	2.91
12	3.44	4	1.57	26	1.77
13	2.83	5	1.40	27	1.61
14	1.76	6	1.51	30	1.25
15	2.00	9	1.08	Jan. 3	1.15
18	2.09	10	1.27	6	1.37
19	1.89	11	1.18	7	1.79
20	1.80	12	1.39	8	1.68
21	1.25	13	1.42	9	1.78
22	1.58	16	2.08	10	1.84

Thus on November 4, $c = 114$, $u = 114/95 = 1.20$, etc. Figure 6.6 shows all 45 u values plotted.

Now what shall we do for center line and limits? There are as usual the two cases.

Standard u' given:

$$\text{¢}_u = u' \tag{6.32}$$

$$\text{Limits}_u = u' \pm 3\sigma_u.$$

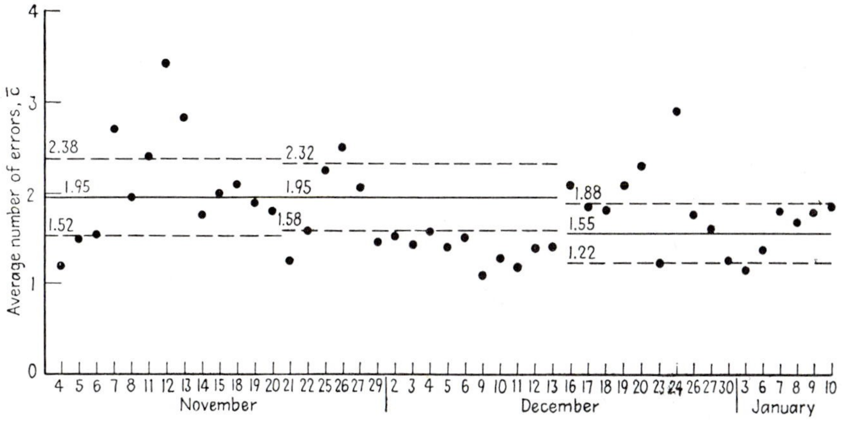

FIG. 6.6. Control chart for average number of errors per truck. Average daily production at station: 95 Nov. 4 to 20; 130 Nov. 21 to Jan. 10. Data in Table 6.6. Reproduced with permission from I. W. Burr, "Engineering Statistics and Quality Control", McGraw-Hill, New York, 1953, p. 240.

But $\sigma_u = \sigma_{c/n} = \sigma_c/n$. But $u' = c'/n$ where c' is the population average <u>total number</u> of defects on the samples of n units. Thus $c' = nu'$. Hence $\sigma_c = \sqrt{c'} = \sqrt{nu'}$, so that $\sigma_u = \sqrt{nu'}/n = \sqrt{u'/n}$. Therefore

$$\text{Limits}_u = u' \pm 3\sqrt{u'/n}. \tag{6.33}$$

For the case analysis of past data, we would merely substitute $\bar{u}$ for u' having

$$\not{C}_u = \bar{u} \tag{6.34}$$

$$\text{Limits}_u = \bar{u} \pm 3\sqrt{\bar{u}/n}. \tag{6.35}$$

Hence for the data at hand we use

$$\not{C}_u = 1.95$$

$$\text{Limits}_u = 1.95 \pm 3\sqrt{1.95/\ 95} = 1.95 \pm .43 = 1.52,\ 2.38.$$

These limits are shown on Figure 6.6 through November 20. Control is quite erratic. This could easily be in some degree due to non-independence of errors. (If an error occurs on a truck the next few may well have the same error.) On November 21, production was increased to 130, so that limits were revised to

$$1.95 \pm 3\sqrt{1.95/130} = 1.58,\ 2.32.$$

Because of the obvious improvement in late November and early December, new control limits were calculated for production after December 16, based upon data of November 21 to December 13, as follows:

$$\bar{u} = 24.87/16 = 1.55$$

$$\text{Limits}_u = 1.55 \pm 3\sqrt{1.55/130} = 1.22,\ 1.88.$$

Considerable trouble was encountered for a while in meeting the new control limits. Particularly note the high point for Christmas Eve. In January the quality performance has settled fairly well into the lower control band.

The average number of errors per truck was driven steadily downward by the psychological effect of the charts and by following through on the various kinds of trouble occurring. A sizable saving in man-hours, delays, supervision and material was achieved.

It is interesting to compare the last limits for u = average defects per truck with limits for defects on an individual truck. Such limits are $1.55 \pm 3\sqrt{1.55} = -, 5.3$. Thus it takes 6 errors for a single truck to be out of control, but only an average of 1.88 errors per truck for the day to be out of control.

There is a variation on the foregoing which may seem different to the reader but which uses in reality just the same principle. It is that case in which we do not have n separate discrete units in a sample, but instead some bulk product by volume or area. To use the standard c chart, the only requirement would be that the area or volume be always the same, so that the field of opportunity for defects is constant. When this is not the case, then we must make some allowance for the changing conditions.

Suppose, for example, that in our original problem we are testing the insulation on 5,000-, 10,000-, and 12,000-ft lengths. We first choose a standard length, say 10,000 ft. Then the other cases are reduced to this basis. The number of defects on any 5,000-ft length is doubled to give a number of defects comparable with that for 10,000 ft, while any number of defects observed on a 12,000-ft length would be divided by 1.2 for the same reason. Then we plot the resulting data, which are thus defects per unit (1 unit = 10,000 ft).

Now to get the center line, we want the total average defects per unit, where each such figure is given the proper weight. Suppose that there are 18 breakdowns of insulation in ten 5,000-ft lengths; then this is 18 defects in 50,000 ft, which is the equivalent of five 10,000-ft lengths. In five 12,000-ft lengths suppose that there are 15 breakdowns. Since we have 60,000 ft in the five, this is equivalent to 15 defects in six 10,000-ft lengths. Finally suppose that there were 43 defects in twelve of the standard lengths. Then altogether we have 18 + 15 + 43, or 76 defects in 5 + 6 + 12, or 23 units of wire. Dividing gives the weighted average

$$\bar{u} = 3.30$$

To systematize the foregoing we may set up the accompanying tabulation.

No. pieces (1)	Total defects (2)	Individual length, ft (3)	Length 10,000 ft (4)	Number Units (1) x (4) (5)
10	18	5,000	.5	5
5	15	12,000	1.2	6
12	43	10,000	1.0	12
	76			23

Now for the control limits we are planning to plot defects per unit u, that is, defects on the piece divided by the item in column (4) (or multiplied by its reciprocal), and hence all points will have the same center line. The width of the control band, however, will vary. The item in column (4) corresponds to n in (6.35), and hence we have

$$\text{5000 ft:} \quad 3.30 \pm 3\sqrt{\frac{3.30}{.5}} = \text{——},\ 11.0$$

$$\text{12,000 ft:} \quad 3.30 \pm 3\sqrt{\frac{3.30}{1.2}} = \text{——},\ 8.3$$

$$\text{10,000 ft:} \quad 3.30 \pm 3\sqrt{\frac{3.30}{1.0}} = \text{——},\ 8.7$$

In this way each point has its own control limits. Of course, as with the p chart, if the variation in area of opportunity for defects is not great no adjustment need be made.

6.4. General Comments on Control Charts for Attributes. Both general types of control charts for attributes involve inspection or testing for defects. It is therefore vital that each defect be clearly defined. For example a very small blemish may not be a defect at all, but above some size it becomes a defect. In fact sometimes there may be three classes as to size (a) not a defect, (b) minor defect (c) major defect. The same can be true of a dimension. For visual defects a collection of limit samples at the dividing line can be a great aid to inspection. Unless each defect is unequivocally defined and agreed upon, inspection is quite ineffective and charts meaningless. The author recalls an actual case in which 60 caps were passed around to many key people in a plant and each one asked which caps had functional defects. The numbers of caps so designated by the people ran all the way from 3 to 60! A similar case was observed on castings. If key people cannot get together on defects, what hope is there for inspection?

6.4. General Comments on Control Charts for Attributes

Some of the steps to be taken in attribute charting are:

1. Decision on what is the purpose of the study. Is it worthwhile? Is there enough at stake?
2. Where in the production process should the samples be drawn for inspection? Usually it is desirable to be as close as possible to the place where trouble is presumed to arise.
   Often it has been found that the first rough operation (for example a grinding) is the key to the trouble.
3. What characteristics would it be desirable to inspect for, and which can be so inspected? What characteristics should be included in a class of defects or defectives to go on one chart?
4. Choice can often be made between charting defectives or fraction defective vs. charting defects. Which should be used? Or would $\bar{x}$,R charts have a better chance of getting at the trouble or making improvement?
5. Choice between p or np charts, c or u charts.
6. Selection of samples should be done so as to be a rational subgroup, that is, all conditions relatively constant (men, material, machine, measuring). This may require rather small samples.
7. Forms can be devised so that a record is made of which defects occurred. A code can be used. Then the more prevalent or costly ones can be followed up on. This form may include a log of the production conditions and personnel involved, or this may be kept separately. Potential assignable causes should be kept track of.
8. Past data may be available. But conditions surrounding such data should be known, or the data can easily be meaningless.

9. As soon as reliable data are available preliminary analysis by control lines should be made and appropriate action taken.

10. A list of all the potential assignable causes which may affect a process could well be made, contributed to by everyone involved.

The probability of a point out on a p or np chart for given p', or on a c chart for given c' can vary quite a bit. This is because the plotted data only go by jumps (whole numbers for d or c), and also because the binomial or Poisson models give unsymmetrical distributions commonly. Therefore the α risk of a point outside the control band varies quite a little, although it is always small. See [2]. Sometimes an executive or foreman wants to be more sure of catching an assignable cause when one comes in, than is done by using three-sigma limits. Then, at the expense of an increased α risk of a false signal, one can use two-sigma limits. This does require careful explanation.

6.5. A Method of Rating Product Quality. There is a useful method for describing quality performance, which was developed in the Bell Telephone Laboratories [3], [4]. It is especially useful for complicated assemblies. If one uses a c chart for all the defects in such assemblies, then each defect will be given a weight of one. The system begun in the Bell System was to give demerit points for each defect noted. They used four classes:

A - Very Serious (Demerit value 100 points)

B - Serious (Demerit value, 50 points)

C - Moderately Serious (Demerit value 10 points)

D - Not Serious (Demerit value 1 point)

For a quite complete description of what defects to include in each such class, see [4], which also covers the general subject thoroughly.

We may define the "demerits" for a unit of product by

$$D = w_A c_A + w_B c_B + w_C c_C + w_D c_D \tag{6.36}$$

where the w's are the demerit values 100, 50, 10, and 1, and the c's are the numbers of defects noted in each class. As such we then have a demerit rating for each sample, which takes account of the relative seriousness of each defect.

Probably the most common application is in a monthly or weekly rating of all product inspected over that time period. For the month, say, then we have

$$\sum_{i=1}^{n} D_i = w_A \Sigma c_A + w_B \Sigma c_B + w_C \Sigma c_C + w_D \Sigma c_D \qquad (6.37)$$

where the Σ's are over 1 to n also, and give the total defects found for the month in each class.

There may be standard values $c'_A, \ldots, c'_D$, of defects per unit, or we may use $\bar{c}_A, \ldots, \bar{c}_D$ for the previous month or several months. In the latter case we may eliminate some data before figuring the averages, if defects were produced from known and eliminated assignable causes.

Since the number of units inspected may well vary each month, the gross demerit sum in (6.37) is not as meaningful as the demerits per unit, which is found by

$$\bar{U} = \frac{\Sigma_{i=1}^{n} D_i}{n} = \frac{w_A \Sigma c_A + w_B \Sigma c_B + w_C \Sigma c_C + w_D \Sigma c_D}{n} \qquad (6.38)$$

$$= w_A \bar{c}_A + w_B \bar{c}_B + w_C \bar{c}_C + w_D \bar{c}_D$$

Now c_A is a count of defects, distributed by the Poisson law, if the conditions are fulfilled, and hence

$$E(c_A) = c'_A, \quad \sigma_{c_A} = \sqrt{c'_A}, \text{ then } \sigma_{\bar{c}_A} = \sqrt{c'_A/n}.$$

Then

$$E(U) = w_A c'_A + w_B c'_B + w_C c'_C + w_D c'_D \qquad (6.39)$$

$$\sigma_U = \sqrt{\frac{w_A^2 c'_A + w_B^2 c'_B + w_C^2 c'_C + w_D^2 c'_D}{n}} \qquad (6.40)$$

since $\sigma^2_{aX+bY} = a^2\sigma_X^2 + b^2\sigma_Y^2$.

Hence

$$₵_U = E(U)$$

$$\text{Limits}_U = E(U) \pm 3\sigma_U \qquad (6.41)$$

If n is small and we use weights for U of 100, 50, 10 and 1, the chart will be quite meaningless, because a single 100 or 50 point defect will throw it out of control, otherwise it will be in control. But for sizable n's the chart can be quite useful for giving the broad picture.

6.6. Summary. We hereby present the basic formulas for control charts for attributes for convenience and to show the essential unity.

I. Charts involving defectives

A. p chart, fraction defective

1. Analysis of past data, use $\bar{p}$

a. $₵_p = \bar{p} = \Sigma d/\Sigma n$

b. $\text{Limits}_p = \bar{p} \pm 3\sqrt{\bar{p}\bar{q}/n}$

2. Standard given, p'

a. $₵_p = p'$

b. $\text{Limits}_p = p' \pm 3\sqrt{p'q'/n}$

B. np or d chart, defectives. Change of scale from p chart

1. Analysis of past data, use $\bar{p} = \Sigma d/\Sigma n$

a. $₵_{np} = n\bar{p}$

b. $\text{Limits}_{np} = n\bar{p} \pm 3\sqrt{n\bar{p}\bar{q}}$

2. Standard given, p'

a. $₵_{np} = np'$

b. $\text{Limits}_{np} = np' \pm 3\sqrt{np'q'}$

II. Charts involving defects

A. c chart, defects on a unit or sample

1. Analysis of past data, k samples, use $\bar{c}$

a. $\text{¢}_c = \bar{c} = \Sigma c_i/k$

b. $\text{Limits}_c = \bar{c} \pm 3\sqrt{\bar{c}}$

2. Standard given, c'

a. $\text{¢}_c = c'$

b. $\text{Limits}_c = c' \pm 3\sqrt{c'}$

B. u chart, average defects per unit = c/n

1. Analysis of past data, k samples, of n each, use

$\bar{u} = \Sigma_{i=1}^{k} u_i/k = \Sigma_{i=1}^{k} c_i/nk$

a. $\text{¢}_u = \bar{u}$

b. $\text{Limits}_u = \bar{u} \pm 3\sqrt{\bar{u}/n}$

2. Standard given u' = population average defects per unit

a. $\text{¢}_u = u'$

b. $\text{Limits}_u = u' \pm 3\sqrt{u'/n}$

C. Quality rating, demerits, see (6.39), (6.40).

III. A relation between np or d charts and c charts: If p' or $\bar{p}$ is small then $1-p' = q'$ or $\bar{q}$ is nearly 1, and may be neglected. Then $n\bar{p} \pm 3\sqrt{n\bar{p}}$ is like $\bar{c} \pm 3\sqrt{\bar{c}}$ and $np' \pm 3\sqrt{np'}$ is like $c' \pm 3\sqrt{c'}$.

IV. Unequal sample sizes

A. Defectives charts

1. d chart: may well use fixed control lines if n's do not differ over about 5 per cent from $\bar{n}$.

2. p chart: may well use fixed control lines if n's do not differ over about 30 per cent from $\bar{n}$.

3. Otherwise use different control lines for each grouping of reasonably homogeneous n's.

B. Defects charts - adjustment of calculation and limits relative to some chosen sample size. Could use criteria similar to A1 for c chart and A2 for u chart.

Again we emphasize the need for sound workable definitions of defects, independence of defects, and the use of rational subgroups, not letting conditions vary too much during the production and inspection of a sample. Also a thorough log of conditions is helpful in tracking down causes.

PROBLEMS

6.1. Data from an optical company are given below, each article, 39 per box, inspected being either good or broken. Make an appropriate control chart for the data and comment. Why do you think the sample size of n = 39 was used? Using a Poisson approximation and $n\bar{p}$ as c', estimate the probability of a box having <u>less</u> than 36 good articles.

Table P6.1

Sample	No. Broken	Sample	No. Broken	Sample	No. Broken
1	2	16	1	31	6
2	1	17	0	32	1
3	1	18	1	33	2
4	0	19	0	34	2
5	2	20	1	35	2
6	3	21	0	36	1
7	2	22	2	37	1
8	1	23	0	38	0
9	1	24	1	39	2
10	2	25	1	40	2
11	1	26	0	41	2
12	1	27	1	42	5
13	1	28	0	43	2
14	2	29	7		
15	1	30	10		

6.2. The following data give results on beer cans inspected for visual defects, in lots of n = 312 cans. Use an appropriate control chart analysis and comment. What would you suggest for a $\bar{p}$ for the next samples?

Table P6.2

Lot	n	Defectives	Lot	n	Defectives	Lot	n	Defectives
1	312	6	11	312	7	21	312	9
2		7	12		7	22		5
3		5	13		6	23		11
4		7	14		6	24		15
5		5	15		6	25	312	10
6		5	16		6			196
7		4	17		23			
8		5	18		10			
9		12	19		8			
10	312	6	20	312	5			

6.3. The following data are on 5-gal pails for paints and varnish. Draw an appropriate control chart. What do you conclude? Is $\bar{p}$ $\Sigma p/11$? Why not?

Table P6.3

Date	No. inspected n	No. defective d	Fraction defective p
June 3	1,024	30	.0293
5	2,056	48	.0233
6	16,835	86	.0051
7	16,069	108	.0067
12	7,365	54	.0073
13	8,738	29	.0033
14	1,220	10	.0082
17	16,242	112	.0069
18	15,145	76	.0050
24	5,845	8	.0014
25	13,582	154	.0113
	104,121	715	

6.4. Lots of bushing screws from a single supplier were inspected 100 per cent with the following results. Draw an appropriate control chart for the data. What can you conclude? Under what circumstances would good control be desirable here if it existed? If the sum of the p's (last column) is divided by 8, do we get $\bar{p}$ = 2.3 per cent? Why not?

Table P6.4

Lot No.	Amount in lot	No. rejected	Percent defective
1	6,000	183	3.0
2	6,000	131	2.2
3	10,000	184	1.8
4	1,783	23	1.3
5	3,089	171	5.5
6	5,774	65	1.1
7	3,000	46	1.5
8	2,724	85	3.1
	38,370	888	2.3

6.5. The following data are from [1], Chapter 4. The data are on an oxygen pressure switch, which has a bourdon tube. Depending on the pressure, the tube opens or closes the electrical circuit through contact points. The notes in the last column refer to p charts on the separate types of defects.

Treating all the data as one set of preliminary data, draw the control chart. How many different sets of control limits are needed? What conclusions can you draw? Is the simple average of the p's (their sum divided by 28) equal to $\bar{p}$? Why not?

Table P6.5

Date produced	No. items inspected	No. defective	Fraction defective	Other information
Feb. 1	2,500	300	.120	Points off and defective receptacle
2	2,314	214	.093	
3	2,435	418	.172	Tube defective, close bars
4	1,217	117	.096	
5	1,348	148	.110	Defective receptacle
6	1,229	142	.115	Miscellaneous trouble
8	1,384	134	.097	
9	1,278	125	.098	

Table P6.5 (continued)

10	1,197	100	.084	
11	1,128	67	.059	
12	1,135	85	.075	
13	1,391	105	.076	
15	1,212	85	.070	
16	1,222	82	.067	
17	1,167	80	.069	
18	1,292	84	.065	
19	1,082	71	.066	
20	1,185	103	.087	Points off running high
22	1,260	110	.087	Points off running high
23	1,286	70	.054	
24	1,204	54	.045	
25	1,142	72	.063	
26	1,216	66	.054	
27	1,204	92	.076	
Mar. 1	1,265	118	.093	
2	844	124	.147	Defective receptacle, out of control
3	759	89	.117	Defective receptacle, out of control
4	745	75	.101	Defective receptacle, out of control
Total	36,641	3,330		

6.6. Data on final inspection of carburetors on a line, preceding those in Table 6.1 by 12 days. (In the check-sheet form for inspectors, there were listed 32 defects, each with three degrees of severity. Other defects could also occur.) Two columns are shown for defectives d and two for per cent defective. They represent (a) observed results and (b) results with deletions for a discovered and removed assignable cause. Plot an appropriate chart (a) for data as observed, and (b) after deletion of defectives from the assignable causes. Comment.

Table P6.6

CARBURETOR NO. EV1 LINE #7 DATE RUN Jan. 15

Sample Size	No. of Defectives		% Defective	
	*	#	*	#
100	10	9	10%	9%
100	4	3	4	3
100	6	5	6	5
100	12	6	12	6
100	6	3	6	3
100	8	4	8	4

Table P6.6 (continued)

100	10	3	10	3
100	12	6	12	6
100	8	2	8	2
100	7	3	7	3
100	3	0	3	0
100	4	1	4	1
100	3	1	3	1
100	4	1	4	1
100	4	0	4	0
100	10	3	10	3
100	7	2	7	2
100	5	1	5	1
100	3	0	3	0
100	6	1	6	1
100	8	1	8	1
100	10	6	10	6
100	4	3	4	3
52	2	2	3.8%	3.8%
2352	156	66		

*Before Removal of Assignable Causes

#After Removal of Assignable Causes

6.7. An interesting example from [1], Chapter 4 is the following. With what factors might an assignable cause be associated? Make an appropriate chart and offer a solution.

"Notes on Unprimed Case Visual Data"
(Small Arms Ammunition)

"These data represent visual defects overlooked by regular inspectors in a 100% inspection operation at an intermediate stage in the manufacture of cartridge cases. The purpose of this inspection operation was to get rid of as much of the imperfect material as possible at an early stage, rather than spend time and effort working it up only to throw it out at the final inspection. The data cover the output of a single station, working

3 shifts a day, over a two-week period. There was a shift change at the end of the first week.

"In the inspection operation the cases were fed mechanically to a track, along which they were moved by two screws. Mirrors were placed parallel to the track, affording the operator a view of the inside and outside, including the head. The chief source of illumination was a Dazor unit with two tubes; during the day there was additional light from large windows nearby. It was the operator's duty to pick off defective cases and drop them down a chute. Unless so picked off, the cases rolled to the end of the track and fell off into tote boxes; these were dumped on the conveyor from time to time by a spot checker.

"At the time these data were obtained much trouble was being encountered due to scale, a condition caused by faulty brass. Sometimes the scale was conspicuous, with metal actually peeling away from the case; often, however, it appeared only as a brownish speck the size of a pinhead, or sometimes as a rippled or watery appearance of the brass.

"Twice during each shift, at times varying from day to day, a sample of 200 cases was taken from the accepted material off each twin screw machine. This sample was visualled by a specially trained inspector, who recorded the number of defectives found. (If the initial inspection had been perfect there would of course have been no defectives in the sample).

"The accompanying data are a typical set, and have the additional interest that from them and the information contained in these remarks it is possible to deduce an assignable cause of variation which was later removed."

Table P6.7

CONTROL CHARTS FOR ATTRIBUTES

Operation: Unprimed Case Visual — TSM No. 6
Data by SDB — Sample size 200

Week Ending May 15

	Mon	Tue	Wed	Thu	Fri	Sat
A shift						
Foreman: Pat						
1st sample	3	3	5	3	5	2
2nd sample	2	0	0	3	1	3
B shift						
Foreman: Rex						
1st sample	2	3	5	5	3	9
2nd sample	3	6	5	10	6	3
C shift						
Foreman: George						
1st sample	4	5	3	5	2	5
2nd sample	1	3	2	1	5	3

Week Ending May 22
(shift change over week end)

	Mon	Tue	Wed	Thu	Fri	Sat
A shift						
Foreman: Rex						
1st sample	2	2	4	2	4	0
2nd sample	5	5	2	3	2	6
B shift						
Foreman: George						
1st sample	4	8	5	4	2	11
2nd sample	6	7	5	6	3	2
C shift						
Foreman: Pat						
1st sample	2	5	5	2	3	6
2nd sample	4	4	3	3	3	1

Total number of samples (2 weeks), 72
Total defectives found, 270
A shift 11 P.M. to 7 A.M.
B shift 7 A.M. to 3 P.M.
C shift 3 P.M. to 11 P.M.

6.8. The data below cover categories of defects found on inspection of large aircraft. Using all data in a category of defects (as assigned) as preliminary data run a c chart and comment. What $\bar{c}$ might you suggest for future production and why?

Table P6.8

	50	51	52	53	54	55	56	57	58	59	60	61	62	63
Alignment	7	6	6	7	4	7	8	12	9	9	8	5	5	9
Adjust	2				1	1	1	4	3	2	2	4	2	
Workmanship	111	145	123	131	147	158	127	157	263	272	118	272	310	145
Tighten	169	155	146	139	134	160	138	161	69	62	135	41	79	149
Cello-Seals	16	23	19	11	16	25	20	25	7	25	17	10	11	10
Mislocations	4		6			3	2	3	1	4	3	6	2	1
Incomplete	299	313	317	360	300	396	378	304	263	266	342	235	289	388
Clearance	108	91	115	116	93	111	108	86	97	125	101	123	110	127
Damage-Mutilations	16	35	23	25	23	37	36	48	18	28	47	25	28	28
Foreign Matter	13	19	18	18	17	16	22	8	9	8	17	8	10	14
Missing Rivets	8	16	14	19	11	15	8	11	21	12	23	16	9	25
Defective Rivets	4	3		3	1	1		1			5	2	2	
Oxygen Leaks									1					
Hyd. System Leaks	18	12	32	5	17	20	15	19	13	30	18	15	15	20
Incorrect Air Pressure														
Replace	2	11	12	13	11	19	11	10	18	21	18	23	20	7

Table P6.8 (continued)

Plug Holes	2	3	5	6	5	6	5	4	6	9	9	6	5	9
Drawing Changes Not Incorporated	1		3	1										
TOTAL	780	832	839	854	780	975	879	853	798	873	863	791	897	932
	<u>64</u>	<u>65</u>	<u>66</u>	<u>67</u>	<u>68</u>	<u>69</u>	<u>70</u>	<u>71</u>	<u>72</u>	<u>75</u>	<u>77</u>	<u>78</u>	<u>79</u>	<u>80</u>
Alignment	8	15	6	4	13	7	8	15	6	6	10	7	13	4
Adjust	1	1		3	2		11	5	1	5	2	6	4	
Workmanship	176	295	114	247	125	239	161	234	141	268	261	95	96	116
Tighten	150	127	98	96	171	108	184	189	192	141	134	161	120	81
Cello Seals	11	14	15	4	6	10	8	9	7	9	14	5	10	1
Mislocation	3	2	1	4	4	6	2	4		1	3		6	2
Incomplete	369	265	312	245	363	243	249	267	264	219	160	296	304	237
Clearance	106	109	75	88	107	76	96	119	103	94	71	79	80	64
Damage-Mutilation	44	17	23	20	30	18	20	10	14	22	18	25	22	23
Foreign Matter	18	10	17	14	10	14	15	20	19	10	12	11	4	25
Missing Rivets	15	9	9	14	11	9	10	22	7	28	9	12	8	14
Defective Rivets		1	1	4	1				2		1			2

Oxygen Leaks							1					1		2
Hyd. System Leaks	29	24	18	7	14	34	5	30	12	17	9	17	16	16
Incorrect Air Pressure														
Replace	14	23	10	25	13	32	27	40	12	24	13	14	18	5
Plug Holes	9	8	7	6	11	12	12	8	8	7	5	5	4	7
Drawing Changes Not Incorporated	2			1		11							1	
TOTAL	955	920	706	782	881	819	809	972	788	851	722	734	706	599

6.9. The following data may be appropriately analyzed by a c chart, if the conditions are met. Review the conditions needed and proceed to chart if you think they might be met. Comment on control.

From each of thirty bales of dry bleached hemlock pulp, a 9" x 12" sheet was taken. The following table gives the number of dirt specks greater than 0.1 sq. m.m. found on these sheets.

Table P6.9

Sample	No. of Dirt Specks	Sample	No. of Dirt Specks	Sample	No. of Dirt Specks
1	12	11	32	21	15
2	24	12	24	22	32
3	21	13	14	23	10
4	30	14	33	24	28
5	16	15	22	25	13
6	22	16	14	26	18
7	12	17	16	27	12
8	15	18	21	28	21
9	31	19	17	29	16
10	18	20	13	30	8
					580

6.10. The following data are from Burns [5]. The data are on outlet leaks in auto radiators found in the first assembling together of the two radiator parts. Make a u chart, using as few different classes of control lines as possible for the varying sample sizes. What conclusions can you draw?

Table P6.10

Date	No. tested n	No. outlet leaks c	Average number per radiator u
June 3	39	14	.36
4	45	4	.09
5	46	5	.11
6	48	13	.27
7	40	6	.15
10	58	2	.03
11	50	4	.08
12	50	11	.22

Table P6.10 (continued)

13	50	8	.16
14	50	10	.20
17	32	3	.09
18	50	11	.22
19	33	1	.03
20	50	3	.06
24	50	6	.12
25	50	8	.16
26	50	5	.10
27	50	2	.04
	841	116	

6.11. The data shown below are on three consecutive months for daily samples of 30 farm implements. All types of defects included in final inspection of implements. For example May 1, 84 defects on 30 implements gave 2.80 for u per implement. Plot the May data and comment. What $\bar{u}$ might you carry on?

Table P6.11

May	u	June	u	July	u
1	2.80	2	3.13	1	2.85
2	4.57	3	3.55	2	3.10
5	4.13	4	4.30	3	1.66
6	2.96	5	3.52	7	2.91
7	2.65	6	3.63	8	**2.65
8	4.54	9	3.38	9	3.17
9	3.35	10	3.83	10	3.01
12	3.95	11	3.95	11	4.15
13	4.42	12	2.82	14	3.58
14	4.34	13	Down	15	2.49
15	4.02	16	3.78	16	2.92
16	4.39	17	3.35	17	3.80
19	4.02	18	2.76	18	No work
20	4.03	19	2.86	21	No work
21	2.86	20	3.59	22	3.03
22	3.37	23	2.96	23	2.51
23	3.08	24	3.17	24	2.82
26	3.19	25	3.75	25	2.44
27	†4.96	26	3.41		
28	2.88	27	*4.56		
29	2.94	30	3.43		
Total	77.45	Total	69.73	Total	47.09

† Section IV out of control

* Production reduced, jobs changed.

** Altered production

6.12. Similarly to 6.11, but using the June data as preliminary data.

6.13. Similarly to 6.11, but using the July data as preliminary data.

6.14. If for a c chart $c' = 2.5$, is the probability of a point above the center line equal to that for a point below? Does this have any bearing on interpreting runs?

6.15. If on a p chart $\bar{p}$ is indicated as .150 and the upper control limit as .180, what is the approximate sample size?

6.16. How might you show an engineer that a sample having $p = .06$, provides no reliable evidence that $p' > .03$?

References

1. American Society for Quality Control, Standard A1-1971 (ANSI Standard Z1.5-1971). Milwaukee, WI.
2. L. A. Aroian, What makes a control chart tick? *Indust. Quality Control* 10 (No. 6), 38-43 (1954).
3. H. F. Dodge, A method of rating manufactured product. *Bell System Tech. J.* 7, 350-368 (1928). Bell Telephone Laboratories Reprint B315.
4. H. F. Dodge and M. N. Torrey, A check inspection and demerit rating plan. *Indust. Quality Control* 13 (No. 1), 5-12 (1956).
5. W. L. Burns, Quality control proves itself in assembly. *Indust. Quality Control* 4 (No. 4), 12-17 (1948).

CHAPTER 7

MISCELLANEOUS TOPICS IN CONTROL CHARTS

7.1. Control Charts for Individual Measurements, Moving Ranges. In over-all data for management's use, a single measurement x may represent one whole shift, day, week or month, or even a whole year. Consequently we can have but few such values available, and moreover we do not want to have to wait until a sample of four or five such results has accumulated before plotting an $\bar{x}$ point. Often a decision is wanted "yesterday" on the new result. In research too, each measurement obtained may be so expensive or time-consuming that we cannot have many of them in our study. Hence the need for the x chart, which we now describe.

The foregoing must not be allowed to give the reader the idea that with 20 measurements and an x chart we can get as good results as we can with 100 measurements, in 20 samples of n=5, using $\bar{x}$ and R charts. The x chart is merely one way to try to get at least "something" from quite meager data.

Naturally, limits for x's require knowledge of what mean μ and σ_x to use. If such standards are available we may easily proceed to set up control charts, but if not available, as in the case analysis of past data then we must use estimates. This is usually done through the use of "moving ranges." For the <u>first</u> moving range we use $R_1 = |x_1-x_2|$, that is, an ordinary range for a sample of n=2. Then for the <u>second</u> moving range, we again use x_2 but now in association with x_3, thusly $R_2 = |x_2-x_3|$.

Proceeding in this manner k observations x, provide k-1 R's.* Then in the analysis of past data we will use $\bar{x}$ for k x's and $\bar{R}$ for k-1 R's.

Let us consider an example. Table 7.1 gives 36 monthly per cents of rejection of steel. Assuming as we do provisionally that all data come from the same (normal) population, we then estimate μ by

$$\bar{x} = \Sigma_{i=1}^{k} x_i/k. \tag{7.1}$$

Then we also compute $\bar{R}$ for as rational subgroups as are available, n=2:

$$\bar{R} = \Sigma_{i=1}^{k-1} R_i/(k-1), \tag{7.2}$$

and use this to estimate σ_x by

$$\hat{\sigma}_x = \bar{R}/d_2 = \bar{R}/1.128, \tag{7.3}$$

using d_2 for n=2.

We now come to the question of what limits to use. Recalling that we are proceeding on a minimum of data, it seems to the author desirable to use "two-sigma" limits for the limits for x's. This is because, with points occurring rather far apart in point of time, we can afford to increase the risk of looking for an assignable cause when none is present. This will then give us less risk of failing to look for an assignable cause when one is present. Hence the author suggests use of the following from (7.1) and (7.2)

$$\text{₵}_x = \bar{x} \tag{7.4}$$

$$\text{Limits}_x = \bar{x} \pm 2\cdot(\bar{R}/1.128). \tag{7.5}$$

---

*Some writers obtain k R's by using $R_k = |x_k - x_1|$ but x_1, x_k is not in any sense a rational subgroup. It can cause serious inflation of $\bar{R}$ if there is something of a trend up or down in the x's.

TABLE 7.1. Monthly Per Cents of Rejection of a Grade of Steel. "Steel" Defects Found by Customer after Cold Drawing of Bars.

Month	Per Cent rejection x	Range R	Month	Per Cent rejection x	Range R
January	4.4		July	4.1	
		1.5			.2
February	2.9		August	4.3	
		1.3			1.0
March	4.2		September	5.3	
		1.2			.0
April	5.4		October	5.3	
		1.2			4.7
May	6.6		November	.6	
		3.0			.2
June	3.6		December	.8	
		2.6			.1
July	1.0		January	.9	
		.3			2.2
August	1.3		February	3.1	
		.1			1.5
September	1.2		March	1.6	
		.0			.4
October	1.2		April	2.0	
		.2			1.5
November	1.0		May	.5	
		1.8			2.0
December	2.8		June	2.5	
		.8			.8
January	2.0		July	3.3	
		1.0			1.1
February	3.0		August	2.2	
		1.2			2.4
March	1.8		September	4.6	
		.5			.5
April	2.3		October	4.1	
		1.0			.3
May	1.3		November	3.8	
		3.9			.4
June	5.2		December	3.4	
		1.1		103.6	42.0

Then we use, as usual, for the R chart:

$$\text{₵}_R = \bar{R}, \quad UCL_R = 3.267\ \bar{R}, \tag{7.6}$$

D_4 being 3.267 for n=2. (One could use two-sigma limits here too, if desired. How?)

Now from Table 7.1 we find

$$\text{₵}_x = \bar{x} = 103.6/36 = 2.88$$

$$\bar{R} = 42.0/35 = 1.20$$

$$\text{Limits}_x = 2.88 \pm 2(1.20/1.128) = .75,\ 5.01$$

$$UCL_R = 3.267(1.20) = 3.92$$

The charts are plotted in Figure 7.1. Considerable lack of control shows up on the x chart and even some on the moving range chart. There the highest point comes from the sharp drop from October to November.

Some disadvantages of the x chart relative to the $\bar{x}$ chart are the following: (1) The x chart is less efficient than the $\bar{x}$ chart, in the sense that it takes more individual measurements on the average to get an indication of the presence of an assignable cause, unless the shift in μ is extremely large, say, 3σ. (See Section 7.7.) (2) The x value does not summarize a set of conditions as well as an $\bar{x}$ does, it being an average of several x's. (3) If the distribution of x's is nonnormal, this will tend to distort the interpretation of the control limits, on the x chart. On the other hand by the central limit theorem the distribution of $\bar{x}$'s is likely to be quite normal even for small n's.

The x chart does have the advantages (1) of being up-to-date at the time the last x is plotted, (2) of having no arithmetic needed before plotting each x, and (3) of being very easy for shopmen to understand. Also comparison of an x chart with specifications, if any, is direct.

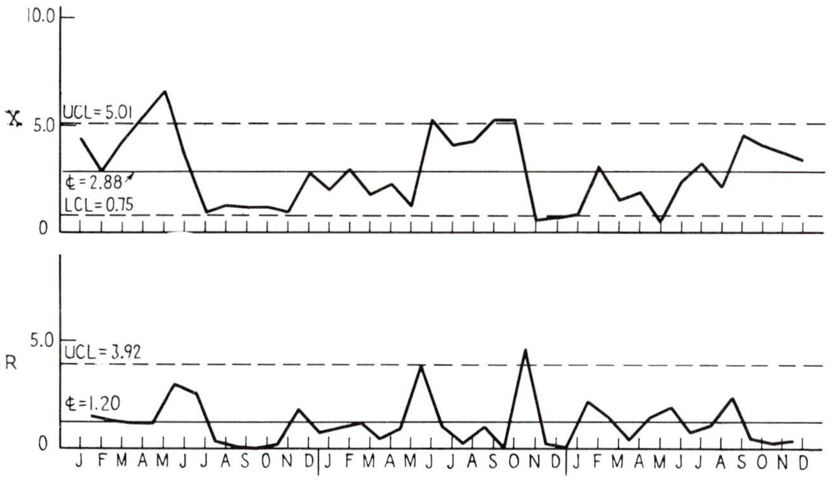

FIG. 7.1. Control charts for individuals x's and for moving ranges R. Monthly per cent of rejection of a grade of steel. Data of Table 7.1. Note that the point for the moving range is plotted between the two months involved. Reproduced with permission from I. W. Burr, "Engineering Statistics and Quality Control", McGraw-Hill, New York, 1953, p. 266.

7.2. Tool Wear and Other Gradual Process Changes. Let us consider the following adaptation of $\bar{x}$ control charts. When we take a sample of product, periodically from the most recent production, these x's may be expected to contain random variation only (unless there has been a very recent resetting of the process average μ). Hence ranges R from such samples and their average, $\bar{R}$, reflect chance causes only. When the R chart shows good control we may estimate the <u>short term</u> σ_x by $\bar{R}/d_2$. Then $\bar{x} \pm 3(\bar{R}/d_2) = \bar{x} \pm 3\hat{\sigma}_x$ are approximate limits to the x distribution for <u>the most recent production</u>. Now unless the process average μ is quite tightly controlled, $\bar{\bar{x}} \pm 3(\bar{R}/d_2)$ will not be the x limits for the <u>total</u> production over, say, a shift. There will be in fact some $\sigma_{(\text{long term})}$ which exceeds $\sigma_{(\text{short term})}$. The former of these will include shifts in the process mean as well as random variation. The whole set of x's for the shift probably will lie within $\bar{\bar{x}} \pm 3\sigma_{(\text{long term})}$.

It is the last set of limits which must meet specifications for x's. Now, again, if $6\sigma_{(\text{short term})}$ = specified tolerance = U-L, then the process mean must be tightly controlled and conventional $\bar{x}$ and R charts are in order. However, cases do occur where $U-L \geq 8\sigma_{(\text{short term})}$, and it would be possible to let the process mean μ vary around a bit. In fact if $U-L=8\sigma_{(\text{short term})}$, then as long as μ lies between $U-3\sigma_{(\text{short term})}$ and $L+3\sigma_{(\text{short term})}=U-5\sigma_{(\text{short ter}}$ we will still be having at least 99.9% of pieces between the limits. Thus there is a band $2\sigma_{(\text{short term})}$ wide within which μ may be permitted to vary , in this case. See Figure 7.2.

Let us in the following understand that σ or $\hat{\sigma} = \bar{R}/d_2$ refer to <u>short term</u> variation, as usual. Now when the above situation of $U-L \geq 8\sigma$ is present, how may we proceed, if we do not wish to maintain tight control on the $\bar{x}$ chart? Let us proceed to illustrate by an example.

Consider the data given in Table 7.2. Looking at the $\bar{x}$ column we can see quite a definite trend of $\bar{x}$'s with a sharp break downward after 12:30 July 27.

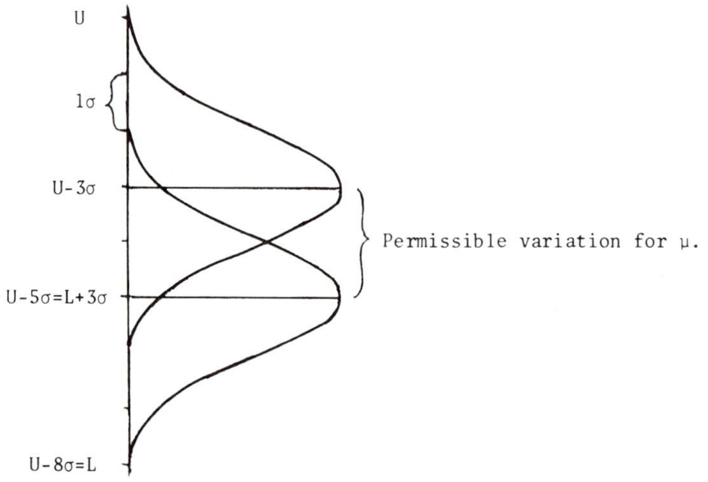

FIG. 7.2. Two normal curves drawn to show the permissible variation in process mean μ, when $\sigma_{(\text{short term})} = \sigma$ is (U-L)/8.

Table 7.2. Outside Diameters, of Spacers, Samples of Five Every 30 Min. From an Automatic Machine. Specifications: .1250 In., + .0000 In., - .0015 In. Readings in .0001 In. Below .1250 In.

Date	Time	x_1	x_2	x_3	x_4	x_5	$\bar{x}$	R
July 27	8:00	- 8	- 9	- 7	- 8	- 8	- 8.0	2
	8:30	- 8	- 9	- 7	- 7	- 6	- 7.4	3
	9:00	- 6	- 6	- 8	- 7	- 7	- 6.8	2
	9:30	- 4	- 6	- 5	- 6	- 6	- 5.4	2
	10:00	- 7	- 5	- 4	- 5	- 7	- 5.6	3
	11:00	- 5	- 6	- 2	- 3	- 4	- 4.0	4
	12:30	- 2	- 4	- 2	- 4	- 5	- 3.4	3
	1:15	- 10	- 11	- 12	- 11	- 10	-10.8	2
	2:00	- 10	- 12	- 13	- 12	- 11	-11.6	3
	2:45	- 11	- 11	- 10	- 11	- 9	-10.4	2
	3:15	- 8	- 10	- 9	- 10	- 11	- 9.6	3
	4:00	- 9	- 10	- 11	- 10	- 9	- 9.8	2
	4:30	- 10	- 11	- 7	- 11	- 8	- 9.4	4
July 30	9:00	- 7	- 8	- 8	- 5	- 9	- 7.4	4
	9:30	- 9	- 10	- 7	- 6	- 7	- 7.8	4
	10:00	- 8	- 6	- 8	- 7	- 8	- 7.4	2
	11:00	- 5	- 4	- 5	- 6	- 7	- 5.4	3
	12:15	- 7	- 8	- 6	- 7	- 7	- 7.0	2
	12:45	- 7	- 6	- 7	- 4	- 5	- 5.8	3
	1:15	- 3	- 3	- 4	- 5	- 4	- 3.8	2
	1:45	- 3	- 4	- 3	- 7	- 6	- 4.6	4
	2:15	- 6	- 3	- 4	- 2	- 6	- 4.2	4
	2:45	- 1	- 3	- 2	- 4	- 4	- 2.8	3
	3:30	- 9	- 6	- 5	- 4	- 7	- 6.2	5
	4:00	- 4	- 7	- 8	- 4	- 5	- 5.6	4
	4:30	- 5	- 7	- 6	- 7	- 6	- 6.2	2
July 31	8:00	- 9	- 10	- 7	- 11	- 10	- 9.4	4
							- 185.8	81

The quality control man, having been recently trained in a short course was quite surprised when he ran an ordinary $\bar{x}$ chart and an R chart. These had

$$\bar{\bar{x}} = -185.8/27 = -6.88 \quad \bar{R} = 81/27 = 3.00$$

$$\text{Limits}_{\bar{x}} = -6.88 \pm .577(3.00) = -8.61, -5.15, \quad UCL_R = 2.115(3.00) = 6.34.$$

He was gratified to find the R chart well in control, but the $\bar{x}$ chart had 13 points out of control. There must be assignable causes. At first he floundered around quite a bit, but inquiry and study revealed that it was due to tool wear. (As a tool, cutting an outside diameter, gradually wears, its point grows farther from the center line, giving slightly larger diameters.) Thus the assignable cause was known. Was it necessary to eliminate this assignable cause by very frequent resettings of the tool?

This last is the basic question. Let us see. Short term variation is estimated by

$$\hat{\sigma}_x = \bar{R}/d_2 = 3.00/2.326 = 1.29$$

But in the coded units U=0, L=-15, so the tolerance is U-L=15. Thus $(U-L)/\hat{\sigma}_x = 11.6$, so we have a lot of latitude for μ to vary within, in fact at least from $-15 + 3(1.29) = -11.13$ up to $0-3(1.29) = -3.87$. (As long as μ lies between, there will be a very small fraction defective.) Therefore quite a bit of tool wear was permissible in this example, because specifications were lenient.

7.2.1 Slanting Control Lines for $\bar{x}$'s. An adaptation for $\bar{x}$'s which is of much use in such a situation is the use of slanting control lines. We plot the $\bar{x}$'s and R's as in Fig. 7.3, analyzing the R's as usual. If they show good control we are justified in estimating σ_x. Next we look at each trend of $\bar{x}$ points and try to draw the best straight line through them.* This represents our guess as to the way the true μ runs with the $\bar{x}$'s varying around this trend line. To test as to whether the tool wear is proceeding along in control, we put in control limits at $3\hat{\sigma}_{\bar{x}} = A_2\bar{R}$ above and below our slanting trend line, measured vertically. If the $\bar{x}$ points

*Fitting such a line by the objective least-squares method is hardly justified.

lie between these control limits then the tool is "wearing in control." This is the case in Fig. 7.3. There, $3\hat{\sigma}_{\bar{x}} = A_2\bar{R} = .577(3.00) = 1.73$. (Compare this with $3\hat{\sigma}_x = 3\bar{R}/d_2 = 3.87$.)

Now at what a level for μ should we aim to start a run? This is

$$L + 3\hat{\sigma}_x = \text{minimum safe process average} \tag{7.7}$$

Then as $\bar{x}$ points accumulate we can draw in the trend line. At what level should we plan to reset the process mean back down again? This should be done just as soon as the trend line hits

$$U - 3\hat{\sigma}_x = \text{maximum safe process average} \tag{7.8}$$

In Figure 7.3, the trends seem to have been handled quite well, perhaps

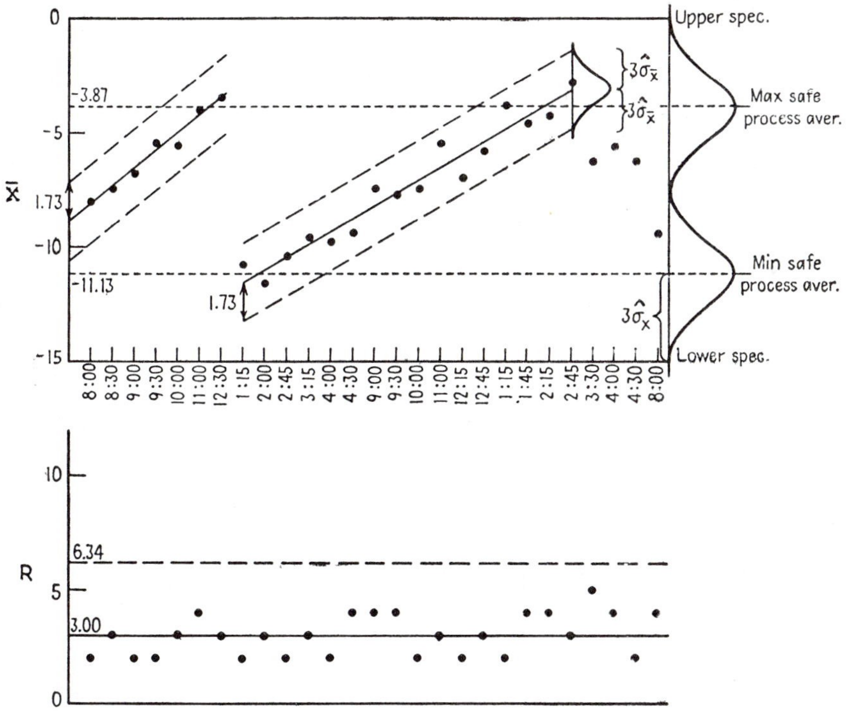

FIG. 7.3. Control charts for outside diameter of a spacer. Samples of five from an automatic. Measurements in .0001 in. from .1250 in. Specifications .1235 in. to .1250 in. Trend control lines for $\bar{x}$'s. Reproduced with permission from I. W. Burr, "Engineering Statistics and Quality Control", McGraw-Hill, New York, 1953, p. 269.

being let run a bit longer than to $3\hat{\sigma}_x$ from U. (But it must be remembered that keeping μ at $3\hat{\sigma}_x$ away from the specification limit is being really quite conservative. Perhaps one can run within $2.5\hat{\sigma}_x$ or even $2\hat{\sigma}_x$ of U occasionally.)

For comparison the author calculated s=2.64 for the 135 x's in the whole Table 7.2. Thus we have

$$\hat{\sigma}_{(\text{long term})} = 2.64, \quad \hat{\sigma}_{(\text{short term})} = 1.29$$

the latter being only a half of the former. We see that the tolerance of 15 is about $6\hat{\sigma}_{(\text{long term})}$.

Careful following up of $\bar{x}$ points outside of the slanting control limits can be fruitful in finding assignable causes in the process.

Besides resetting when the trend line hits the maximum safe process average, we should also reset if a single x goes out of specifications.

Trends can also be downward, as for example for an <u>inside</u> diameter. Other examples of trends are (1) die wear, (2) grinder or polisher wear, (3) human fatigue, (4) room temperature or humidity, whether or not air-conditioned, (5) viscosity in a chemical process, (6) moisture content of lumber or a plastic. The author recalls hearing one executive claim that such trend charts had doubled the tool and die life in his division!

7.2.2.* Modified Limits for $\bar{x}$'s. A practice associated with the approach just described is the use of "modified limits" for $\bar{x}$'s. The general idea is to place modified limits for $\bar{x}$'s on the chart instead of ordinary control limits, when the tolerance U-L is substantially greater than $6\sigma_{(\text{short term})}$, that is, when we can afford to let μ vary around a bit and still meet the tolerance limits for x's, L and U. Such usage gives up on trying to find assignable causes affecting μ.

Let us reconsider the example of Sections 7.2 and 7.2.1. When we are near the upper specification limit U, we have to be concerned with the "maximum safe process average"= - 3.87. This is as high as μ can be permitted to go. Now if μ= - 3.87, where will the $\bar{x}$'s lie? They will be between - 3.87 $\pm$ $A_2\bar{R}$ = - 3.87 $\pm$ 1.73 = - 5.60,- 2.14. Suppose now that an $\bar{x}$ point goes <u>above</u> -2.14. See Figure 7.4. Then surely μ is above -3.87. On the other hand, if an $\bar{x}$ point lies <u>below</u> -2.14 does this give assurance that μ is at or below -3.87? Not at all. In fact μ might be at -2.14 or even as high as -1 and still yield such an $\bar{x}$ point. Thus the -2.14 line, which we call "modified limit A", gives rather unsymmetrical protection against the two errors: stopping and resetting the process when we should not, and not stopping to reset when we should.

The author recommends that a better balance is obtained by "modified limit B" found by again coming down from U by $3\hat{\sigma}_X$ (to -3.87), then back up by <u>only</u> $2\hat{\sigma}_{\bar{X}}$ (from -3.87 to -2.72). We then have

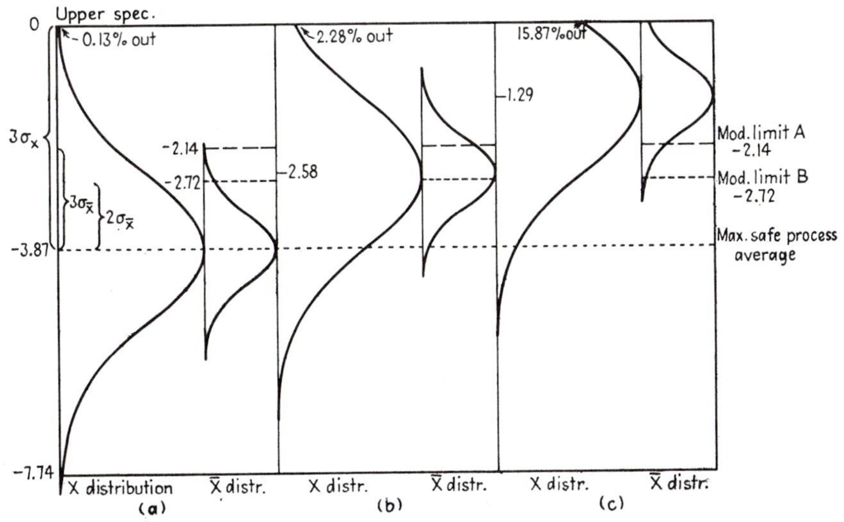

FIG. 7.4. Three process levels to illustrate the operating characteristics of two different modified limits. This is in relation to the data of Table 7.2. Reproduced with permission from I. W. Burr, "Engineering Statistics and Quality Control", McGraw-Hill, New York, 1953, p. 272.

$$\text{Modified limit A} = U - 3\bar{R}/d_2 + 3[\bar{R}/(d_2\sqrt{n})] \tag{7.9}$$

$$\text{Modified limit B} = U - 3\bar{R}/d_2 + 2[\bar{R}/(d_2\sqrt{n})] \tag{7.10}$$

Figure 7.4 shows in section (a) the x distribution if μ is at -3.87 and the $\bar{x}$ distribution centered there too. Almost no $\bar{x}$'s will be above limit A and only about 2 % above limit B, in each case giving a false warning that μ is too high. Now in section (b) both distributions center at -2.58 giving 2.28% of x's outside U. Under this condition much less than half of the $\bar{x}$'s will exceed limit A, providing a warning; but over half are above limit B. Finally in section (c) where nearly 16% of x's are beyond U, about 7% of the time $\bar{x}$ will <u>not</u> be above limit A under this alarming situation, while only .7% of the time will an $\bar{x}$ <u>not</u> be above limit B. Thus limit B provides much better protection than limit A, at little expense on the α risk. See Fig. 7.5.

Modified limit B is much like those used in England, whereas limit A is often used in the USA. Neither is safe unless the R chart shows control. Limit B can be useful in short runs for an immediate check of setup.

Reference [1] gives quite a complete discussion of this subject, along with operating characteristic curves for various lines.

7.3. Charts for $\bar{x}$ and s, Large Samples. If it is desired to run control charts for $\bar{x}$ and s for sample sizes larger than those given in Table VI, then we can treat c_4 as 1, which it is approaching. Also

$$\sigma_s \doteq \sigma_x/\sqrt{2n} \text{ or } c_5 = 1/\sqrt{2n} \tag{7.11}$$

Then using these in (5.9) and (5.16) gives

$$B_3 \doteq B_5 \doteq 1 - (3/\sqrt{2n}) \qquad B_4 \doteq B_6 \doteq 1 + (3/\sqrt{2n}\,) \tag{7.12}$$

Also by (5.11)

$$A_3 \doteq 3/\sqrt{n}. \tag{7.13}$$

Then we may use these results in the formulas in Section 4.10 for the charts.

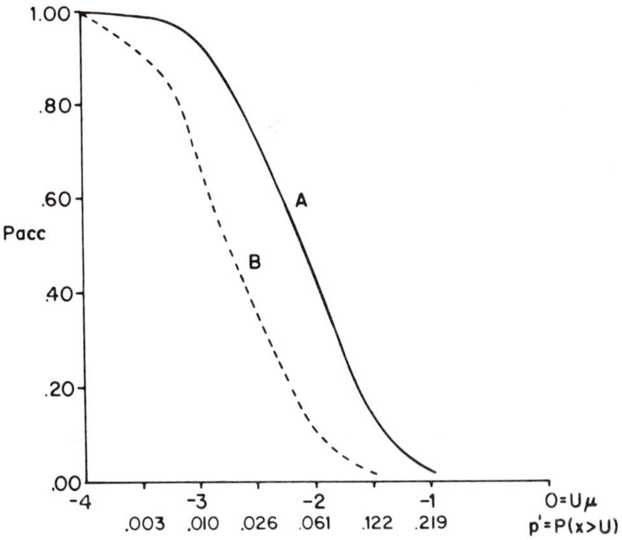

FIG. 7.5. Operating characteristic curves for Type A and Type B modified limits. Probability of acceptance versus process average μ and fraction defective p'.

As was true for p charts, we may let the n's vary from their average $\bar{n}$, by about 30% of $\bar{n}$. Then for any points close to a limit we can use the actual n for more accurate comparison.

7.4. Charts for $\bar{x}$ and R or s, with Varying Sample Sizes. Sometimes in laboratory or research data we wish to test for homogeneity of level and of variability by control charts. This is the equivalent of testing the hypothesis that all samples come from the same (normal) population with unknown μ and σ_x. In analyzing such data for control we seek the best estimates of μ and σ_x, assuming the hypothesis is true. For this we may proceed as shown in [2]. For μ we use a simple weighted average

$$\hat{\mu} = \bar{\bar{x}} = \Sigma_{i=1}^{k} n_i \bar{x}_i / \Sigma_{i=1}^{k} n_i = (\text{sum all x's})/(\text{sum n's}) \qquad (7.14)$$

For σ_x we use the weighted average

$$\hat{\sigma}_x = \frac{e_1R_1 + e_2R_2 + \ldots + e_kR_k}{f_1 + f_2 + \ldots + f_k} \tag{7.15}$$

where e_i, f_i are for R_i's sample size n_i, in Table 7.3. Or, if the sample variabilities are measured by s, we use

$$\hat{\sigma}_x = \frac{g_1s_1 + g_2s_2 + \ldots + g_ks_k}{h_1 + h_2 + \ldots + h_k} \tag{7.16}$$

Table 7.3. Factors for Best Linear Estimates of Population Standard Deviation σ for Sample R's and s's. [2]

Sample Size	For Ranges Equation (7.15)		For Standard Deviations Equation (7.16)	
n	e	f	g	h
2	1.55	1.75	2.20	1.75
3	2.14	3.63	4.13	3.66
4	2.66	5.48	6.09	5.61
5	3.12	7.25	8.07	7.59
6	3.52	8.93	10.06	9.57
7	3.90	10.53	12.05	11.56
8	4.24	12.06	14.04	13.55
9	4.55	13.52	16.04	15.55
10	4.84	14.91	18.03	17.54
11			20.03	19.54
12	With sample sizes		22.03	21.53
13	above 10 divide		24.03	23.53
14	into subgroups.		26.02	25.53
15			28.02	27.53
16			30.02	29.52
17			32.02	31.52
18			34.02	33.52
19			36.02	35.52
20			38.02	37.52

Then the next step is to treat our $\hat{\mu}$ and $\hat{\sigma}_x$ as though they were given standards μ and σ_x, and use appropriate formulas (4.5) to (4.9). In this latter work we can use an $\bar{n}$ if the n's are close together. For example use n=5 if $\bar{n}$ is near 5, and the n's are 4, 5 or 6.

Let us give an example from [2], see Table 7.4, for the x's, $\bar{x}$'s, R's for five sets of lamps.

$\bar{\bar{x}} = 158.82/16 = 9.93 = \hat{\mu}$

Then by (7.15)

$$\hat{\sigma}_x = \frac{3.52(.47) + 2.14(.48) +...+ 1.55(.92)}{8.93 + 3.63 +...+ 1.75} = \frac{4.9032}{19.69} = .249$$

Now the center line for all $\bar{x}$'s is $\bar{\bar{x}} = 9.93$, the limits vary as n varies using $\bar{\bar{x}} \pm A\hat{\sigma}_x$. Meanwhile both the center line and limits for R vary with n, using $d_2\hat{\sigma}_x$ and $D_2\hat{\sigma}_x$ respectively. (D_1=0, up through n=6.) We then have

		n=2	n=3	n=6
Center	$\bar{x}$:	9.93	9.93	9.93
Lines	R:	.281	.422	.631

Table 7.4. Samples of Lamps for Luminous Flux Measurements, in Lumens per Watt. From [2]. Various Companies.

Company	Measurements	n	$\bar{x}$	R	e	f
1	9.47, 9.00, 9.12, 9.27, 9.27, 9.25	6	9.23	.47	3.52	8.93
2	10.80, 11.28, 11.15	3	11.08	.48	2.14	3.63
3	10.37, 10.42, 10.28	3	10.36	.14	2.14	3.63
4	10.65, 10.33	2	10.49	.32	1.55	1.75
1*	9.54, 8.62	2	9.08	.92	1.55	1.75
	Total 158.82	16				

* Held in stock for one year.

Control	$\bar{x}$:	9.40,10.46	9.50, 10.36	9.63,10.23
Limits	R:	——, .918	——, 1.085	——, 1.264

Looking at Figure 7.6, we see all $\bar{x}$ points on a control limit or outside the limits indicating marked differences between companies. There is also one high range point indicating excessive variability for the lamps held for one year.

7.5. Different Types of Samples. The success of anyone who uses statistics in science or industry will depend in large measure upon how he collects his samples. This section describes some types of samples.

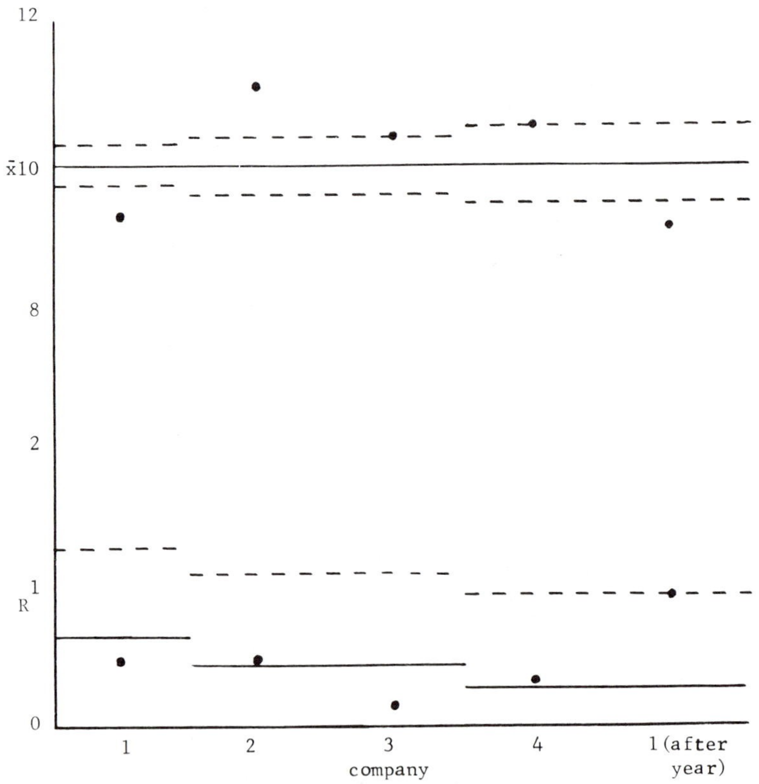

FIG. 7.6. Control charts for laboratory tests on luminous flux (lumens per watt) for samples of lamps from various companies. Data in Table 7.4.

7.5.1. Rational Sample. The ideal way to sample is to take a rational sample or subgroup. By this we mean that all of the pieces or measurements have been taken under the same conditions, so that what variation there is, within the sample, is all attributable to chance causes only. Thus the pieces or tests were made on the same machine (under one setting), by the same people, with the same methods, and at approximately the same time. Succeeding samples will be taken under differing sets of conditions. Whenever a point goes out of the control band, we try to find what condition was different for the out-of-control sample than for the other in-control samples. In this way one can hope to track down the assignable cause of the off performance. The success of the program depends in part upon careful records of conditions in the production process and measuring techniques.

An illustrative experiment often used in short courses in statistical quality control, is to use five bowls, which represent five machines, or five spindles, heads, stations or orifices for one machine. Suppose that specifications are -5 to +5. Then distributions of Table 4.1 are used in five bowls as follows: three of A(μ=0, σ=1.7), one of B(μ=2, σ=1.7) and one of C(μ=0, σ=3.5). A sample of n=5 is taken from each bowl at the first "time," a record being kept of which bowl is used for each sample, and $\bar{x}$'s and R's plotted, in such a way as to distinguish the sources of the data. Then at the next "time" five more samples, each of five are taken. The procedure continues until five or six sets of samples are taken, all being plotted on the same charts. Then $\bar{\bar{x}}$, $\bar{R}$ and control limits are found for all 25 or 30 samples taken together. If in control, the lines are extended and new samples drawn. But usually either distribution B or C or both will provide at least one point out of control. Then appropriate "action" is taken to "reset the process" or to "overhaul" the process. Data from the out of control bowl are eliminated, center lines recalculated, etc.

Such an experiment can be quite realistic and a stimulus to thinking, illustrating many of the principles of control chart operation.

7.5.2. Random Sample from Mixed Product. A very common type of sampling in industry is to take a sample of n at random from the total output of a machine or process. Such a sample is often taken from a conveyor belt or a tote pan of parts. These parts have been made from different heads, spindles, mold cavities, etc. But when such a sample is drawn we cannot tell from which head we have each part. We have simply sampled the overall output which includes not only chance variation but also any assignable-cause differences between heads.

Now if such samples show good control <u>and</u> the calculation of $\bar{\bar{x}} \pm 3\bar{R}/d_2$ shows the natural process limits to lie about on or within specification limits for x's, then we are on quite good ground and do not need to dig more deeply into the process. But if the natural process limits from the x's do not meet specifications, then we need to analyze more deeply into the process and to try to compare the output from the different heads or spindles, etc.

The author recalls being asked by a man from a can company what the probability is for a random sample of 12 cans, from the total output of a can-making machine with four heads, to contain at least one can from each of the four heads. He said many answers were arrived at by his associates, but all were different! The probability proves to be .87345; rather lower than was hoped for.

7.5.3. Stratified Sample. Another type of sample often used is the stratified sample. In it we include one part from each head or spindle on the machine. For example, on a five-spindle automatic machine, we can take the next five pieces to be produced. Then we are sure to have one part from each, but unless we are very careful, we cannot tell which part was made by spindle number one, which by number two, etc. Now such a sample does even better at estimating the overall process mean, than does the random sample from mixed product. But it has the same weakness of lumping into one sample, parts and therefore x's from potentially different populations (each spindle). Thus again the ranges, R, will include potential assignable causes as well as chance variation. Hence, again $\bar{R}$ will be inflated. Moreover such sampling does not permit comparison of spindles.

But now if we do make a record of which spindle produced each part, then we can plot rational samples for each spindle. Then using such rational samples we have from $\bar{R}$ a valid estimate of chance variation, and can moreover distinguish between spindles.

Let us illustrate some of these principles with experiments. In Table 7.5 are shown the results of drawing 30 stratified samples. The x_1 measurement is of a part from spindle 1, x_2 from spindle 2, etc. Then at, say, 8:15, each spindle contributes an x to the sample. Thus we have as rows in the table, 30 stratified samples. Figure 7.7(a) shows the $\bar{x}$ and R charts. For these samples we have

$\bar{\bar{x}} = -3.4/30 = -.11 \quad \bar{R} = 263/30 = 8.77 \quad UCL_R = 2.115(8.77) = 18.5$

$\text{Limits}_{\bar{x}} = -.11 \pm .577(8.77) = -.11 \pm 5.06 = -5.17, +4.95$

As seen in the figure, control is perfect. But in point of fact it is entirely too good. On the $\bar{x}$ chart there is no $\bar{x}$ closer than a third of the way from center line to limit, that is, all $\bar{x}$'s lie within $\pm 1\hat{\sigma}_{\bar{x}}$. This results from the ranges having been substantially inflated by including spindle-to-spindle differences as well as chance causes. Also the R chart exhibits rather too perfect control. This is because we always have an x from the highest average spindle and one from the lowest in the same sample thus forcing rather unnatural uniformity in the R's.

Next in Figure 7.7(b) we see the same data plotted in rational samples. These samples are shown in columns of Table 7.5 with the $\bar{x}$ and R given. Thus we find 6 $\bar{x}$'s and 6 R's in the x_1 column for spindle 1, etc. Now

$\bar{\bar{x}} = -.11 \quad \bar{R} = 119/30 = 3.97 \quad UCL_R = 2.115(3.97) = 8.40$

Table 7.5. Samples of n = 5. Five Populations of Table 4.1, Used for x_1 to x_5 Respectively. Stratified Samples Horizontally, and Rational Samples Vertically in Each Block. "Specifications" are ± 5.

Time	Date	x_1	x_2	x_3	x_4	x_5	$\bar{x}$	R
8:15	2/11/73	+ 2	- 5	+ 1	0	+ 2	0.0	7
17		0	- 1	+ 1	- 1	+ 6	+1.0	7
19		+ 1	- 5	0	- 4	+ 6	-0.4	11
21		+ 2	+ 1	+ 1	- 1	+ 5	+1.6	6
23		+ 2	- 6	- 2	+ 2	+ 4	0.0	10
	$\bar{x}$	+1.4	-3.2	+0.2	-0.8	+4.6		
	R	2	7	3	6	4		
9:20		+ 2	- 5	- 3	0	+ 4	-0.4	9
22		+ 1	- 5	- 2	- 3	+ 4	-1.0	9
24		+ 2	- 4	0	0	+ 3	+0.2	7
26		+ 1	- 5	+ 3	+ 1	+ 4	+0.8	9
28		0	- 4	- 1	- 1	+ 6	0.0	10
	$\bar{x}$	+1.2	-4.6	-0.6	-0.6	+4.2		
	R	2	1	6	4	3		
10:09		+ 2	- 5	+ 1	- 1	+ 5	+0.4	10
11		0	- 5	+ 1	- 2	+ 2	-0.8	7
13		+ 4	- 2	+ 2	- 2	+ 5	+1.4	7
15		+ 2	- 6	+ 1	- 3	+ 7	+0.2	13
17		+ 2	- 4	+ 2	- 5	+ 2	-0.6	7
	$\bar{x}$	+2.0	-4.4	+1.4	-2.6	+4.2		
	R	4	4	1	4	5		
10:57		+ 3	- 7	+ 1	- 2	+ 6	+0.2	13
59		+ 4	- 4	+ 2	0	+ 2	+0.8	8
11:01		+ 2	- 5	- 1	- 2	+ 6	0.0	11
03		+ 2	- 5	0	- 5	+ 2	-1.2	7
05		0	- 5	0	- 5	+ 4	-1.2	9
	$\bar{x}$	+2.2	-5.2	+0.4	-2.8	+4.0		
	R	4	3	3	5	4		
11:47		0	- 7	+ 2	- 2	+ 5	-0.4	12
49		+ 3	- 3	0	- 3	+ 1	-0.4	6
51		+ 2	- 4	0	+ 2	+ 6	+1.2	10
53		0	- 2	- 4	+ 1	+ 1	-0.8	5
55		+ 2	- 6	- 1	- 1	+ 3	-0.6	9
	$\bar{x}$	+1.4	-4.4	-0.6	-0.6	+3.2		
	R	3	5	6	5	5		
1:03	2/11/73	0	- 8	- 1	- 3	+ 6	-1.2	14
05		+ 2	- 4	+ 2	- 2	+ 1	-0.2	6
07		0	- 7	+ 2	- 4	+ 2	-1.4	9
09		+ 2	- 4	+ 3	- 4	0	-0.6	7
11		+ 1	- 4	- 1	0	+ 4	0.0	8
	$\bar{x}$	+1.0	-5.4	+1.0	-2.6	+2.6		263
	R	2	4	4	4	6		
	$\Sigma\,\bar{x}$	+9.2	-27.2	+1.8	-10.0	+22.8	-3.4	
	Σ R	17	24	23	28	27	119	

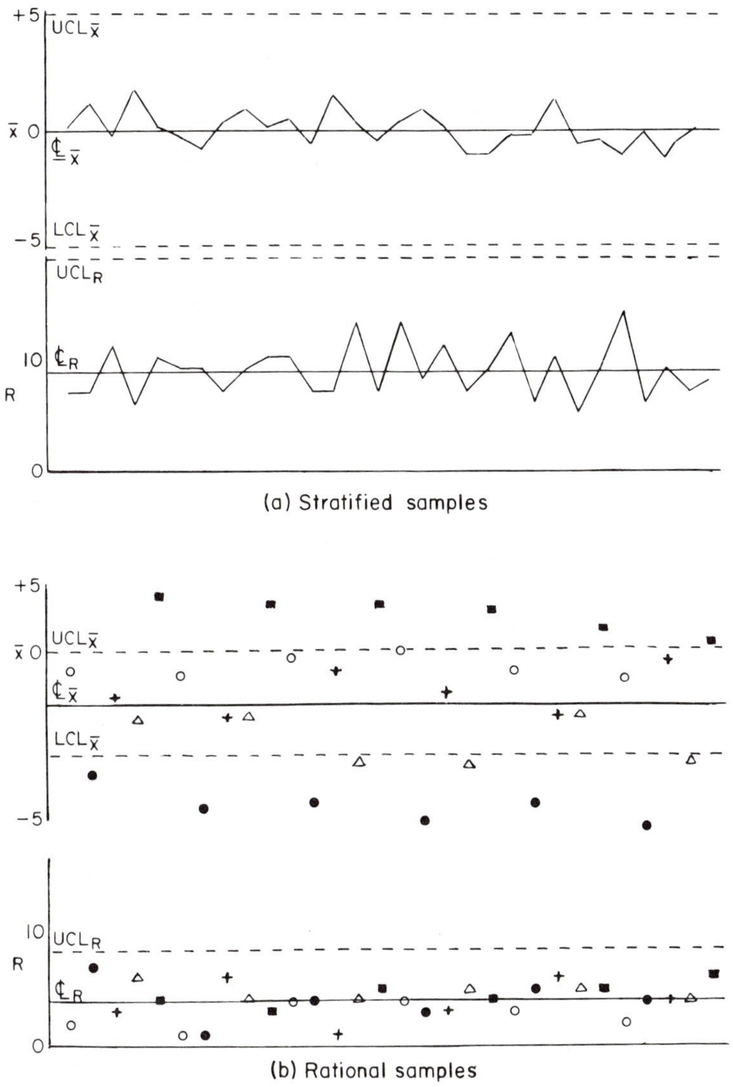

FIG. 7.7. Two analyses of data of Table 7.5. In (b) the rational samples have each spindle pictured by a different symbol.

$\text{Limits}_{\bar{x}} = -.11 \pm .577(3.97) = -.11 \pm 2.29 = -2.40, +2.18.$

The $\bar{x}$ and R points are plotted with different point symbols in rotation, a set of five at each basic time. The R chart shows perfect control with typical appearance. But the $\bar{x}$ chart shows bad control, with points outside the control band in both directions. Examining more closely, we find all six $\bar{x}_2$'s below the $LCL_{\bar{x}}$ and also three of $\bar{x}_4$'s. Also all six $\bar{x}_5$'s are above $UCL_{\bar{x}}$. Thus we can readily perceive substantial differences in spindles as to level (but the variabilities are apparently homogeneous). Action can be taken to adjust the spindles so as to meet specifications of $\pm$ 5. But rather tight control of μ must be exercised, since $\hat{\sigma}$ is close to 1.7 and thus is about a sixth of the total tolerance of 10.

Actually the x_1's were from population B; x_2's: G; x_3's: A; x_4's: F and x_5's: E, of Table 4.1.

The author ran a companion experiment in which all five populations were dumped into a single bowl and thoroughly mixed. Then random samples were taken from this "conveyor belt of mixed product." The naive person might expect to have one from each spindle in such a random sample. Actually none of the 30 samples had such a balanced composition. One even had four of the five from the same spindle. We shall not list the data, but did have:

$\bar{\bar{x}} = +6.4/30 = +.21 \quad \bar{R} = 221/30 = 7.37 \quad UCL_R = 15.6$

Limits $_{\bar{x}} = -4.04, + 4.46.$

The $\bar{x}$'s ran -2.6 to +3.4 and were well in control as were the R's, 3 to 12.

This would in practice justify estimation of the process population:

$\hat{\mu} = +.21 \quad \hat{\sigma}_x = 7.37/2.326 = 3.17$

yielding x limits of

$+.21 \pm 3(3.17) = -9.30, + 9.72$

If one combines the five populations A, B, E, F and G of Table 4.1, the overall distribution has μ=0, σ=3.32, and it runs from -9 to + 9. If these had been the specifications we would not have to analyze the individual spindles and adjust them. But with specifications of $\pm$ 5, we do need to analyze the individual spindles.

Other plottings of the rational samples of Table 7.5 could be made. We could plot all six samples of x_1's first, then the six of x_2's etc. This gives a fine picture and is quite feasible when analyzing past data. But plotting in rotation may be better when extending into the future.

Or one can make a "group control chart." For it one plots for a given time the highest and lowest $\bar{x}$ points and highest and lowest R points. So instead of plotting five $\bar{x}$'s and five R's at, say, 8:19 a.m. we only plot two $\bar{x}$'s and two R's, the highest and lowest. Then near the highest $\bar{x}$ point at 8:19 we place a little number to indicate which spindle had the highest $\bar{x}$, and similarly for the lowest. In Table 7.5 we would plot $\bar{x}_5 = +4.6$ and $\bar{x}_2 = -3.2$ and put respectively a "5" and "2" near the two points. Similarly the largest R was R_2 and lowest R_1. Only these two R's are plotted, with the "2" and "1". Now for the $\bar{x}$'s there would be a string of 5's by the high points indicating that spindle 5 is set high, and similarly a string of 2's by the low $\bar{x}$'s. But for the R's, the spindles "take turns" being high and low indicating no particular differences in variabilities. Group control charts are often used with greater numbers of spindles, heads, etc. and n=2 at each.

7.5.4.* Stratification Control Charts. There is another approach which can be used on stratified samples, if a record is kept of which observation goes with which mold cavity or head, etc. The basic approach is to find the average per cavity, over, say, 20 or 25 stratified samples. Then the next step is to substract from all of the cavity 1 results, the amount $\bar{x}_1 - \bar{\bar{x}}$, where $\bar{x}_1$ is the average of all cavity 1 x's i.e. the x_1's. The adjusted x_1 values then average $\bar{\bar{x}}$, (the whole set of adjusted x_1's), not $\bar{x}_1$. Adjust all the x_2's by substracting $\bar{x}_2-\bar{\bar{x}}$, etc. Then we use an ordinary $\bar{x}$ and R chart set-up for the adjusted data. Since the sum of the adjustments is 0, the row means (means at each time) are unaffected. But the ranges will in general be much smaller, since they are no longer inflated by cavity-to-cavity differences, these having been removed by the adjustments. This approach was developed in [3] and [4].

7.6.* Relative Efficiency of Different Types of Charts. One of the reasons for using an $\bar{x}$ chart rather than an x chart is that the former will average fewer measurements before giving an indication of a moderate shift in the process average. (But if the shift is extremely large, the x chart will be quicker.) Suppose, for example, that a process which has been in control is suddenly affected by an assignable cause which increases μ by $1\sigma_x$. How soon can we expect an indication of this on an x chart, and how soon on an $\bar{x}$ chart with n=4? See Figure 7.8. In the left hand section showing the original process and limits for x and $\bar{x}$, we see that in either case the

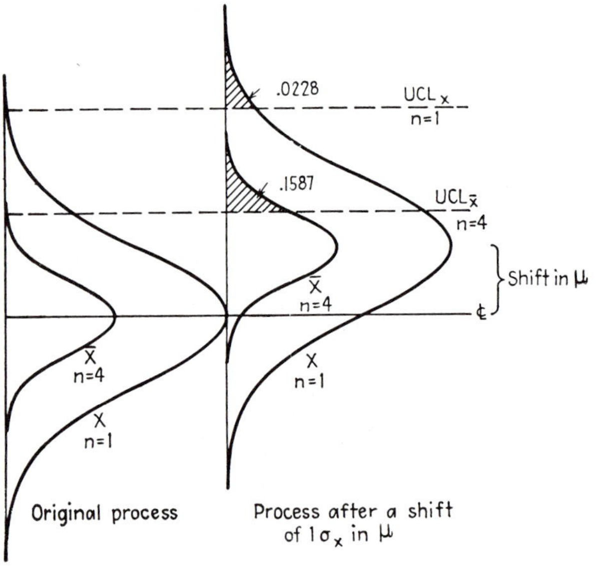

FIG. 7.8. Distributions of x and $\bar{x}$ (n=4), with given μ and σ, and 3σ upper control limits, in left hand section. Right hand section shows distributions after μ increases by $1\sigma_x$. Probabilities of a point out of the respective upper control limits are .0228 and .1587. Reproduced with permission from I. W. Burr, "Engineering Statistics and Quality Control", McGraw-Hill, New York, 1953, p. 280.

probability of a point outside is .0026 (3 sigma limits, standards given). In the right hand section, we see the distributions after μ is increased by $1\sigma_x$, with the original upper control limit lines extended. The probability of an x point lying above the original UCL_x is now .0228, while that for an $\bar{x}$ point above the original $UCL_{\bar{x}}$ is .1587. These are from respectively

$$z = \frac{\mu + 3\sigma_x - (\mu + \sigma_x)}{\sigma_x} = +2 \qquad z = \frac{\mu + 3(\sigma_x/\sqrt{4}) - (\mu + \sigma_x)}{\sigma_x/\sqrt{4}} = +1$$

Thus there is a much greater chance for an $\bar{x}$ point to show out of control, than for an x point. But remember that the x chart has four times as many opportunities to have a point out, than does the $\bar{x}$ chart, because the latter requires four x's to make one $\bar{x}$. How can we obtain a fair comparison?

The solution is to find the "average run length" or ARL, in terms of the average number of x's before a warning. Let us consider the x's first, and find the average number till the first point goes outside the limits. Let the probability be p' and its complement q'. Then the probability of the first warning on the first point is

$$P(1) = p',$$

and the probability of the first point out occurring on the second point is

$$P(2) = q'p',$$

and in general the probability of the first point out occurring on the i'th point is

$$P(i) = (q')^{i-1}p' \qquad (7.17)$$

(This distribution of number of trials is called the "geometric distribution.") Hence to find the average number till the first point goes out we need

$$E(n) = 1P(1) + 2P(2) + 3P(3) + \ldots + iP(i) + \ldots, \qquad (7.18)$$

or, say,

$$\mu_n = 1p' + 2q'p' + 3q'^2p' + 4q'^3p' + \ldots . \qquad (7.19)$$

Multiplying (7.19) by q' gives

$$q'\mu_n = \quad q'p' + 2q'^2p' + 3q'^3p' + \ldots . \qquad (7.20)$$

Subtracting (7.20) from (7.19) yields

$$(1-q')\mu_n = p' + q'p' + q'^2p' + q'^3p' + \ldots$$
$$= p'(1 + q' + q'^2 + q'^3 + \ldots)$$

But by division $1/(1-q') = 1 + q' + q'^2 + q'^3 + \ldots$, and hence

$$(1-q')\mu_n = p'[1/(1-q')].$$

Since $(1-q') = p'$ we have

$$p'\mu_n = p'/p'.$$

Thus

$$\mu_n = 1/p' \qquad (7.21)$$

Hence the average number of x's to receive a warning for the case in Figure 7.8, is

$$\mu_n = ARL_x = 1/.0228 = 44.$$

On the other hand the average number of $\bar{x}$'s to receive a warning on the $\bar{x}$ chart is

$$ARL_{\bar{x}} = 1/.1587 = 6.3.$$

But this is 6.3 sets of 4 x's each or

$$4ARL_{\bar{x}} = 4(6.3) = 25$$

Thus on the average, it takes about 25 x's made up into $\bar{x}$'s to receive a warning point using $\bar{x}$'s, but 44 x's for the x chart.

Note that this is for x vs. $\bar{x}$ charts for the case "standards given" and when $n=4$ and the shift is $1\sigma_x$.

If the jump is $k\sigma_x$ and the sample size is n, then generalization of the foregoing gives

$$nARL_{\bar{x}} = \frac{n}{\int_{3-k\sqrt{n}}^{\infty} \phi(z)\,dz} \qquad (7.22)$$

for the average number of x's till the first point out.

Note that use of (7.22) with $k=0$, gives ARL's to a false warning, when process is actually in control. P(point out of ± limits) = .0026 in either case. Then $ARL_x = 385$, while $ARL_{\bar{x}} = 4/.0026 = 1540$.

Comparison may also be made between $\bar{x}$ and p charts when the criterion for defectives is a measurable characteristic, normally distributed. Then if some jump in μ of an amount $k\sigma_x$ occurs, we may use (7.22) for the average number of x's till a warning is obtained. Moreover, the jump of $k\sigma_x$ in μ involves some increased fraction defective relative to the upper specification U, say from p' to p". Then we can find by the binomial distribution (2.30) the probability of a point above the upper control limit for a p chart of sample size n. Finally, (7.21) is used by n/P(p point above UCL).

7.7*. Artificial "Specifications" for Greater Efficiency in Attribute Charts. This section is basically a continuation of the preceding section. The idea is to increase artificially the fraction "defective" by using artificial "specification limits." Consider a somewhat extreme case, where σ_x of the process is precisely a sixth of the tolerance U-L. Then assuming normality and μ right at (U+L)/2, there will be a fraction defective relative to L,U of only .0026. As shown in Section 7.6, if we use $\bar{x}$'s for n=4, the probability of a point outside if μ increases by $1\sigma_x$ is .1587. See Figure 7.8. Hence the average run length to the first $\bar{x}$ point out is 4/.1587, or about 25 x's.

Now what if we are using a snap gage or go — not — go* gage instead of a measurement gage, just to tell whether or not each piece is within L,U? (This costs much less time per piece than measuring.) Take n=200, say. As shown above, p'=.0026, np'=.52. For an np chart we may neglect q'=.9974, and use limits like a c chart as follows

$$np' \pm 3\sqrt{np'} = .52 \pm 3\sqrt{.52} = -,\ 2.68.$$

Thus 3 defectives in 200 would be out of control. Using the Poisson Table VII, we find by interpolation P(2 or less|.52) = .984 or P(point out|.52) = .016. Hence the average number of attribute inspections: good or defective, is

*Two gages: the part should enter the "go" gage and not enter the "not-go" gage.

200/.016 = 12,500. This ARL compares favorably with $ARL_x = 380$ and $ARL_{\bar{x}} = 1540$ for x and $\bar{x}$ charts respectively under this case of controlled conditions. The point is that we like long ARL's when the process is in control, but short ARL's when the process goes out of control by some specified amount.

Next what is the effect on our assumed p chart if μ were to jump from $(U+L)/2$ to $(U+L)/2 + \sigma_x$?
Now the fraction defective p' jumps from .0026 to .0228. Hence now $np' = 200\ (.0228) = 4.56$, giving, by Table VII, a probability of .833 for an occurrence of three or more defectives in the 200, thus signaling lack of control. Then $ARL_{np} = 200/.833 = 240$. This compares with $ARL_x = 44$, $ARL_{\bar{x}} = 25$; in fact, unfavorably.

Now suppose that we artificially use "specifications" at $L' = L + .25(U-L) = (3L + U)/4$ and $U' = U - .25(U-L) = (L + 3U)/4$. This gives a centered "tolerance range" L' to U', just half as wide as the actual tolerance range. Also under our assumptions L' is at $\mu - 1.5\sigma_x$ and U' at $\mu + 1.5\sigma_x$. Thus by Table I the fraction "defective" has been artificially raised to $p' = 2(.0668) = .1336$. Also suppose we try n=20 for an np chart:

$$np' = 20(.1336) = 2.67$$

$$\text{Limits}_{np} = 2.67 \pm 3\sqrt{2.67(.8664)} = -,\ 7.23.$$

Now what is the probability of a warning signal, that is, of finding eight or more defectives in the sample, if p' = .1336. From appropriate binomial tables, this proves to be only about .0029, giving an ARL of 20/.0029 = 690 under conditions of control. Again this is compared to $ARL_x = 385$, $ARL_{\bar{x}} = 1540$. Now suppose as before that μ jumps by $1\sigma_x$ to $(U+L)/2 + \sigma_x$. Then the probability of a piece above $U' = (U+L)/2 + 3\sigma_x/2$ is found by $z = [(U+L)/2 + 3\sigma/2 - (U+L)/2 - \sigma]/\sigma = .5$, yielding .3085. Therefore $ARL_{np} = 20/.3085 = 65$.
This is quite comparable to the ARL's after a $1\sigma_x$ jump of 44 for x's and 25 for $\bar{x}$'s.

We must point out that to make a really fair comparison in these cases of efficiency of charts, the ARL's should be approximately equal when the process is in control at the satisfactory condition. Moreover we should compare the charts with various conditions such as varying k in a $k\sigma_x$ jump. Also we could vary the n's, and the multiple of sigma for control limits, and so on. See [5] - [7].

Such an approach can be useful. Other terms for this type of artificial limits are tests of "increased severity" or "narrow limit gaging."

7.8*. Summary Charts for $\bar{\bar{x}}$ and $\bar{\bar{R}}$. From time to time the author hears someone mention an "$\bar{\bar{R}}$ chart" and he immediately listens intently. Nearly always it is someone speaking loosely about an R chart. (True: the centerline is at $\bar{R}$.) But occasionally it is a real "$\bar{\bar{R}}$ chart."

For summarization purposes, we may sometimes wish to summarize a collection of $\bar{x}$'s and R's taken during a shift. Suppose, for example, that eight samples of five pieces each are measured about* every hour during a shift. Then we can let k=8, n=5 and find a pair of points to represent the <u>whole shift</u>:

$$\bar{\bar{x}} = \sum_{i=1}^{8} \bar{x}_i/8, \quad \bar{R} = \sum_{i=1}^{8} R_i/8$$

These are $\bar{\bar{x}}$ and $\bar{\bar{R}}$ points for charts so named.

To set centerlines and limits let us suppose we have 15 shifts in a week. Then

$$\text{₵}_{\bar{\bar{x}}} = \sum\bar{\bar{x}}/15 = \bar{\bar{\bar{x}}} \qquad \text{₵}_{\bar{\bar{R}}} = \sum \bar{R}/15 = \bar{\bar{R}}. \tag{7.23}$$

Now $\bar{\bar{R}}$ is the average of 15(8)=120 R's for the week, each of a sample of n=5. Hence we have

$$\hat{\sigma}_x = \bar{\bar{R}}/d_2 = \bar{\bar{R}}/2.326$$

---

*It is rather desirable not to make the interval exactly the same, for several reasons.

the d_2 value being for n=5. Then we have

$\hat{\sigma}_R = d_3\hat{\sigma}_x = .864\hat{\sigma}_x$

and since there are eight R's per shift

$\hat{\sigma}_{\bar{R}} = .864\hat{\sigma}_x/\sqrt{8}.$

Likewise for $\bar{\bar{x}}$ from 8(5) = 40 x's (or 8 $\bar{x}$'s):

$\hat{\sigma}_{\bar{\bar{x}}} = \hat{\sigma}_x/\sqrt{40}.$

Then these are used for "three-sigma" limits:

$$\text{Limits}_{\bar{\bar{x}}} = \bar{\bar{\bar{x}}} \pm 3\hat{\sigma}_x/\sqrt{kn} \quad (7.24)$$

$$\text{Limits}_{\bar{R}} = \bar{\bar{R}} \pm 3d_3\hat{\sigma}_x/\sqrt{k}. \quad (7.25)$$

Such limits are not especially good at pointing out specific assignable causes, but are good for giving general trends and summarizing. A single point may be summarizing a considerable variety of conditions. It is true, however, that an operator or a gage may be "off" level for one whole shift, giving an $\bar{\bar{x}}$ point out of control.

7.9.* Variations of Control Charts. Let us now discuss quite briefly some other adaptations and related literature on control charts. In the next chapter, we present many references on applications along with appropriate discussion.

1. Tool Wear and Trends. A quite simple approach to control of processes subject to consistent change in level has already been given in Section 7.2. An early paper is by Tippett [8] presenting the general approach, while an applicational one is by Manuele [9]. Somewhat more sophisticated approaches using mathematically fitted trend lines are Mandel [10] and Behnken [11]. Gibra [12] gives a least cost approach to such problems using necessary input.

2. Trend Charts and Moving Averages. We presented control charts for individual measurements x and moving ranges R, in Section 7.1. It is possible to use this same approach with moving averages, that is, $\bar{R}$ is used for estimating σ_x, then we may plot $\bar{x}$ for the last two, last three or the last

four x's, as each new x comes along. Such an approach has the highly desirable element of simplicity. It can be used along with x's for expensive and/or rather infrequent measurements, such as, octane numbers for gasoline. A variation of considerable merit is to give the most recent x the maximum weight, and earlier x's successively less weight. These weights can be on a geometrically decreasing basis. Amplification on the approach is given by Hammer [13], Beall [14] and Roberts [15]. Divers [16] discusses this approach and time-related control charts. Freund [17] gives a useful presentation of several approaches. Two other recent papers are Wortham and Heinrich [18] and Turner [19].

3. Limits Other than Three-sigma. Some workers have experimented with control limit lines other than those set at plus and minus three standard deviations (estimated) of the characteristic being plotted. As a matter of fact Shewart [3.1] and his associates did a great amount of such experimentation before settling upon three-sigma limits. See Section 3.6. Two variations are (1) to use some other multiple than three, of the estimated or standard sigma for the characteristic, and (2) to provide "warning limits" in addition to three-sigma "action limits."

Freund in [1] developes the former variation by setting an alpha (α) risk of signaling an assignable cause of change of level when none is present and setting a beta (β) risk of failing to call for action when the level is off by a specified amount. This leads to a requisite sample size and the needed limits. We shall elaborate on this type of significance test in Chapter 11.

Another related approach is that of modified limits, which may be used when tight control of process level μ is not needed and σ is known, so that we may set modified limits for $\bar{x}$'s inside of specifications for x's. See Section 7.2.2. Also see Hill [21].

Often with warning limits, two $\bar{x}$ points in succession outside of warning limits is cause for action. Some workers use two-sigma limits for warning. Early work on this general approach was done by Page [22], [23]. See also Mitten and Sanoh [24] and Shahani [25]. Other multiple warning criteria are given in Moore [26], Satterthwaite [27], Weiler [28], Goetz [29] and Gibra [30]. These approaches buy a more favorable operating characteristic curve for the test on the process for a given n, but at a cost of considerable loss in simplicity and/or in required input, such as, known σ and a desire to guard against a particular size of process jump. Hence the author is not inclined to urge their use, although they do have their place.

4. Multiple Measurements. An applicational paper on group control charts, described here in Section 7.5.3, is by Boyd [31]. Multiple measurements on a single piece are discussed by Hammer [32].

5. Miscellaneous Graphical Procedures. A variety of graphical procedures are discussed in Freund [33] and Noble [34]. Further analyses of means are presented by Ott [35] and Small [36]. Useful approaches are given by Seder [37], [38]. Traver and Davis [39], describe how to determine process capabilities from short runs. Rhodes [40] discusses the sources of variation in lots of bulk material, and how they may be estimated. When a process is in good control but with too much spread to fit between the specifications L to U, one may find the most economical level to minimize scrap plus rework costs, Springer [41].

6. Adjustments for Small Initial Run. When but few measurements are initially available, the control limits can be adjusted slightly to take this into account, so as to preserve the usual probabilities. See Proschan and Savage [42] and Yang and Hillier [43].

7. Power Characteristics of Control Charts. The operating characteristic curve (probability of acceptance) or its inverse, the power curve (probability of rejection) are of importance in yielding risks of wrong decisions. For the

case of standards given (especially σ), this is not a difficult problem. But when analyzing past data it is more difficult. Several papers on the subject are Olds [44], Page [45], [46], Roberts [47] (in which he gives average run lengths), and Enrick [48]. Hilliard and Lasater [49] consider the rather difficult problem of risks when more than one test is used for rejection criteria for measurement charts.

8. Adaptations Using Other Sample Statistics. In the cases of ranges and standard deviations when σ is known, we use as center lines $d_2\sigma$ and $c_4\sigma$, that is, the expected values respectively. Or if a standard σ is unknown we use $\bar{R}$ or $\bar{s}$. But the distributions of R and s are not symmetrical (the amount of asymmetry depending upon n), and there is a greater probability of a value below the center line than above. As a consequence a long run of points below the center line is much more likely than a long run above it. To equalize the probability of long runs by chance we may use the median of R's or s's instead of their mean, Eisenhart [50].

Some workers suggest the use of medians rather than $\bar{x}$'s. Or mid-ranges, $[x_{(1)} + x_{(n)}]/2$, where $x_{(i)}$ is the i'th smallest, may be used. Such statistics can be plotted on a plotting of x's. See Ferrell [51], [52], Clifford [53], von Osinski [54] and Shahani [55]. When the sample size varies we may plot a standardized variable $z = \dfrac{\bar{x} - \mu}{\sigma/\sqrt{n}}$ if σ is known, with μ set, perhaps at (L+U)/2. See Duncan [56].

Another adaptation is to plot $x_{1i} - x_{2i} = y_i$ $i = 1,\ldots,n$, where each pair is on a single part or individual, Grubbs [57] and Eichelberger [58]. Or, one may plot the maximum $x_{(n)}$ and minimum $x_{(1)}$ in each sample, and set an upper limit on the former and a lower limit for the latter, Howell [59] and Madison and Ostle [60].

9. Use of Computers. An optimum procedure for adjusting the process level, readily adapted to machine control is given by Grubbs [61]. A variety of papers on computer applications follows in chronological order:

Amber and Amber [62], Roberts [63], Williams and Johnson [64] a resume for process industries, Kallet [65] and Lieberman [66].

7.10. Summary. It seems pointless to try to summarize here the diversity of problems described in the preceding sections. These adaptations and special cases can be usefully applied when occasions occur. Moreover careful study of them will surely aid the student in consolidating his understanding of control charting principles, the distinctions between attribute and variable charts and the interplay of charts with specifications.

PROBLEMS

7.1. The following data give the number of .01 per cent of moisture in a polymer (Polysar S). The maximum desired is .50 per cent. Measurements once every 2 hr., starting at 4 p.m., Sept. 25: 36, 20, 16, 21, 32, 34, 32, 34, 23, 25, 12, 31, 25, 31, 34, 38, 26, 29, 45, 27, 29, 26, 33, 33, 38, 45, 42, 47, 45, 35, 44, 37. Draw the x and moving range charts. How many x's would have been outside three standard deviation x limits?

7.2. The following data represent one analysis per heat on each of 30 consecutive heats of 1045 steel. The specifications are .70 to .90 per cent for manganese. Analyses in .01 per cent of manganese: 74, 79, 77, 81, 72, 66, 75, 80, 76, 86, 84, 70, 80, 62, 74, 71, 68, 79, 81, 76, 79, 79, 84, 78, 74, 88, 71, 80, 79, 74. Draw the x and moving range charts, and comment. How many x's would have been outside three standard deviation x limits?

7.3. Below are given for 40 consecutive reels the average basis weight x (pounds per 1000 sq. ft.) of paperboard before the start of a quality control system. Make an x and moving range chart and draw any conclusions you can. [67].

Table P7.3

Reel	x	Reel	x	Reel	x	Reel	x	Reel	x
1	53.6	9	52.2	17	49.8	25	50.0	33	52.8
2	51.6	10	50.6	18	50.6	26	49.8	34	51.8
3	51.6	11	49.6	19	50.0	27	48.8	35	50.6
4	54.4	12	49.0	20	49.8	28	50.8	36	50.6
5	51.2	13	48.2	21	50.2	29	52.0	37	50.0
6	50.8	14	49.2	22	50.0	30	52.8	38	52.6
7	50.6	15	47.6	23	50.0	31	53.4	39	50.4
8	51.6	16	49.6	24	49.4	32	55.2	40	49.2
									2032.0

7.4. For the data in Problem 7.3, form ten samples of $n=4$ from the consecutive x's, and run $\bar{x}$ and R charts, comparing with the x and moving range chart results.

7.5. Make an x and moving range chart for the following data taken while the author was listening to a lecture. Time in seconds for a large "pill bug" to pass across consecutive 9¼" tiles: 35, 17, 18, 19, 23, 20, 25, 25, 24, 25, 20, 29, 19, 23, 26. Comment.

7.6. A description of a job by a quality-control man follows (see Table 7.6):

"Three different vendors were supplying forgings. These forgings were machined on a Gridley automatic. There were constant tool changes and tool resets. As a result of these changes tool life was short and pieces were often out of specification. It was the opinion that the difference among vendors was responsible for pieces being out of specification. However, a study, such as attached, was made on each vendor's pieces and there was no significant difference in the trend of any of them. A chart was placed on the machine and the operator was instructed to set the machine so that the average of 5 pieces would approximate the Lower Modified Control Limit and then to let the machine run until the Upper Modified Control Limit was approached. At this time he was to reset the tool. This chart indicated that approximately 75 pieces could be produced before a tool needed to be reset.

"By using this type of chart, what used to be a difficult operation with short tool life is now running without defects, and 100% inspection is reduced to 5 pieces each 30 minutes (checked at the machine). Tool life has increased considerably."

Table P7.6. Outside Diameter at Base of Stem of an Exhaust-Valve Bridge. Specifications: 1.156 In., + .000 In., - .001 In. Samples of Five, all Pieces Measured. Data in .0001 In. Above 1.154 In.

x_1	x_2	x_3	x_4	x_5	$\bar{x}$	R
12	13	13	14	13	13.0	2
13	14	14	15	14	14.0	2
14	13	15	15	15	14.4	2
15	14	16	14	14	14.6	2
15	15	16	14	15	15.0	2
17	15	16	15	15	15.6	2
17	17	17	17	16	16.8	1
17	17	18	16	17	17.0	2

Table P7.6 (continued)

18	18	18	17	17	17.6	1
17	18	17	18	19	17.8	2
18	17	19	18	18	18.0	2
18	19	19	18	19	18.6	1
19	20	19	18	19	19.0	2
12	13	11	11	12	11.8	2
12	13	12	12	13	12.4	1
11	13	13	13	13	12.6	2
13	13	12	13	13	12.8	1
14	12	13	12	14	13.0	2
13	13	14	13	14	13.4	1
13	13	15	14	13	13.6	2

The "modified limits" used were analogous to what we called modified limit A in Sec. 7.2.2. Draw the charts, slanting control limits, and modified limits, and comment on this job.

7.7. For the data of Table 7.7, first draw the R chart for all 44 points. Is there evidence of an assignable cause? What is it? If you were in charge of this operation, could you remove it? Let us eliminate the two points from the data and recompute $\bar{R}$ and the limits. Is there control? Next analyze the data by slanting control lines on the $\bar{x}$ chart. What can you conclude?

Table 7.7. Outside Diameter of a Spacer. .0001 In. below .1250 In. Specifications: .1235 to .1250 In. Another Run of Data Subsequent to Table 7.2.

Sample	x_1	x_2	x_3	x_4	x_5	$\bar{x}$	R
1	- 10	- 3	- 9	- 2	- 3	- 5.4	8
2	- 12	- 13	- 13	- 11	- 10	- 11.8	3
3	- 12	- 8	- 13	- 13	- 11	- 11.4	5
4	- 11	- 10	- 8	- 12	- 11	- 10.4	4
5	- 11	- 10	- 13	- 10	- 12	- 11.2	3
6	- 11	- 9	- 13	- 11	- 13	- 11.4	4
7	- 10	- 12	- 11	- 10	- 9	- 10.4	3
8	- 12	- 12	- 11	- 8	- 11	- 10.8	4
9	- 8	- 10	- 9	- 10	- 11	- 9.6	3
10	- 11	- 10	- 8	- 9	- 8	- 9.2	3
11	- 10	- 8	- 10	- 9	- 9	- 9.2	2
12	- 9	- 10	- 9	- 8	- 10	- 9.2	2
13	- 10	- 10	- 8	- 9	- 8	- 9.0	2
14	- 8	- 7	- 9	- 9	- 8	- 8.2	2
15	- 7	- 7	- 6	- 8	- 10	- 7.6	4
16	- 7	- 6	- 9	- 8	- 7	- 7.4	3
17	- 6	- 6	- 7	- 7	- 8	- 6.8	2
18	- 4	- 5	- 6	- 7	- 4	- 5.2	3
19	- 4	- 2	- 4	- 3	- 3	- 3.2	2
20	- 11	- 2	- 10	- 9	- 10	- 8.4	9
21	- 10	- 9	- 8	- 11	- 9	- 9.4	3
22	- 11	- 10	- 11	- 9	- 9	- 10.0	2
23	- 9	- 7	- 11	- 11	- 9	- 9.4	4
24	- 9	- 7	- 9	- 10	- 8	- 8.6	3
25	- 8	- 10	- 9	- 8	- 7	- 8.4	3
26	- 10	- 11	- 8	- 8	- 7	- 8.8	4
27	- 7	- 8	- 8	- 7	- 6	- 7.2	2
28	- 6	- 8	- 7	- 9	- 9	- 7.8	3
29	- 6	- 7	- 7	- 8	- 8	- 7.2	2
30	- 7	- 9	- 5	- 7	- 7	- 7.0	4
31	- 5	- 7	- 7	- 6	- 5	- 6.0	2
32	- 5	- 8	- 7	- 6	- 7	- 6.6	3
33	- 7	- 6	- 8	- 8	- 7	- 7.2	2
34	- 8	- 8	- 7	- 9	- 7	- 7.8	2
35	- 7	- 8	- 7	- 6	- 7	- 7.0	2
36	- 6	- 5	- 6	- 5	- 5	- 5.4	1
37	- 5	- 8	- 5	- 6	- 4	- 5.6	4
38	- 5	- 6	- 4	- 6	- 6	- 5.4	2
39	- 4	- 4	- 5	- 5	- 4	- 4.4	1
40	- 11	- 12	- 10	- 11	- 12	- 11.2	2
41	- 9	- 8	- 11	- 10	- 10	- 9.6	3
42	- 9	- 9	- 8	- 10	- 8	- 8.8	2
43	- 7	- 8	- 9	- 10	- 8	- 8.4	3
44	- 8	- 9	- 8	- 7	- 10	- 8.4	3

7.8. The following data are for edge widths (thicknesses) of samples of piston rings following the preliminary disc grind. Specifications .1070 ± .0007 in. Data in table are for the number of .0001 in. from .1070 in.

Day	1	2	3	4	5	6
$\bar{x}$	+ 6.6	+ 3.2	+ 4.6	+ 3.8	+ 7.5	+ 4.7
s	10.8	5.0	6.1	4.5	8.4	5.1
n	100	120	135	105	125	105

What conclusions can you draw?

7.9. The following data are for edge widths (thicknesses) of samples of piston rings after "clear disc grind." Specifications .0957 ± .0005 in. Data in table are for the number of .0001 in. from .0957 ± .0005 in. Same part number as in Problem 7.8. Data are in .0001 in. from .0957 in.

Day	1	2	3	4	5	6
$\bar{x}$	+ .6	+ .4	+ .07	+ .2	+ .2	- .05
s	3.0	1.6	1.8	1.6	1.5	2.5
n	140	90	135	105	135	100

What conclusions can you draw?

7.10. The data below are from [68] Use control charts to test the data for any differences between types of packing. Chemical engineering research data.

Type of Packing	Pressure Drops for Individual Runs
Random B	46, 29, 48
Random C	37, 35, 50, 39
Random D	29, 16, 5, 24
Dense E	36, 27, 25, 28, 23
Dense F	16, 23, 29
Random H	21, 29

7.11. For a grinding operation, samples of n=5 are taken about every half hour, 16 per shift, with three shifts per day, five days per week. At the end of a week, we find $\bar{\bar{x}}$ = .22510 in., and $\bar{\bar{R}}$ = .00071 in. Set control limits for a shift $\bar{\bar{x}}$ point and $\bar{R}$ point. Compare to specifications of .2250 $\pm$ .0010 in., making and stating what assumptions you need.

7.12. Compare the average run length ARL_x before the first indication of out-of-controlness for the following two charts for the case μ, σ_x given, when μ goes to $\mu + 1.5\sigma_x$:

(a). $\bar{x}$ chart, n=5

(b). x chart with $\pm 2\sigma_x$ limits.

REFERENCES

1. R. A. Freund, Acceptance control charts. _Indust. Quality Control_ 14 (No. 4), 13-23 (1957).

2. I. W. Burr, Control charts for measurements with varying sample sizes. _J. Quality Techn._ 1, 163-167 (1969).

3. A. E. R. Westman and B. H. Lloyd, Quality control charts for $\bar{x}$ and R adjusted for within - subgroup pattern. _Indust. Quality Control_ 5 (No. 5), 5-10 (1949).

4. I. W. Burr and W. R. Weaver, Stratification control charts. _Indust. Quality Control_ 5 (No. 5), 10-15 (1949).

5. W. L. Stevens, Control by gauging. _J. Roy. Statist. Soc. Suppl._ 10, 54-98 (1948).

6. A. E. Mace, The use of limit gages in process control. _Indust. Quality Control_ 8 (No. 4), 24, 28-31 (1952).

7. E. R. Ott and A. B. Mundel, Narrow-limit gaging. _Indust. Quality Control_ 10 (No. 5), 21-28 (1954).

8. L. H. C. Tippett, The control of industrial processes subject to trends in quality. Biometrika 33, 163-172 (1944).

9. J. Manuele, Control chart for determining tool wear. Indust. Quality Control 1 (No. 6), 7-10 (1945).

10. B. J. Mandel, The regression control chart. J. Quality Techn. 1, 1-9 (1969).

11. D. W. Behnken, Moving trend control charts. Indust. Quality Control 14 (No. 5), 7-11 (1957).

12. I. N. Gibra, Optimal control of processes subject to linear trends. J. Indust. Engin. 18, 35-41 (1967).

13. P. C. Hammer, Limits for control charts. Indust. Quality Control 2 (No. 3), 9-11 (1945).

14. G. Beall, Control of basis weight in the machine direction. TAPPI 39, 26-29 (1956).

15. S. W. Roberts, Control charts based on geometric moving averages. Technometrics 1, 239-250 (1959).

16. C. K. Divers, Some unique but highly successful control charts. Annual Tech. Conf. Trans. Amer. Soc. Quality Control, 589-597 (1966).

17. R. A. Freund, Control charts eliminate disturbance factors. Chem. Engin. 73 (No. 3), 70-76 (1966).

18. A. W. Wortham and G. F. Heinrich, A computer program for plotting exponentially smoothed average control charts. J. Quality Techn. 5, 84-90 (1973).

19. T. F. Turner, Smoothing data trend ratios. Quality Progress 4 (No. 4), 20-22 (1971).

20. R. A. Freund, A reconsideration of the variables control chart, with special reference to the chemical industries. Indust. Quality Control 16, (No. 11), 35-41 (1960).

21. D. Hill, Modified control limits. Appl. Statcs. 5, 12-19 (1956).

22. E. S. Page, Control charts with warning lines. Biometrika 42, 243-257 (1955).

23. E. S. Page, A modified control chart with warning lines. Biometrika 49, 171-176 (1962).

24. L. G. Mitten and A. Sanoh, The $\bar{x}$ warning limit chart. Indust. Quality Control 18 (No. 2), 15-19 (1961).

25. A. K. Shahani, A three-zone control chart. Qual. Engin. 37 (No. 4), 95-99 (1973).

26. P. G. Moore, Some properties of runs in quality control procedures. Biometrika 45, 89-95 (1958).

27. F. E. Satterthwaite, Pre-control for supervisors. Quality Progress 6 (No. 2), 26-28 (1973).

28. H. Weiler, A new type of control chart limits for means, ranges and sequential runs. J. Amer. Statist. Assoc. 49, 298-314 (1954).

29. B. E. Goetz, 3σ? Adv. Mgt. 23, 23 (1958).

30. I. N. Gibra, Economically optimal determination of the parameters of $\bar{X}$-control chart. Management Science 17, 635-646 (1971).

31. D. F. Boyd, Applying the group chart for $\bar{X}$ and R. Indust. Quality Control 7 (No. 3), 22-25 (1950).

32. P. C. Hammer, Limits for control charts, second article. Indust. Quality Control 2 (No. 6), 20-22 (1946).

33. R. A. Freund, Graphical process control. Indust. Quality Control 18 (No. 7), 15-22 (1962).

34. C. E. Noble, Variations in conventional control charts. Indust. Quality Control 8 (No. 3), 17-22 (1951).

35. E. R. Ott, Analysis of means - a graphical procedure. Indust. Quality Control 24 (No. 2), 101-109 (1967).

36. B. B. Small, Simplified method of performing an analysis of variance with control charts. Proc. Fourteenth Midwest Qual. Control Conf., 223-231 (1959).

37. L. A. Seder, Diagnosis with diagrams. Indust. Quality Control, Part I 6 (No. 4), 11-19, Part II 6 (No. 5), 7-10 (1950).

38. L. A. Seder, The technique of preventing defects. Indust. Quality Control 9 (No. 6), 23-30 (1953).

39. R. W. Traver and J. M. Davis, How to determine process capabilities in a developmental shop. Indust. Quality Control 18 (No. 9), 26-29 (1962).

40. R. C. Rhodes, Lot structure and sampling bulk materials. Quality Progress 5 (No. 11), 24-27 (1972).

41. C. H. Springer, A method for determining the most economic position of a process mean. Indust. Quality Control 8 (No. 1), 36-39 (1951).

42. F. Proschan and I. R. Savage, Starting a control chart. Indust. Quality Control 17 (No. 3), 12-13 (1960).

43. C. H. Yang and F. S. Hillier, Mean and variance control chart limits based on a small number of subgroups. J. Quality Techn. 2, 9-16 (1970).

44. E. G. Olds, Power characteristics of control charts. Indust. Quality Control 18 (No. 1), 4-10 (1961).

45. E. S. Page, Control charts for the mean of a normal population. J. Roy Statist. Soc. Suppl. 16, 131-135 (1954).

46. E. S. Page, Comparison of process inspection schemes. Indust. Quality Control 21, 245-249 (1964).

47. S. W. Roberts, A comparison of some control chart procedures. Technometrics 8, 411-430 (1966).

48. N. L. Enrick, Operating characteristics of reject limits for measurements. Indust. Quality Control, 2 (No. 2), 9-10 (1945).

49. J. E. Hilliard and H. A. Lasater, Type I risks when several tests are used together on control charts for means and ranges, no standard given. Indust. Quality Control 23 (No. 2), 56-61 (1966).

50. C. Eisenhart, Probability center lines for standard deviation and range charts. Indust. Quality Control 6 (No. 1), 24-26 (1949).

51. E. B. Ferrell, Control charts using midranges and medians. Indust. Quality Control 9 (No. 5), 30-34 (1953).

52. E. B. Ferrell, A median, mid-range chart using run-size subgroups. Indust. Quality Control 20 (No. 10), 22-25 (1964).

53. P. C. Clifford, Control charts without calculations: some modifications and some extensions. Indust. Quality Control 15 (No. 11), 40-44 (1959).

54. R. von Osinski, Use of median charts in the rubber industry. Indust. Quality Control 19 (No. 2), 5-8 (1962).

55. A. K. Shahani, A control chart based on sample median. Qual. Engin. 35 (No. 7), 7-9 (1971).

56. A. J. Duncan, Detection of non-random variation when sample size varies. Indust. Quality Control 4 (No. 4), 9-12 (1948).

57. F. E. Grubbs, The difference control chart with an example of its use, Indust. Quality Control 3 (No. 1) 22-25 (1946).

58. L. S. Eichelberger, Statistical quality control in a press shop. Indust. Quality Control 13 (No. 2), 12-17 (1956).

59. J. M. Howell, Control chart for largest and smallest values. Ann. Math. Statist. 20, 305-309 (1949). Errata: 21, 615-616 (1950).

60. R. L. Madison and B. Ostle, An evaluation of the Maytag double-line chart. Indust. Quality Control 14 (No. 3), 11-13 (1957).

61. F. E. Grubbs, An optimum procedure for setting machines or adjusting processes. Indust. Quality Control 11 (No. 1), 6-9 (1954).

62. G. H. Amber and P. S. Amber, QC computers for machine control. Electrical Mfg. I 58 (No. 1), 80-88, II 58 (No. 2), 78-85, 298 (1956).

63. T. W. Roberts, Jr., Automatic quality control: pipe dream or reality? Indust. Quality Control 14 (No. 6), 14-18 (1957).

64. T. J. Williams and R. C. Johnson, Computers in the process industries. Instruments and Automation 31, 90-94 (1958).

65. F. T. Kallet, The computer-another tool for quality control. Indust. Quality Control 22 (No. 2), 89-90 (1965).

66. G. J. Lieberman, Statistical process control and the impact of automatic process control. Technometrics 7, 283-292 (1965).

67. C. A. Bicking, The fundamentals of control charts. TAPPI 36, 544-549 (1953).

68. W. B. Seefeldt, Pressure drop in packed columns. M.S. Thesis, Purdue University (1948).

CHAPTER 8

APPLICATIONS OF CONTROL CHARTS

8.1. What Manufacturing Problem to Work On. The time of a Quality Control Department, or an individual person, is divided between (1) routine controls, (2) "putting out fires" on an emergency basis, and (3) problem-solving. Also some time should be available for long-range planning. Too often nearly all of the time is spent on routine matters and in meeting emergencies as they arise. Keeping in mind the profit and loss statement, it is of high importance to reserve a generous portion of time beyond day-to-day matters, to seek improvements in quality and costs. For this we need to choose processes and operations to work on.

Each part number has many characteristics, each of which is determined by one or more processes or operations. Ideally a study should be made, perhaps by Accounting, on the money costs of bad quality for each operation. Such costs may be classified under (1) prevention (planning, problem solving, improvement, process control), (2) detection (inspection, after the fact) and (3) failure (scrap, rework, salvage, field failure, warranties and claims). Often too little effort is put on the prevention of defective or non-conforming material.

Studies have shown that if one finds the costs for each operation, a very few will be found to contribute a completely disproprotionate share of the total. (Such a maldistribution is often said to follow a "Pareto law.") Accordingly we ought to try to seek out those operations which are the most costly, and bend our efforts to markedly improve them, that is, to work where there is the potential for big financial gains. Not a few companies have obtained reliable quality costs on each of their various operations, and placed strong effort on the most costly, to their great advantage. Even without such a complete study, this philosophy can be used with subjective judgement and such objective information as may be available.

In summary then we may say "Seek out the operations where the biggest losses for bad quality occur, and place substantial effort and problem-solving talent on them." Do not let day-to-day matters prevent such effort.

8.2. Attacking an Industrial Problem. "Now go out and apply these methods." Easier said than done! A practical problem often looks entirely different from a book exercise, with the data all collected and neatly tabulated. Hence we give the following suggestions:

In the first place, finding a solution to an industrial problem is a team job. Secondly, appropriate action must be taken at the proper time. This is usually the function of management at some level. Finally, follow-through is necessary in order to avoid having the original trouble return, and also to be on guard against future trouble.

With these general observations in mind let us proceed with the following outline of steps:

1. The problem naturally involves some product or material which is not satisfactory. Determine what quality characteristic is causing the most trouble.

2. If for the characteristic in question, an appropriate measuring device is available, test it to make sure that it gives repeatable results. This can be tested by using an R chart as follows: choose several (say 8 or 10) samples of material or individual pieces, which may well differ from each other, but each such sample, by itself, is as homogeneous as possible. Present to the technician the several samples or pieces to measure or analyze. Then give him the same samples or pieces in a different order to be measured again, preferably not telling him they are the same. Repeat until we have n measurements

on each sample or piece, where n may be 3, 4 or 5. This gives 8 or 10 samples, each of n repeated measurements. Naturally an $\bar{x}$ chart would show poor control, but the R(or an s) chart should show good control, if the measurement technique is repeatable. If good control is demonstrated then $\bar{R}/d_2$ (or $\bar{s}/c_4$) is an estimate of the measurement error σ_e. If control is poor take steps to improve the technique.

3. If a measuring device for the characteristic is not available, it may well be desirable to devise one, since measurement control charts are more efficient at finding trouble than are attribute charts.

4. If attribute charts are to be used (or must be), make sure that the definition of a defect or a defective is clear and specific. This can be checked in the same sort of way as in 2. A larger number of samples or pieces should be offered for attribute inspection than for repeated measurements.

5. Go back in the plant operations on the product to where it is believed that the trouble begins. (In machining or grinding operations, for example, this is surprisingly often the first rough work.) Samples should be drawn as near to that operation as possible, and the testing or measurement done as soon after the operation was performed on the pieces or material as possible.

6. Make a list of the potential assignable causes which we wish to correct. Everyone associated with the operation should be consulted in this: the operator, setup man, inspector, foreman, engineer, designer, chemist, or whoever may have some ideas on it.

7. There may be some useful data already available. Usually this is not the case, because the results may not be in a useful sample size, conditions were not adequately controlled, or a record was not made of the conditions under which the pieces were made and tested. Attribute data more often are of use, however. In any case it is worth looking for past data, at least for a comparison.

8. Arrange for the collection of new data. This involves the choice of sample size. For measurements, four or five are standard, while for a p chart the sample size should be small enough so as to make it as easy as possible to keep conditions uniform, but not so small that $n\bar{p}$ is much below 1. As far as possible, the testing or measuring should be done by the same person on the same test set, so that any out-of-control indications are due to product variation and not to the testing or personnel.

9. Careful records should be made of the conditions under which each sample was made. Ideally the situation with regard to all the potential assignable causes in (6) should be recorded. But it is obviously impossible in any practical problem to record them all.

10. If past data are not available, then a new run of preliminary data must be collected. This is ordinarily done by taking samples about as often as they can be handled, rather than taking them as far apart as they will routinely be taken. In this way a quicker start can be made. Also, it is often advantageous to see how a run of most, if not all, of the pieces does behave.

11. Analyze the preliminary data. Show them to all the people involved, and explain with great care and simplicity the interpretation of the charts.

12. The hunt for assignable causes is now on, and this is where teamwork counts. Basically the approach is to find what conditions or potential assignable causes were different for the out-of-control samples. Was there any one condition present for all of the points out on the high side and not present on many, if any, of the other points? Were there any trends? (Be sure that any resettings or retoolings are marked on the chart.) If the process is in control and still not meeting specifications, this will require a rather drastic change in the process.

13. Corrective action may be tried on any suspected assignable cause. If no further points go out for a while, it might even be worth while to

bring in the apparent assignable cause again to see whether or not any points go out. Finding and proving an assignable cause is an experimental job.

14. Corrective action on a proved assignable cause may require action by management and possibly the expenditure of money. One of the surest ways to kill off a promising quality-control man (or department) is for management repeatedly to fail to take action when the need is clearly proved and when it is also economically sound.

15. A brief, clear, and factual report should present the case to the appropriate persons in the organization. Interesting graphs help. Money saved is always of interest to management. Also improved quality assurance.

16. Follow-through is essential. Control charts should usually be maintained even after the problem is completely solved. The samples may be taken at rather long time intervals, and they can still give warning when the process is getting out of hand. This is especially true when 100 per cent sorting has been replaced by sampling. Operations have a tendency to get out of hand when not kept track of.

8.3. List of Typical Applications of Control Charts. The following list is given to stimulate your interest and imagination, and as a check list of ideas for you to consider in your own organization. They are all based on actual work in a wide variety of organizations. Many applications are written up in the references in Section 8.6.

1. Saving time in looking for assignable causes when none is present. In particular avoiding resetting or tampering with the process average when it is actually just about right. Often such excessive resetting is done when the process is simply incapable of holding the specified tolerance (maximum minus minimum). When this is the case, we are simply increasing variability and making matters worse.

2. Knowing when a machine or process is doing the best that can be expected of it. If in control, but the variability is too great, a fundamental process change is needed.

3. Help in finding causes of trouble by telling when to look, and by rational sampling, where. Careful records of conditions under which each sample was taken can be of much help in tracking down the causes.

4. Decreasing product variability. Careful watch of R or s chart and improving control of these charts and the $\bar{x}$ chart usually greatly decreases variability.

5. Determining repeatability of a measuring or analytical technique and its error. First step is to gain control on R or s chart for homogeneous material. Then $\bar{R}/d_2$ or $\bar{s}/c_4$ estimates σ_e. See Burr [26].

6. Saving on scrap and rework costs. Control charts can give aid in running at a safe level from both specification limits to minimize scrap relative to one limit and rework relative to the other. Also review 4 in this outline.

7. Increasing tool and die life, determining when to retool or reset. See Section 7.2 again.

8. Decreasing inspection for processes in control at satisfactory level. Process control samples can then be spaced farther apart and/or final inspection or customer's receiving inspection reduced. Example 4 of Section 4.7 gives a case where the sampling was reduced to once in eight hours.

9. Safer guaranteeing of product, reducing customer complaints. Processes in control are the basic ingredient. Enables Sales to know what to promise customers.

10. Obtaining warning of impending trouble. In a woolen processing line a control chart would have given a warning right at the time of process change, whereas without such a chart, trouble was not noticed for six weeks resulting in fumbling for six months.

11. Improving engineering-production relations. By production obtaining maximum economic performance of processes and reporting capabilities to engineering, more realistic specifications may be set and costs minimized.

12. Improving producer-consumer relations, by more factual knowledge of process capabilities, and by making measurements and defect definitions compatible.

13. Decreasing defects on sub-assemblies. Many cases could be cited of improvement through p, c and u charts.

14. Comparison of several inspectors, machines, processes, heads, spindles, orifices and so forth, by comparing one sample from each. See also Section 7.9, 4, on group control charts.

15. More efficient use of present equipment, knowing when new equipment is needed, checking it for stability and capability when received or even in the builder's plant.

16. Stabilization of metallurgical or chemical processes.

17. Justifying a "pat on the back." Charts give an objective way of telling who is doing an excellent job and when. Can be a great stimulus to quality performance.

18. As an objective, fact-finding tool for deciding between courses of action or proposals.

19. Determining stability and quality level of producer. May be able to safely decrease inspection (100 percent sorting to sampling, or sampling to reduced sampling). Can compare performance of different producers.

20. Meaningful reports from consumer to producer, and vice versa. In latter, producer's records may well justify greatly reduced receiving inspection or even just spot checking, as performance is improved and confidence built up.

21. Convenient, meaningful and compact records of quality performance. Standardized forms help. Records can be absolutely vital in event of a liability suit.

22. Interpretation of the many management figures such as, production, absenteeism, unit cost, budget performance, man-hours per ton, returns, sales, complaints, and so forth.

23. Exploratory work in research and development, comparison of experimental conditions.

24. Savings in loading or filling operations. As better control is obtained it becomes possible to run the average fill closer to the minimum requirement, and to do so safely. A very commonly occurring opportunity. May be able to cut "factors of safety" in similar manner.

25. Improvement of visual inspection by balancing errors of (a) missing defectives, and (b) rejecting good pieces. See Hayes [65].

26. Saving loss from too many meals loaded on an airplane.

27. Decreasing clerical errors, for example, in a mail-order house.

28. Getting an organization "quality minded."

29. Decreasing accidents. Out-of-controlness in hospital calls may foreshadow a lost-time accident, and suggest a safety campaign.

30. Classifying lumber stacks by moisture content for less spoilage and more efficient use of kilns (for drying).

31. Improving packaging of product.

32. Improving tin-plate and printing.

33. Facilitating random assembly so as to avoid the necessity of selective assembly.

34. Control of inventory, alerting as to when to restock by studying demands on inventory.

35. Getting at the causes of trouble rather than expecting the nonconforming pieces to be sorted out and not missed. The objective is to "make it right the first place," and thus to save on inspection, reworking and scrapping.

8.4. Use in the Laboratory. Is there any evidence of nonrandomness in our sample? In the light of our observed data, do we have a right to assume that laboratory conditions were held constant? How much is the repeatability error in our measurements or analyses? Control charts can supply answers to these basic questions in the laboratory.

Most statistical methods for analyzing one or more samples of data such as those in Chapters 9-12, assume randomly chosen samples. When feasible, the randomness should be tested, at least until proper sampling techniques are being used. A simple and useful test of randomness is to form the measurements into a series of small samples of three, four or five each, in the order in which the results occurred.*

The same approach can be used in attribute cases, but the opportunities are less common. Thus from a sample of 1000 pieces or tests, 10 samples of 100 each (or 20 of 50) could be used to form an np chart. Likewise, if the counts or colonies of bacteria are taken from 15 slides or plates, a c chart with 15 points could be run. In these cases, as with measurements, the average performance for the whole sample is meaningful and sound only if there is some positive evidence that the underlying conditions were substantially constant throughout, and that thus we can assume a single population.

---

*What will happen if we mix our large sample and keep drawing small samples from it until it is exhausted? The charts will show good control even if the large sample was obtained with conditions badly controlled. See Section 7.5.2.

The control chart can be used as a tool of research by taking a sample under homogeneous conditions, then varying the conditions in some way and taking another sample, etc. In this way a series of samples may be taken and tested by control charts. This approach is particularly helpful in the exploratory phases of research where it is not yet known which factors are most influential on the variable under study, or, if they are known, at approximately what level they should be held. When the important factors are known, and approximately the best levels (for desired performance, for example production or yield), then it is possible to use a statistical technique known as analysis of variance in efficient experimental designs for answering a number of questions on a minimum of data. It is possible to use control charts to analyze a rectangular table of data for both row-to-row and column-to-column effects, as explained in Section 7.5.4 Although a similar approach can be used for more complex designs, analysis of variance supplies a more direct method.

Whenever a new measuring device or analytical technique is under study, a control chart is a useful tool. If charts are run on a series of measurements and they show control, then we have some evidence of the repeatability of the measurements, and can also estimate the error of measurement by $\bar{R}/d_2$ or $\bar{s}/c_4$. This will help us in interpreting results using this technique.

The following are helpful articles on this subject: A. J. Duncan [41], M. D. Gross [59], R. H. Noel and M. A. Brumbaugh [87], E. R. Ott [94], [95], L. A. Seder [104], G. Wernimont [124], [126].

8.5. The Quality Control Person. Many of the personal qualities which make for success in other activities are important to those in quality control. Since quality control depends so largely upon teamwork, the good quality-control worker cannot be a credit grabber. For the same reason he should like people and be able to get along well with them. They should have confidence in him. The more familiar he is with the processes and products in the plant, the better.

He must be practical and down-to-earth, and able to explain statistical tools and interpretations to those in the plant who have come up "the hard way," and in fact most of the people he works with will have had scarcely any statistical background. Thoroughness and common sense are essential too. Of course some mathematical and analytical ability and background are necessary. Since much of the time, those who are untrained in statistical matters need to trust the quality-control person's word on problems, he should be of complete honesty and integrity. Keeping on studying the quality control literature, past and present, and attending related society meetings is basic too. We have left for the last that most necessary characteristic—imagination. It is the priceless ingredient.

Now, naturally no one is going to be fully equipped in all of the foregoing respects. But they are all worth developing and keeping in mind as ideals toward which to strive. Quality control is an extremely varied, interesting and important work for those who qualify.

8.6. Bibliography on Applications. In this section we shall list a substantial number of references to control charting in particular, and to quality control in general. Many of these are in the journal, *Industrial Quality Control* of the American Society for Quality Control. This journal was split into two in 1968: *Quality Progress*, and the *Journal of Quality Technology*. Abstracts of quality control articles are available in two early volumes*, and in the abstracting periodicals: *International Journal of Abstracts on Statistical Methods in Industry*, 1954-1963, and *Quality Control and Applied Statistics*, Interscience Publishers, May, 1956—. All of these sources are well indexed.

You can seek out from the following, titles of articles that appear of interest in your work.

---

*G. I. Butterbaugh, "A Bibliography of Statistical Quality Control" 1946 and supplement 1951. University of Washington Press, Seattle, Washington.

1. Standard Group of the Standards Committee of Amer. Soc. for Quality Control, "Definitions, Symbols, Formulas, and Tables for Control Charts." Standard A1-1971 (Also Z1.5-1971 of Amer. Natl. Standards Inst.) Milwaukee, Wisconsin (1971).

2. Standards Group of the Standards Committee of Amer. Soc. for Quality Control, "Glossary of General Terms Used in Quality Control." Standard A3-1971 (Also Z1.7-1971 of Amer. Natl. Standards Inst.) Milwaukee, Wisconsin (1971).

3. Standards Committee of Amer. Soc. for Quality Control, "Guide for Quality Control." Standard B1 (Also Z1.1 of Amer. Natl. Standards Inst.) Milwaukee, Wisc. (1969).

4. Standards Committee of Amer. Soc. for Quality Control, "Control Chart Method of Analyzing Data." Standard B2 (Also Z1.2 of Amer. Natl. Standards Inst.) Milwaukee, Wisc. (1969).

5. Standards Committee of Amer. Soc. for Quality Control, "Control Chart Method of Controlling Quality During Production." Standard B3 (Also Z1.3 of Amer. Natl. Standards Inst.) Milwaukee, Wisc. (1969).

6. Statistics Technical Committee of Amer. Soc. for Quality Control, "Glossary and Tables for Statistical Quality Control." Milwaukee, Wisconsin (1973).

7. Amer. Soc. for Testing Materials, "Manual on Quality Control of Materials." ASTM, Philadelphia, Pennsylvania, 1951.

8. Industry's Use of Quality Control. <u>Indust. Quality Control</u> 2 (No. 4A), 2-19 (1946). Twenty-one short articles on applications.

9. W. H. Abbott, Which standard deviation? <u>Tooling and Production</u> 22, 86-88 (1956).

10. P. D. M. Anderson and H. Herne, Application of computers in quality control. <u>Qual. Engin</u>. 33 (No. 5), 16-25 (1969).

11. G. R. Armstrong and P. C. Clarke, Frequency distribution vs. acceptance table. Indust. Quality Control 3 (No. 2), 22-27 (1946).

12. W. M. Armstrong, Foundry applications of quality control. Indust. Quality Control 2 (No. 6), 12-16 (1946).

13. C. A. Bicking, How statistics contributes to quality control in the chemical industry. Indust. Quality Control 3 (No. 4), 17-20 (1947).

14. C. A. Bicking, Statistical techniques for production control and plant scale experimentation. Indust. Quality Control 3 (No. 6), 11-14 (1947).

15. C. A. Bicking, Quality control as an administrative aid. Indust. Quality Control 14 (No. 11), 36-43 (1958).

16. C. A. Bicking, New angles on old problems of measurement and data analysis. Indust. Quality Control 22 (No. 10), 510-515 (1966).

17. C. A. Bicking and R. T. Trelfa, Some influences of paper machine design and operation on variability of paperboard. Indust. Quality Control 10 (No. 2), 17-19 (1953).

18. W. Bidlack, E. R. Close and J. C. Warren, Applications of quality control at Bausch and Lomb Optical Company. Indust. Quality Control, 2 (No. 1), 3-9 (1945).

19. R. S. Bingham, Jr., Practical chemical process control. Indust. Quality Control 13 (No. 11), 46-54 (1957).

20. R. S. Bingham, Jr., Tolerance limits and process capability studies. Indust. Quality Control 19 (No. 1), 36-40 (1962).

21. R. S. Bingham, Jr., Quality control applications in the coated abrasives industry. Indust. Quality Control 19 (No. 5), 5-12 (1962).

22. H. R. Bolton, The use of average and range charts to control material consumption. Indust. Quality Control 10 (No. 2), 50-52 (1953).

23. H. L. Breunig, Some uses of statistical control charts in the pharmaceutical industry. Indust. Quality Control 21 (No. 2), 79-86 (1964).

24. D. G. Browne, Quality control in the forging industry. Indust. Quality Control 7 (No. 3), 25-30 (1950).

25. M. A. Brumbaugh, A report on pre-war quality. Indust. Quality Control 2 (No. 2), 11-12 (1945).

26. I. W. Burr, Confidence limits on process capabilities. Indust. Quality Control 10 (No. 2), 52-55 (1953).

27. I. W. Burr, Short runs. Indust. Quality Control 11 (No. 2), 16-22 (1954).

28. R. Chateauneuf, Modern QC pays off in woodwork. Indust. Quality Control 17 (No. 3), 19-25 (1960).

29. Editors, Chem. Week, How to cash in on process control computers. Chem. Week 102. (Apr. 20), 72-82 (1968).

30. V. J. Clancey, Statistical methods in chemical analysis. Nature 159, 339-340 (1947).

31. B. L. Clark, Statistical methods in the chemical industry. Chem. and Engin. News 27, 1426-1428 (1949).

32. P. C. Clifford, Integrated control of product quality—what it means. Qual. Engin. 28, 90-92 (1964).

33. P. C. Clifford, A process capability study using control charts. J. Quality Techn. 3, 107-111 (1971).

34. A. G. Dalton, Some engineering aspects of quality control. Mech. Engin. 37, 305-307 (1948).

35. J. A. Davies, Quality control in radio tube manufacture. Proc. of Inst. of Radio Engins. 37, 548-556 (1949).

36. W. E. Deming, Some principles of the Shewhart methods of quality control. Mech. Engin. 66, 173-177 (1944).

37. W. E. Deming, Report to management. Quality Progress 5 (No. 7), 2, 3, 41 (1972).

38. H. F. Dodge, Keep it simple. Quality Progress 1 (No. 8), 11-12 (1973).

39. J. W. Dudley, Inventory policy by control chart. Indust. Quality Control 16 (No. 7), 4-8 (1960).

40. D. J. Duffy, A control chart approach to manufacturing expense. J. Indust. Engin. 11, 451-458 (1960).

41. A. J. Duncan, Confidence limits for $\bar{X}'$ when kn is less than 100. Indust. Quality Control 10 (No. 4), 31 (1954).

42. A. R. Eagle, A method for handling errors in testing and measuring. Indust. Quality Control 10 (No. 5), 10-14 (1954).

43. N. M. Ebeoglu and S. P. Phocas--Cosmetatos, How to promote statistical control charts without really trying. Quality, EOQC Jour. 10, 13-17 (1966).

44. L. S. Eichelberger, Statistical quality control in a press shop. Indust. Quality Control 13 (No. 2), 12-16 (1956).

45. S. Eilon and J. R. King, On total quality control. Qual. Engin. 29, 8-15 (1965).

46. N. L. Enrick, Variations flow analysis for process improvement. Indust. Quality Control 19 (No. 1), 23-29 (1962).

47. E. J. Erdman, L. E. Bailey and W. C. Knowler, The production line SQC helped. Quality Progress 2 (No. 8), 20-22 (1969).

48. A. A. Evans, Foundry quality control. Indust. Quality Control. 9 (No. 4), 12-15 (1953).

49. A. V. Feigenbaum, Quality programs for the 1970's. Quality Progress 1 (No. 12), 9-16 (1968).

50. V. Filimon, R. Maggass, D. Frazier and A. Klingel, Some applications of quality control techniques in motor oil can filling. Indust. Quality Control 12 (No. 2), 22-25 (1955).

51. W. C. Frey and W. M. Spencer, Automatic statistical filling control. Indust. Quality Control 14 (No. 8), 13-16 (1958).

52. L. G. Ghering, Refined method of control of cordiness and workability of glass during production. J. Amer. Ceram. Soc. 27, 373-387, (1944).

53. E. F. Gibian, Quality control by statistical methods applied to line production. Indust. Quality Control 3 (No. 2), 7-12 (1946).

54. P. E. Gnaedinger, Quality control of factory mixed batches of rubber. Rubber Age 64, 711 (1949).

55. P. E. Gnaedinger, Attaining uniformity in processing rubber batches. Indust. Quality Control 11 (No. 7), 5-10 (1955).

56. W. P. Goepfert, Some quality control applications in the Aluminum Company of America. Indust. Quality Control 11 (No. 3), 17-25 (1954).

57. R. W. Good, Quality control in relation to time - study. Indust. Quality Control 4(No. 2), 13-16 (1947).

58. W. L. Gore, Statistical methods in plastics research and development. Indust. Quality Control 4 (No. 2), 5-8 (1947).

59. M. D. Gross, Quality control in the synthetic rubber industry. Indust. Quality Control 4 (No. 3), 11-16 (1947).

60. F. E. Grubbs, On estimating precision of measuring instruments and product variability. J. Amer. Statist. Assoc. 43, 243-264 (1948).

61. F. E. Grubbs and H. J. Coon, On setting test limits relative to specification limits. Indust. Quality Control 10 (No. 5), 15-20 (1954).

62. R. L. Grunwald and C. Williams, Quality control system fits short-run jobs. Mill and Factory 61, 120-123 (1957).

63. C. K. Hamlin, Statistical quality control methods at Rome Cable Corporation. Indust. Quality Control 14 (No. 2), 8-16 (1957).

64. H. R. Harrison, Statistical quality control will work on short-run jobs. Indust. Quality Control 13 (No. 3), 8-11 (1956).

65. A. S. Hayes, Control of visual inspection. Indust. Quality Control 6 (No. 6), 73-76 (1950).

66. B. Hecht, Process control methods. Indust. Quality Control 4 (No. 1), 7-11 (1947).

67. G. V. Herrold, The introduction of quality control at Colonial Radio Corporation. Indust. Quality Control 1 (No. 1), 4-9 (1944).

68. D. Hill, Modified control limits. Appl. Statcs. 5 (No. 1) 12-19 (1956).

69. C. E. Hoover, SQC in the manufacture of piston rings. Indust. Quality Control 6 (No. 3), 21-22 (1949).

70. B. W. Jenney, General cost advantages of a quality control system. Qual. Engin. 36 (No. 7), 7-9 (1972).

71. J. M. Juran, Cultural patterns and quality control. Indust. Quality Control 14 (No. 4), 8-13 (1957).

72. W. A. Kerr, Some axioms of quality control. Indust. Quality Control 1 (No. 1), 3, 14-15 (1944).

73. E. P. King, Probability limits for the average chart when process standards are unspecified. Indust. Quality Control 10 (No. 6), 62-64 (1954).

74. J. R. King and R. Miller, Automation and quality control. Indust. Quality Control 17 (No. 4), 13-18 (1960).

75. A. G. Klock and C. W. Carter, Woolen carding meets quality control. Indust. Quality Control 8 (No. 6), 35-38 (1952).

76. G. J. Lieberman, Statistical process control and the impact of automatic process control. Technometrics 7, 283-292 (1965).

77. D. L. Lobsinger, Air transportation finds new and lucrative uses for SQC. Indust. Quality Control 6 (No. 6), 76-78 (1950).

78. J. Mandel, Efficient statistical methods in chemistry. Ind. and Engin. Chem. 17, 201-206 (1945).

79. V. E. McCoun, The case of the perjured control chart. Indust. Quality Control 5 (No. 6), 20-23 (1949).

80. R. A. McLean, V. L. Anderson, H. S. Bayer and G. W. McElrath, A scientific approach to experimentation for consulting statisticians. J. Quality Techn. 5, 1-6 (1973).

81. T. R. Meyer, J. H. Zambone and F. L. Curcio, Application of statistical quality control in glass fabrication. Glass Indust. 34, 539-541, 564 (1953).

82. D. Mills, The influence of materials and processes on quality control costs. Qual. Engin. 36 (No. 7), 11-17 (1972).

83. J. A. Mitchell, Quality control in raw materials acceptance and chemical specifications. Indust. Quality Control 4 (No. 3), 16-20 (1947).

84. V. F. Moore, Control charts to cut costs--a case study. J. Indust. Engin. 16, 288-290 (1965).

85. E. J. Newchurch, J. S. Anderson and E. H. Spencer, Quality control in petroleum research laboratory. Anal. Chem. 28, 154-157 (1956).

86. R. H. Noel, Statistical quality control in the manufacture of pharmaceuticals. Indust. Quality Control 7 (No. 2), 14-18 (1950).

87. R. H. Noel and M. A. Brumbaugh, Applications of statistics to drug manufacture. Indust. Quality Control 7 (No. 2), 7-14 (1950).

88. J. F. Occasione, Quality control as applied to continuous processes. Indust. Quality Control 13 (No. 4), 9-13 (1956).

89. A. Oladko, Modern quality control applied to compounding. Rubber Age 65, 44 (1949).

90. E. G. Olds and L. A. Knowler, Teaching statistical quality control for town and gown. J. Amer. Statist. Assoc. 44, 213-230 (1949).

91. P. S. Olmstead, How to detect the type of an assignable cause. Indust. Quality Control Part I, Clues for particular types of trouble, 8 (No. 3), 32-38 (1952), Part II, Procedure when the probable cause is unknown, 8 (No. 4), 22-32 (1953).

92. R. von Osinski, Use of median charts in the rubber industry. Indust. Quality Control 19 (No. 2), 5-8 (1962).

93. R. von Osinski, Computers and your job. Quality Assur. 5 (No. 3), 20-25 (1966).

94. E. R. Ott, Indirect calibration of an electronic test-set. Indust. Quality Control 3 (No. 4), 11-14 (1947).

95. E. R. Ott, Variables control chart in production research. Indust. Quality Control 6 (No. 3), 30-31 (1949).

96. R. G. Pippitt, More than cost reduction. Quality Progress 2 (No. 6), 18-20 (1969).

97. J. B. Pringle, SQC methods in telephone transmission maintenance. Indust. Quality Control 19 (No. 1), 18-22 (1962).

98. H. W. Reece, Optimising quality costs. Qual. Engin. 36 (No. 8), 15-19 (1972).

99. T. W. Roberts, Jr., Automatic quality control, pipe dream or reality? Indust. Quality Control 14 (No. 6), 14-18 (1957).

100. W. T. Rogers, Quality control of tubular steel products. Indust. Quality Control 13 (No. 2), 6-11 (1956).

101. G. A. T. Saywell, Practical problems in quality control. Qual. Engin. 28, 58-62 (1964).

102. R. Schin, Quality control engineering in process and product control. Indust. Quality Control 16 (No. 10), 11-17 (1960).

103. E. M. Schrock, Matters of misconception concerning the quality control chart. J. Amer. Statist. Assoc. 39, 325-334 (1944).

104. L. A. Seder, The technique of experimenting in the factory. Indust. Quality Control 4 (No. 5), 6-14 (1948).

105. W. A. Shewhart, Contribution of statistics to the science of engineering. Metal Progress 41, 854-858 (1942). Also Bell Telephone System Monograph B 1319.

106. J. Sittig, Defining quality costs. Qual. Engin. 28, 105-113 (1964).

107. B. B. Small, Use of control chart techniques in making a quality audit. Indust. Quality Control Part I, 6 (No. 1), 15-19, Part II, 6 (No. 2), 11-15 (1949).

108. H. M. Smallwood, Quality control in the manufacture of small-arms ammunition. Mech. Engin. 66, 179-182 (1944).

109. H. Smith, Jr., Testing non-randomness. Indust. Quality Control 22 (No. 3), 691-692 (1966).

110. K. M. Smith, Twenty-one ways to fail in quality control. Foundry 92 (No. 8), 46-50 (1964).

111. H. E. Thompson, A talk with the foreman about quality control. Indust. Quality Control 6 (No. 6), 29-34 (1950).

112. W. W. Thompson Jr., Some decision rules for establishing an optimal process mean. AIIE Transactions 2 (No. 2), 118-121 (1970).

113. W. S. Traylor, Use of statistical methods for time study of batch processes. Indust. Quality Control 4 (No. 4), 18-22 (1948).

114. G. Ver Beke, Statistical quality control in the foundry. Indust. Quality Control 7 (No. 6), 82-88 (1951).

115. J. F. Verigan, Quality control at the Weatherhead Company. Indust. Quality Control 6 (No. 6), 36-40 (1950).

116. G. R. Vils and M. D. Gross, Application of statistical methods to the production of synthetic rubber. Rubber Age 57, 551-558 (1945).

117. H. A. Vorhees and J. E. Culbertson, Control charts and automation applied to analysis of field failure data. Proc., Second Natl. Sympos. on Qual. Control and Reliab. in Electronics 18-45 (1956).

118. J. T. Walter, Continuous process control in a petroleum refinery. Indust. Quality Control 12 (No. 6), 5-7 (1955).

119. J. T. Walter, How reliable are lab analyses? Petrol. Refiner. 35, 106-108 (1956).

120. R. V. Ward, SQC applications in the chemical industry. Indust. Quality Control 20 (No. 1), 4-8 (1963).

121. C. B. Way, Statistical quality control applications in the food industry. Indust. Quality Control 17 (No. 11), 30-34 (1961).

122. W. R. Weaver, The foreman's view of quality control. Indust. Quality Control Part I, 5 (No. 2), 6-14, Part II, 5 (No. 3), 19-22 (1948).

123. G. Wernimont, Use of control charts in the analytical laboratory. Indust. and Engin. Chem. 18, 587-592 (1946).

124. G. Wernimont, Statistical quality control in the chemical laboratory. Indust. Quality Control 3 (No. 6), 5-11, (1947).

125. G. Wernimont, Statistical control in chemical production. Chem. Eng. 55, 272 (1948).

126. G. Wernimont, Statistics applied to analysis. Anal. Chem. 21, 115-120 (1949).

CHAPTER 9

ACCEPTANCE SAMPLING FOR ATTRIBUTES

9.1. Why Acceptance Sampling? In the relationship between producer and consumer, whether two separate business concerns, or two departments within a single organization, there is always the problem of acceptance of material (units, pieces, parts, articles or bulk product). Should the consumer accept or reject the lot of material? In the latter case, should it be returned to the producer for rectification or should the lot be 100% sorted or otherwise rectified at the consumer's inspection station?

If each lot received is to be sorted 100 per cent for all characteristics of importance, there are only two problems: arranging for sound inspection and paying for all of it. Such an approach is necessary, if most lots are not of satisfactory quality when submitted. If, however, many of the lots would be satisfactory as submitted, then acceptance sampling becomes feasible. Acceptance sampling is the use of a sample (or samples) from a lot so as to make a decision to accept or reject the lot.

Why acceptance sampling? There are two basic reasons, both good. First, it saves money, and second, in many cases it is absolutely necessary. As to money, one company using a great many small parts cut its receiving inspection costs from $7.00 per 1000 parts to $.28 by using scientific acceptance sampling. Furthermore no more trouble was encountered in assembly than previously. In this way only lots needing such attention were sorted 100 per cent, the other lots being accepted about as received.

The second reason for sampling is that it is often impossible to inspect 100 per cent. Thus, if the test is destructive (as for example in tensile strength or life tests), we must use sampling inspection if we are to have any pieces left to use! Furthermore, even a moderately simple part may have forty or fifty characteristics which are written into the description of the piece. We cannot possibly inspect for each of them on every piece. We can concentrate only on the most important attributes of the part. The less important characteristics of all parts are only sampled, if indeed checked at all.

Since acceptance sampling is desirable and often essential, we ought to make sure that we choose the right plan for our purposes and that we use it properly. Under the latter comes sound definition of a defect, sound method of inspection or testing, careful usage of that method, and random sampling from the lot. The present chapter discusses how to analyze the characteristics of attribute sampling plans for defectives or defects, while Chapter 10 describes some standard sampling plans, and Chapter 11 discusses acceptance sampling using measurements.

9.2. Characteristics of Sampling Plans. In general we need a way of describing lot quality so that we can be objective about it. When lot quality is to be described in terms of defectives, then for the class of defectives in question, each piece is a "good" or a "defective" one. We let

$$p' = \text{lot or process fraction defective.} \tag{9.1}$$

For a process or a series of lots, this is the proportion of pieces which are classified as defectives. Or for one particular lot

$$N = \text{number of pieces in the lot} \tag{9.2}$$

$$D = \text{number of defective pieces in the lot.} \tag{9.3}$$

Then for this one lot

$$p' = D/N. \tag{9.4}$$

Currently, lot quality is often described as defectives per hundred pieces. Let us use P' for this, that is

$$P' = \text{lot or process defectives per 100 pieces} \quad (9.5)$$

$$= \text{per cent defective.}$$

Also when interest is in defects rather than defectives and several "defects" can occur on one piece, then we can use for lots or processes:

$$c' = \text{average number of defects per piece} \quad (9.6)$$

or

$$C' = \text{average number of defects per 100 pieces.} \quad (9.7)$$

Meanwhile for measurable characteristics of product we have the lot or process average μ and standard deviation σ, which together are usually a sufficient description, unless the distribution of x's is markedly non-normal.

Now when sampling a lot or process of any given quality, and using objective acceptance-rejection criteria, there will be some calculable probability of acceptance, Pa.

$$Pa = \text{probability of acceptance.} \quad (9.8)$$

This is always a conditional probability, depending upon the quality level submitted. Now if by some definition the quality submitted is "undesirable" or "bad," then we would like to reject the lot. But we cannot be perfectly sure, so there is always some risk of erroneous acceptance. We call such a risk by the Greek letter β, and also call it a consumer's risk:

$$P(\text{acc.}|\text{rejectable quality}) = \beta = \text{consumer's risk} = \text{CR.} \quad (9.9)$$

Moreover, unless the lot is perfect, we cannot be absolutely sure of the sampling plan accepting a lot of acceptable quality. Thus there is almost always some risk of erroneous rejection. We call such a risk by the Greek letter α, and also call it a producer's risk:

$$P(\text{rej.}|\text{acceptable quality}) = \alpha = \text{producer's risk} = \text{PR.} \quad (9.10)$$

For most sampling plans it is easy to draw a curve showing Pa versus submitted quality. We call this the <u>Operating Characteristic</u> or OC curve for the sampling plan. It vividly shows the risks under the plan. See Section 9.4.

Another characteristic of a sampling plan is the cost of using it. Such costs are directly related to the sample size needed for a decision, and the cost of inspecting or testing each piece and making the decision.

We are interested not only in the risks of wrong decisions and cost of inspection for just one lot, but also on these characteristics for a series of lots: protection vs. cost. Since a high degree of discriminating power is purchased at relatively high costs, we need to seek a balance between (a) low protection at low cost and (b) high protection at high cost.

9.3. Acceptance Criteria for Attribute Acceptance Plans. We shall give here criteria for decisions based upon single and double sampling. In single sampling, just one sample of n pieces is drawn at random and inspected, from which a decision is always made. In double sampling, after the first sample is inspected we may accept, reject, or require a second sample. After the latter is inspected then a decision is always rendered.

The following notations will be used in addition to those in the previous section.

Single sampling:

n = number of pieces in the single sample (9.11)

d = number of defectives or defects found in the sample (9.12)

Ac = acceptance number for d (9.13)

= maximum allowable number of defectives or defects for acceptance

= c (in some systems)

Re = rejection number for d (9.14)

= minimum number of defectives or defects for rejection

= $Ac + 1$

Double sampling:

n_1 = number of pieces in first sample (9.15)

n_2 = number of pieces in second sample (9.16)

d_1 = number of defectives or defects in first sample (9.17)

d_2 = number of defectives or defects in second sample (9.18)

Ac_1 = acceptance number for d_1 (9.19)

= c_1 (in some systems)

Re_1 = rejection number for d_1 (9.20)

$Re_1 > Ac_1 + 1$ (9.21)

Ac_2 = acceptance number for $d_1 + d_2$ (9.22)

= c_2 (in some systems)

Re_2 = rejection number for $d_1 + d_2$ (9.23)

$Re_2 = Ac_2 + 1$ (9.24)

Flow Chart, Single Sampling:

n from lot yielding d defectives or defects

Acc. Lot ← $d \leq Ac$ $d \geq Re$ → Reject Lot

Flow Chart, Double Sampling:

n_1 from lot yielding d_1 defectives or defects

Acc. Lot ← $d_1 \leq Ac_1$ $Ac_1 < d_1 < Re_1$ $d_1 \geq Re_1$ → Reject Lot

↓

n_2 more from lot yielding d_2 defectives or defects

Acc. Lot ← $d_1+d_2 \leq Ac_2$ $d_1+d_2 \geq Re_2$ → Reject Lot

9.4. The Operating Characteristic (OC) Curve. Defectives.

In analyzing the characteristics of sampling plans, the one most basic is the OC curve. It is concerned with the protection provided by the plan: protection against acceptance of "bad" lots if offered, and protection against rejection of "good" lots when offered for acceptance. Or, if used on a process, we have "bad" or "good" process performance. What is the probability of acceptance, Pa, as a function of submitted quality?

For sampling decisions on defectives we have two different types of OC curves, designated as Type A and Type B. [1].

(a). Type A OC Curve. Account is taken of the lot size. The probabilities, Pa, are calculated for a finite lot of size N, which contains D defectives. Historically such probabilities were of direct interest to the consumer, especially regarding lots of undesirable quality. For example, if a consumer regards 1% as a satisfactory P' lot quality, but would regard P' = 4% as distinctly undesirable, then he might ask of a sampling plan: Suppose a lot of 500 should come along containing .04(500) = 20 defectives, then what is the probability it will be accepted? The hypergeometric distribution is exact for such Type A probabilities. See Section 9.4.2 for discussion of this distribution.

(b). Type B OC Curve. No account is taken of the lot size. The probabilities, Pa, are calculated for lots considered to be random samples from a process having a constant fraction defective, p'. In this case the number of defectives D per lot of N will vary from lot to lot, according to the binomial distribution (2.30). Moreover, the probability Pa for acceptance of a lot by the sampling plan is calculated from the binomial distribution, because it can be proven that in this case random samples of n from such a series of lots behave like random samples directly from a process at p', even though they are samples of n chosen from samples of N. (Further it can be shown that in this case of a controlled process at p', the defectives d in n and D-d in N-n are independent.) Historically, such probabilities, Pa, were of direct interest to the producer. Thus if the consumer is going to use a sampling plan n = 100, Ac = 3, he may ask at what level p' must he run his process, so as to have 95% of his lots accepted. The binomial distribution is exact for such Type B probabilities. See Section 2.5.3

We thus see that Type A probabilities make practical sense from the consumer's viewpoint when considering bad quality lots, and that Type B probabilities make practical sense to the producer when considering how good a quality level, at which he must run, to have a high proportion of his lots accepted. Thus the two types were historically associated with CR and PR respectively. As we shall see, however, it makes rather little difference which type of calculation is used. This is fortunate indeed because calculations using the hypergeometric distribution are extremely tedious.

9.4.1. Type B Operating Characteristic Curves. Let us now consider the calculation of Type B OC curves for single sampling plans. First consider a single sampling plan:

$$n = 80,\ Ac = 3,\ Re = 4.$$

Then Pa = P(3 defectives or less in 80). The exact calculation in this case involves the binomial (2.30), namely, $P(d) = C(80,d)\, p'^{d}\, q'^{80-d}$, and thus

$$Pa = \Sigma_{d=0}^{3}\, C(80,d)\, p'^{d}\, q'^{80-d}$$

Direct use of this formula can be quite a chore, but can be simplified a bit by first finding $P(d=0) = q'^{80}$, then using it to find the other needed probabilities as follows:

$$P(1) = P(0)(80/1)\cdot p'/q', \quad P(2) = P(1)(79/2)\cdot p'/q', \text{ etc.}$$

But there are available tables, [2] - [6], which provide for most cases in practice. (In fact it is not too great an undertaking to program a computer for a specific job.)

Now an available further simplification is to use the Poisson distribution (2.35) as an approximation to the binomial. This begins to be a useful apprxoimation when $n \geq 10$ and $p' \leq .10$. To visualize the approach, see Figure 9.1 from [7]. If we make use of the Poisson approximation we may use our Table VII, in the back of the book, interpolating as necessary.

The Pa values for the Type B OC curve for the given single sampling plan are shown in Table 9.1. The exact binomial entries were found in [3] using Pa = P(3 or less) = 1-P(4 or more), the last P being entries in the table used. In order to use the Poisson approximation we form an np' column in Table 9.1 providing the expected number of defectives in the sample. Then these np' are used in the left column of Table VII, and c = 3 for the column heading, and so the entry in the body of the table gives Pa. These are shown in the last column.

Careful inspection of Table 9.1 reveals that for very low p', .00 to .03, the exact Pa is greater than or equal to the approximate Pa, whereas, for worse p''s, .04 on up, the exact Pa is lower than the approximate Pa. Thus the exact Pa gives a slightly more "square-shouldered" OC curve than does the Poisson approximation. Or to put it another way the exact OC curve provides slightly smaller risks, <u>both</u> PR and CR than does the approximated OC curve. This is in general true. Hence using the Poisson approximation overstates the risks of wrong decisions in general.

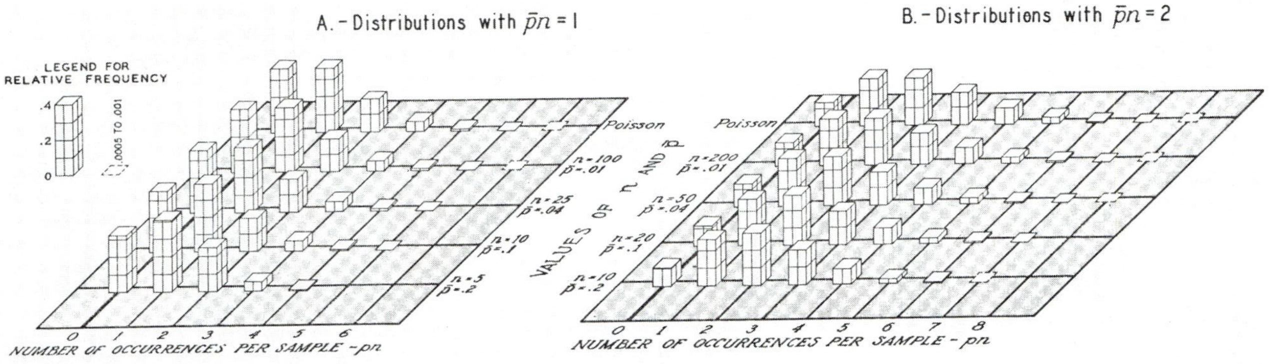

FIG. 9.1. Binomial approach to Poisson distribution [7]. Reproduced by kind permission, from "Holbrook Working, A Guide to Utilization of the Binomial and Poisson Distributions in Industrial Quality Control". Stanford University Press, Stanford University, California, 1943.

Table 9.1. Type B OC Curve for Single Sampling Plan:

n = 80, Ac = 3, Re = 4

Process p'	Exact Pa[3] = P(3 or less)	Expected Defectives in Sample 80·p'	Poisson Approximation P(3 or less). Table VII
.00	1.000	.0	1.000
.01	.991	.8	.991
.02	.923	1.6	.921
.03	.781	2.4	.779
.04	.602	3.2	.603
.05	.428	4.0	.433
.06	.286	4.8	.294
.07	.181	5.6	.191
.08	.109	6.4	.119
.09	.063	7.2	.072
.10	.035	8.0	.042

Figure 9.2 shows the OC curve for the plan together with that for a double sampling plan with quite similar characteristics. We shall shortly be discussing analyses of double sampling plans. Section 9.8.

9.4.2* The Hypergeometric Distribution. For calculations for Type A OC curves, we need the hypergeometric distribution. Since the distribution may well not be included in elementary statistics courses, we find it desirable to discuss it here.

Consider the problem of drawing randomly four motors from a lot of ten motors, of which three are defective somehow (not necessarily unusable). We might now ask what is the probability that the sample contains precisely one defective motor (and three good ones)? To answer this we treat the ten motors as distinguishable, for example, by a serial number.

Now how many distinct samples are possible, drawing at random four out of the ten, without replacement? The answer is the combination

$$C(10,4) = \frac{10\cdot 9\cdot 8\cdot 7}{4\cdot 3\cdot 2\cdot 1} = 210.$$

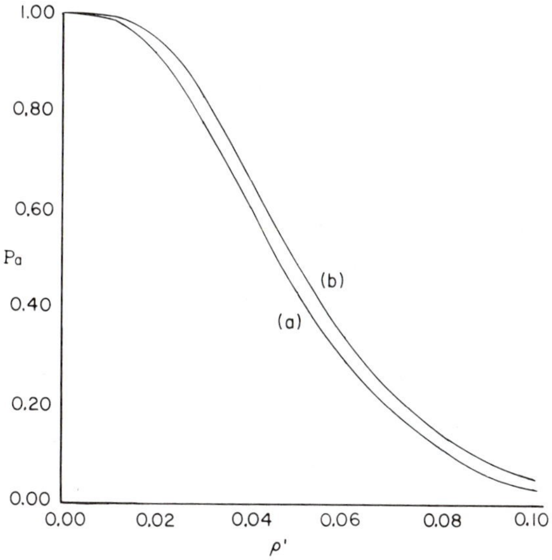

FIG. 9.2. Two Type B OC curves. (a) $n = 80$ $Ac = 3$, $Re = 4$. (b) $n_1 = n_2 = 50$, $Ac_1 = 1$, $Re_1 = 4$, $Ac_2 = 4$, $Re_2 = 5$.

Because of random drawing, all of these 210 are equally likely. Next, of the 210, how many will contain exactly one of the defectives and exactly three of the good ones? We can choose the former in

$$C(3,1) = 3$$

distinct ways, and the latter three out of the seven good ones in the lot in

$$C(7,3) = \frac{7\cdot6\cdot5}{3\cdot2\cdot1} = 35$$

distinct ways. Since each of the 3 ways can be paired with each of the 35 ways, this gives 3(35) = 105 samples with just one defective. Thus

$$P(1) = \frac{105}{210} = .500.$$

Hence half of the time the sample of four will contain exactly one defective. The other probabilities prove to be:

$$P(0) = .1667,\ P(2) = .3000,\ P(3) = .0333.$$

Now let us set down the usual symbols for the problem and obtain the general form. We have in the above

$N = 10,\ D = 3,\ n = 4,\ d = 1$

or in general

$$N = \text{lot or population size} \tag{9.25}$$

$$D = \text{number of "defectives" in lot} \tag{9.26}$$

$$n = \text{number of pieces drawn at random from lot} \tag{9.27}$$

$$d = \text{number of "defectives" in sample} \tag{9.28}$$

With this notation, then, the number of possible samples of n from the lot is C(N,n). Meanwhile we can draw d defectives from the D in the lot in C(D,d) ways, and n-d good from the N-D good in the lot in C(N-D,n-d) ways. Multiplying the last two combinations gives the number of possible samples with d defectives in the n. Thus with all specified samples being equally likely we have:

$$\begin{aligned} P(d|N,D,n) &= \frac{C(D,d)\ C(N-D,n-d)}{C(N,n)} \\ &= \frac{D!(N-D)!n!(N-n)!}{d!(D-d)!(n-d)!(N-D-n+d)!N!}\ . \end{aligned} \tag{9.29}$$

For calculational purposes we can best use the latter form in terms of factorials, by using Table IX, which gives logs of factorials. Since factorials only exist for zero or positive integers, inspection of the denominator shows the range of sample defectives to be

$$\text{Max}(0,D+n-N) \le d \le \text{Min}(D,n). \tag{9.30}$$

In quality control problems, since both D and n are usually much smaller than N, the lower limit for d is almost always 0, not positive.

Calculation of hypergeometric distribution terms is quite a chore. The best approach is to try to find the desired probabilities in a table. About the only one of substantial size is [8]. This is because we must make a quadruple entry table with probabilities listed for d, N, D, n. Hence any semblance of completeness requires a huge number of entries. Fortunately, as pointed out in [8], there are many symmetries, which reduce the number of entries needed. For example, in (9.29) we can interchange D and n without affecting the probability.

It may readily be shown that the average and standard deviation for sample defectives d are

$$E(d) = \frac{nD}{N} = np', \quad \sigma_d = \sqrt{\frac{nD}{N}\,\frac{N-D}{N}\,\frac{N-n}{N-1}} = \sqrt{np'(1-p')\frac{N-n}{N-1}} \tag{9.31}$$

where the second expression uses $p' = D/N$, the lot fraction defective in order to show the resemblance to the binomial characteristics (2.31), (2.32). $E(d)$ is identical to (2.31) and if n is small relative to N, then σ_d is nearly identical to (2.32).

These results suggest that the binomial is a good approximation for the hypergeometric, if n is, say, N/10 or less. Thus by (2.30)

$$P(d|N,D,n) \doteq C(n,d)(p')^d(1-p')^{n-d} \qquad p' = D/N. \tag{9.32}$$

Or, if D is smaller than n, a better approximation is usually given by

$$P(d|N,D,n) \doteq C(D,d)(n/N)^d[1-(n/N)]^{D-d}. \tag{9.33}$$

See [9] on these two approximations. This reference also gives formulas for improving these approximations by provision of a correction term.

Furthermore if n is, say, at least 20, and D/N, say, .05 or less, we can go even further and use the Poisson distribution (2.35) as an approximation as follows:

$$P(d|N,D,n) \doteq e^{-c'}(c')^d/(d!) \qquad c' = nD/N. \tag{9.34}$$

Thus we often can use Table VII to approximate required probabilities, even for Type A operating characteristic curves.

But if we do need to calculate hypergeometric terms from direct use of (9.29), the usual procedure is to find one term, using logarithms of factorials, and then find subsequent terms recursively. Thus from (9.29)

$$\frac{P(d+1|N,D,n)}{P(d|N,D,n)}$$

$$= \frac{D!(N-D)!n!(N-n)!d!(D-d)!(n-d)!(N-D-n+d)!N!}{(d+1)!(D-d-1)!(n-d-1)!(N-D-n+d+1)!N!D!(N-D)!n!(N-n)!}$$

$$= \frac{(D-d)(n-d)}{(d+1)(N-D-n+d+1)}.$$

Therefore

$$P(d+1|N,D,n) = \frac{(D-d)(n-d)}{(d+1)(N-D-n+d+1)} P(d|N,D,n). \qquad (9.35)$$

It is often easier to use the actual integers of N, D, n, d rather than treating (9.35) as a formula to be substituted into.

For purposes of comparison we give Table 9.2, wherein N=800, n=100 and D=16. It is easily noted that, of the two binomial approximations, the second which treats D as n (and n as D) does much better. This is because D of 16 was so

Table 9.2. Comparison of Probabilities from Hypergeometric Distribution with Two Binomials and a Poisson Approximation. N=800, n=100, D=16.

No. defectives d, in Sample	Hypergeometric Distribution	Binomial Approximations n=100,p'=.02	n=16,p'=.125	Poisson c'=2
0	.11553	.13262	.11807	.13534
1	.26985	.27065	.26986	.27067
2	.29207	.27341	.28915	.27067
3	.19443	.18227	.19276	.18045
4	.08909	.09021	.08950	.09022
5	.02979	.03535	.03068	.03609
6	.00752	.01142	.00804	.01203
7	.00146	.00313	.00164	.00344
8	.00022	.00074	.00026	.00086
9	.00003	.00015	.00004	.00019
10	.00000	.00003		.00004
11		.00001		.00001

much smaller than n=100. So in general use as n in (2.30) the smaller of D and n. Also the Poisson approximation is a little further off than either binomial.

9.4.3. Type A Operating Characteristic Curves. For Type A operating characteristic curves, we are interested in the probability of acceptance of a lot of N pieces, which contains exactly D "defectives." This may well be the situation of interest when we are to make a decision on one or two isolated lots, instead of a decision on each of a series of lots. For exact treatment this requires us to take into account the lot size, and thus to make "finite-lot" calculations, using the hypergeometric distribution. A characteristic of such sampling is that the probabilities on the draws does not remain constant. Thus in the example of Table 9.2, if on the first draw a good piece is obtained then on the second draw

$$p' = 16/799$$

while if the first piece was a defective

$$p' = 15/799.$$

Always p' at any draw depends on what happened before this draw. The hypergeometric distribution takes this dependency into account.

We therefore have for a single sampling plan defined by n and Ac, for given N:

$$P(acc|D \text{ in } N) = \sum_{d=0}^{Ac} P(d|N,D,n) \quad \text{Type A OC.} \tag{9.36}$$

The OC curve is then drawn for Pa versus fraction defective D/N. Usually it is entirely satisfactory to use one of the three approximations given in Section 9.4.2. Or one could use a second order approximation from [9], unless [8] gives the needed entries.

9.5. The Average Outgoing Quality Curve, AOQ Curve. There is a second type of curve which pictures another _protection_ characteristic of a sampling plan. When it is feasible to sort 100% the rejected lots, screening them of all defective pieces, we may ask what the average fraction defective is, including both accepted lots and those rejected and screened. This average is called the "average outgoing quality", AOQ. Obviously, the AOQ will depend upon the average submitted quality p', because if p', is relatively very good, then nearly all lots will be accepted and the AOQ fraction defective will be approximately p', (but actually a trifle less). But if p' is relatively poor, many lots will be rejected and screened, and the AOQ will be well below p'. Thus

AOQ = average outgoing quality (9.37)

= average fraction defective in all lots, after the rejected lots have been sorted and cleared of defectives; an average quality for practically perfect lots (those sorted) and lots still approximately at p' (those accepted).

Also an important number, descriptive of a plan's protection is:

AOQL = average outgoing-quality limit (9.38)

= maximum value of the AOQ's for all values of p'.

This is the worst long-run average quality with which we will have to work, no matter what p' or p''s come in for acceptance.

We shall assume Type B calculation, that is, a series of lots produced from a process at p', so that the binomial distribution is exact.

There are several slightly different interpretations possible for AOQ's, depending upon what one does with defectives which have been found. Three are:

1. The defectives found in the samples may be retained with the accepted lots and not replaced with good ones. They are presumably reworked or salvaged. Also all the defectives in the rejected lots are found and replaced by good ones. This is the simplest assumption. It may be practical in some vendor-certification programs.

2. The defectives found in the samples and the sorting of rejected lots are all removed, but not replaced by good pieces. This is the usual assumption where sampling inspection is done at some considerable distance from the place where the pieces were produced. Hence it is not very feasible to replace the defectives. This is typical of much receiving inspection.

3. The defectives found in the samples and sorting are all removed and replaced by good pieces. This is the usual assumption where sampling inspection is done close to the place where the pieces are produced, so that defectives may readily be replaced. This is typical of acceptance sampling of lots at final inspection.

In all three cases, as we do throughout the chapter, we assume that the inspectors find all the defectives in the samples and in the sorting of the lots, and that they do not reject any pieces which are actually good ones. Although this is never quite fulfilled in practice, it does provide a reasonable first approximation. Adjustments can be made if one knows in what way an inspector's work is inaccurate, whether too lenient or harsh, or inaccurate in both types of error.

Suppose for the moment we assume interpretation 1. Then we have two kinds of lots with which to work: (a) accepted lots still at an average of just about p', and (b) perfect lots. Moreover all lot sizes are still at N. Now suppose that 1000 lots each of 2000 pieces with an average fraction defective of p' = .02 are submitted for inspection under the single plan: n=80, Ac=3. We can then <u>expect</u> that 921 will be accepted, Table 9.1, and 79 will be rejected. Thus AOQ is

$$\text{AOQ} = \frac{921(.02)2000 + 79(.00)2000}{921(2000) + 79(2000)} = .921(.02) = .0184$$

In general it is seen that the proper formula for interpretation 1 is

$AOQ_{p'} = p'Pa$ (interpretation 1). (9.39)

This is also very nearly correct for the other interpretations.

Under interpretation 3 when using single sampling, the calculation of AOQ may be slightly modified to again attain full accuracy. In 3, we note that we have in the accepted lots n pieces which are all good (following any needed replacements), and N-n still at about p'. Hence the average fraction defective in the accepted lots in p'(N-n)/N, so that under interpretation 3:

$AOQ_{p'} = p'Pa[1-(n/N)]$ (interpretation 3). (9.40)

This formula may also be used in double sampling by using the average number of pieces used for a decision instead of the single n.*

Interpretation 2 gives an AOQ somewhere between those of the other two interpretations, because it does not have the improving benefit of the removed defectives being replaced by good ones.

Table 9.3 lists the AOQ's for both plans, along with other characteristics yet to be discussed. Figure 9.3 shows the two AOQ curves for the plans we are analyzing, using (9.39). The AOQL for the single sampling plan may be read off from the AOQ curve, Figure 9.3 or from inspection of the AOQ entries as in Table 9.3. It is about .024. The AOQL for the slightly more lenient double sampling plan is about .027. Such AOQL's will only occur in practice, however, if p' remains at just about .035. If the incoming fraction defective is either lower or higher than .035 then the outgoing fraction defective AOQ will be lower than the AOQL. Thus in practice the actual average outgoing quality is likely to be considerably better than the AOQL of a sampling plan.

*Since n/N is usually small, this correction is likely to be of little practical importance.

Table 9.3. Comparison of Characteristics (Type B) of Two Sampling Plans with N=2000: (a) Single n=80, Ac=3, Re=4; (b) Double $n_1=n_2=50$, $Ac_1=1$, $Re_1=4$, $Ac_2=4$, $Re_2=5$. Calculations by Poisson Distribution. Interpretation 1 for AOQ's.

Process Fraction	Single				Double			
Defective p'	Pa	AOQ	ASN	ATI	Pa	AOQ	ASN	ATI
.00	1.000	.0000	80	80	1.000	.0000	50.0	50
.01	.991	.0099	80	97	.996	.0100	54.4	62
.02	.921	.0184	80	232	.950	.0190	62.2	168
.03	.779	.0234	80	504	.831	.0249	68.8	393
.04	.603	.0241	80	842	.662	.0265	72.6	722
.05	.433	.0216	80	1169	.488	.0244	73.6	1058
.06	.294	.0176	80	1436	.339	.0203	72.4	1346
.07	.191	.0134	80	1633	.224	.0157	70.0	1567
.08	.119	.0095	80	1772	.145	.0116	67.0	1720
.09	.072	.0065	80	1862	.091	.0082	64.0	1824
.10	.042	.0042	80	1899	.057	.0057	61.2	1890

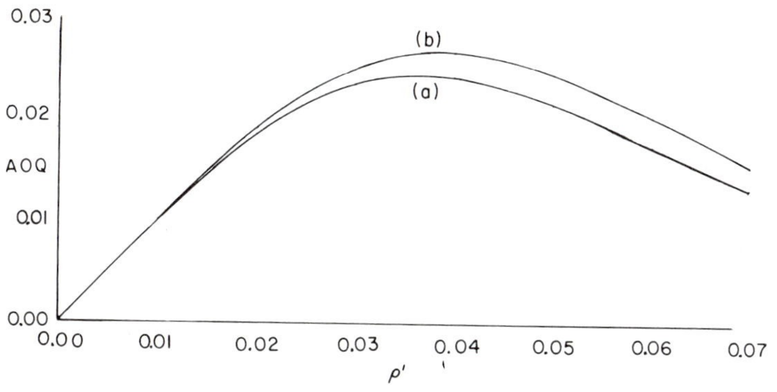

FIG. 9.3. AOQ curves for (a) single sampling plan: n=80, Ac=3, (b) double plan: $n_1=n_2=50$, $Ac_1=1$, $Re_1=4$, $Ac_2=4$, $Re_2=5$.

9.6. The Average Sample Number Curve, ASN Curve. The average sample number for a sampling plan is the average number of pieces needed to be inspected from a lot, in order to reach a decision, whether for acceptance or rejection. In single sampling, after inspection of n pieces we always reach a decision.

Now it could well occur for a lot of relatively poor quality that we reach the rejection number Re, well before inspection of n pieces. Since by then d=Re, we could reject at once without completing the sampling. Indeed this is what we would likely do if we have an isolated lot. But if we are sample inspecting a series of lots we will in general complete the inspection of all n in each lot for the record, so as to study the performance of the producer as to quality level and consistency, via a p or an np chart.

Thus for single sampling the ASN curve is in fact a straight line at height n, for every incoming p'. (As a matter of practical expediency however, if most pieces or all proved defective, by some mischance, we should probably not complete the entire sample.)

For double sampling, however, the ASN curve starts off at n_1, if p'=0, because all pieces in the sample are good and acceptance with $d_1=0$ always occurs. But as p' increases from zero the ASN curve rises, reaches a maximum and then begins to decline, eventually dropping to n_1 again. See Table 9.3.

9.7. The Average Total Inspection Curve, ATI Curve. A second characteristic directly related to inspection cost is the average total inspection, ATI, as a function of submitted quality. This is the average number of pieces inspected per lot, including those required for the decision, plus those additionally required in 100% inspection of rejected lots.

For single sampling a formula for the ATI is easily obtained. First, we must always inspect n pieces to reach a decision. Whenever this decision is for rejection, inspection of the remainder of the lot is called for, namely N-n pieces. This additional inspection occurs in a proportion of lots, 1-Pa, which of course is dependent upon the incoming fraction defective p'. Thus we have the formula

$$ATI_{p'} = n + (N-n)(1-Pa) \quad \text{Single sampling.} \tag{9.41}$$

For our example, the ATI values are given in Table 9.3. There we see that the ATI starts at n for p'=0, and steadily rises to N, when p' is high enough that virtually all lots are rejected.

The calculations for the AOQ, ASN and ATI are done in general only for Type B probabilities of acceptance Pa, since they are averages and only meaningful over a series of lots. Type A probabilities Pa are usually only used for one or a few isolated lots, if used at all.

9.8. Characteristics of Double Sampling Plans. We now consider the analysis for the characteristics of a double sampling plan. We have already presented the decision making process in Section 9.3. In Figure 9.2 and Table 9.3 we have some of the characteristics of a double sampling plan in comparison to the single sampling plan exemplified. It had n_1=50, n_2=50, Ac_1=1, Re_1=4, Ac_2=4, Re_2=5. Such decision criteria are often placed in a box like the following:

n	Σn	Acc. No.	Rej. No.
n_1 = 50	50	1	4
n_2 = 50	100	4	5

Thus if $d_1 \leq 1$ there is acceptance on the first sample, or if $d_1 \geq 4$, there is rejection on the first sample; while if d_1=2 or 3, we must draw a second random sample of 50. Inspection of the second sample yields d_2 defectives. Then if $d_1 + d_2 \leq 4$, the lot is accepted, while if $d_1 + d_2 \geq 5$ the lot is rejected.

Therefore, for Pa we must calculate the probabilities for the various "avenues" by which acceptance can occur:

d_1 = 0 or 1

d_1 = 2 followed by $d_2 \leq 2$, so that $d_1+d_2 \leq 4$

d_1 = 3 followed by $d_2 \leq 1$, so that $d_1+d_2 \leq 4$.

The calculations are made for each avenue of acceptance and then added, for Pa, since these avenues are mutually exclusive.

We could make the calculations for Type A OC curves, if necessary. But this would be very tedious, because of the necessity of using hypergeometric probabilities, and because the composition of the remainder of the lot, after the first sample is drawn, varies with the results on that sample. If the lot size is 100 or less, one can use Lieberman and Owen [8]. Otherwise log(n!) tables are necessary for (9.29). The usual recourse, however, is to use the binomial or Poisson distributions for an approximation.

More often, however, we are interested in Type B calculations of Pa, for which the binomial is exactly correct (for lots from a process with an assumed constant fraction defective p', and independence piece to piece). Then binomial tables [2] to [6], are exact and may be used. However, since p' is usually small in practical applications and $n \geq 20$, we can use the Poisson distribution as an approximation, and thus Table VII.

Let us now illustrate the calculation of Pa for our double sampling plan, using Table VII which gives

$P(c \text{ or less} \mid c')$

that is, cumulative probabilities. We take the case of p' = .04. A convenient tabular form follows:

p' = .04	$n_1=50$		$n_2=50$	Contrib. to Pa
Expectation:	$n_1p'=2.0$		$n_2p'=2.0$	
	$P(1 \text{ or less} \mid 2)=.406$			.406
		$P(2 \mid 2)=.271$	$P(2 \text{ or less} \mid 2)=.677$	.677(.271)=.183
	$P(2 \text{ or less} \mid 2)=.677$			
		$P(3 \mid 2)=.180$	$P(1 \text{ or less} \mid 2)=.406$	.406(.180)=.073
	$P(3 \text{ or less} \mid 2)=.857$			
				Pa=.662

The rationale is quite simple as discussed previously. Out of say 1000 lots incoming from a process at p' = .04, 406 of these are expected to yield none or one defective in the first sample of 50, and thus all 406 are accepted at once. Likewise 271 of these lots are expected to yield 2 defectives on the first sample because P(2) = P(0-2) - P(0,1). Judgment is withheld till a second sample is inspected. P(0-2) = .677, so 67.7% of these 271 lots can be expected to be passed by having 2 or less defectives for d_2, so that $d_1 + d_2 \leq Ac_2 = 4$. This gives 183 lots of 1000 being accepted via this route. Similarly for those with $d_1 = 3$, of which we expect 180, 40.6% of them are expected to be accepted. Then we total the probabilities for Pa.

For the exact binomial calculation at p' = .04 we use [3] obtaining the following

$p' = .04$, $n_1 = 50$, $n_1p'=2$		$n_2 = 50$, $n_2p'=2$		Contr. to Pa
P(0,1) = .40048.				.40048
	P(2) = .27623	P(0-2) = .67671	.67671(.27623) =	.18693
P(0-2) = .67671	P(3) = .18416	P(0,1) = .40048	.40048(.18416) =	.07375
P(0-3) = .86087			Pa =	.66116

This result turns out to be very close to that for the Poisson approximation.

Let us do one more, using the Poisson approximation, for comparison. Take p' = .08.

$p' = .08$ $n_1 = 50$		$n_2 = 50$		Contrib. to Pa
Expectation: $n_1p'=4.0$		$n_2p' = 4.0$		
P(0,1) = .092.				.092
	P(2) = .146	P(0-2) = .238	.238(.146) =	.035
P(0-2) = .238	P(3) = .195	P(0,1) = .092	.092(.195) =	.018
P(0-3) = .433			Pa =	.145

Both of these points on the OC curve are somewhat above the corresponding points for the single sampling plan, indicating the double plan to be more lenient at p' = .04 and .08. Other points are shown in Table 9.3. Upon comparing entries in Table 9.3 we see that the double sampling plan has a higher Pa at each p'(except at 0), and it is therefore somewhat more lenient. The OC curves are both shown in Figure 9.2, that for the double plan lying above the single plan throughout.

Next consider the average outgoing quality, AOQ, curve. We again might consider the three alternate interpretations given in Section 9.5, and could work with either Type A or Type B probabilities. Interpretation 1 leads to the same equation

$$AOQ_{p'} = p'Pa \text{ (interpretation 1, double sampling)} \tag{9.42}$$

using the same reasoning as for single sampling.

If we use interpretation 3, then in the accepted lots we have either n_1 pieces which we know are all good, or if a second sample is drawn resulting in acceptance, then n_1+n_2 pieces are known to be good. Taking account of the probability of acceptance on (a) the first sample and (b) the second sample, we have an average number of pieces known to be good as being the ASN of the plan. See below. Thus we have the following

$$AOQ_{p'} \doteq p'Pa[1 - (ASN/N)] \text{ (interpretation 3, double sampling).} \tag{9.43}$$

This formula is not quite exact. Once again, (9.42) and (9.43) are likely to give very similar results, and also interpretation 2 will give results lying between. See Table 9.3 and Figure 9.3 for the example.

For the average sample number, ASN, curve, under double sampling we really do have a "curve." In order to reach a decision we will sometimes require only n_1 pieces in a sample and other times n_1+n_2 pieces. The probability of the former is

$$P(\text{decision on first sample}) = P(d_1 \leq Ac_1) + P(d_1 \geq Re_1)$$

while the probability of the second case is

$$P(\text{decision on second sample}) = P(Ac_1 < d_1 < Re_1).$$

Since we always look at n_1 pieces, we can use the latter as the probability of having to look at n_1+n_2. Hence we have

$$ASN_{p'} = n_1 + P(Ac_1 < d_1 < Re_1)n_2 \text{ (double sampling)} \tag{9.44}$$

In order to calculate ASN's we can easily use our OC calculation tables as illustrated earlier in this section. For example if p'=.04, then

$$\text{ATI} = 50 + (.857-.406)50 = 50 + .451(50) = 72.6$$

or if $p'=.08$

$$\text{ATI} = 50 + (.433-.092)50 = 50 + .341(50) = 67.0.$$

The ASN curve for this double sampling plan is shown in Figure 9.4 from Table 9.3. Note that the average amount of inspection is everywhere less than that for the single plan, even though they have quite similar OC curves. (Of course part of the time the sample size required for the double sampling plan to reach a decision is 100, exceeding n=80 for the single plan.)

Finally let us take up the calculation for the average total inspection, ATI for double sampling. The overall inspection load is compounded of three possibilities: n_1 and n_1+n_2 under acceptance and N under rejection, since in the last case all pieces must be inspected. We have but to multiply these three amounts of inspection by the corresponding probabilities to obtain the average amount. The respective probabilities are seen to be

$$P(\text{acceptance on first sample}) = P(d_1 \leq Ac_1)$$

$$P(\text{acceptance on second sample}) = Pa - P(d_1 \leq Ac_1)$$

$$P(\text{rejection}) = 1-Pa.$$

Therefore

$$\text{ATI}_{p'} = n_1 P(d_1 \leq Ac_1) + (n_1+n_2)[Pa - P(d_1 \leq Ac_1)] + N(1-Pa). \tag{9.45}$$

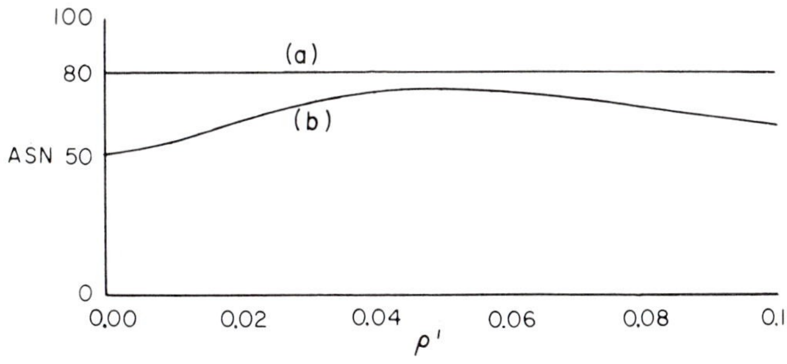

FIG. 9.4. ASN curves for (a) single plan: n=80, Ac=3 and (b) double plan: $n_1=n_2=50$, $Ac_1=1$, $Re_1=4$, $Ac_2=4$, $Re_2=5$.

For double sampling plan when p'=.04 we thus have

$$ATI = 50P(d_1 \le 1) + 100[Pa - P(d_1 \le 1)] + 2000[1-Pa]$$

$$= 50(.406) + 100(.662-.406) + 2000(1-.662) = 722$$

Note that the sum of the weights is one.

The ATI curves are drawn in Figure 9.5 from the results given in Table 9.3. The curves start at the minimum sample sizes, and steadily rise to N=2000 "asymptotically."

9.8.1.* Truncation of Inspection. When using single sampling on a series of lots, ordinarily the full sample of n pieces is inspected on each lot so as to complete the record. (An exception is made, however, if the great majority of the pieces should prove to be defectives.) By completing the entire sample we thereby make it easy to run an np chart (or a c chart if concern is with defects), and also we can easily give each lot the same weight in determining $\bar{p}$ for the producer's process.

On the other hand, in double sampling we always complete the first sample for the record as in single sampling, but since a second sample is not always required for a decision, we are at liberty to curtail the inspection as soon

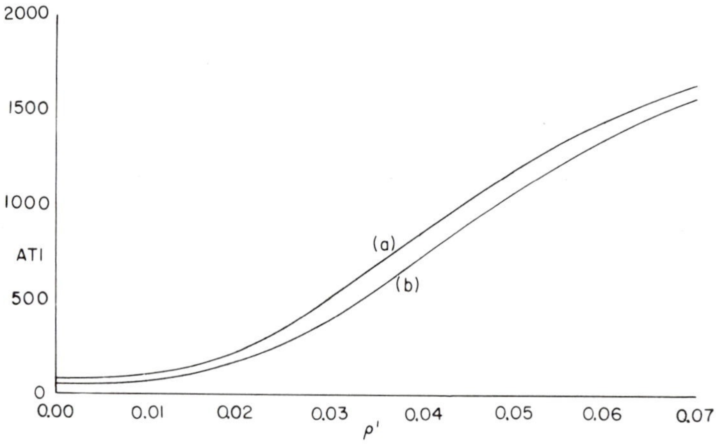

FIG. 9.5. ATI curves for (a) single sampling plan: n=80, Ac_1=3, (b) double plan: $n_1=n_2=50$, $Ac_1=1$, $Re_1=4$, $Ac_2=4$, $Re_2=5$.

as a decision is reached. Thus in our example of a double sampling plan, suppose that we find $d_1=2$ on the first sample. We are then allowed two additional defectives on the second sample to permit acceptance. Suppose we find none in the first 48. We would know then that we will have an acceptance and could stop inspection. Why? Or, if we had one additional defective in the first 49 of the second sample we would again have a sure acceptance, even if the last one were a defective. But these possible gains of Ac_2-d_1 are so relatively small that we commonly never curtail or truncate sampling in the case of an acceptance decision.

But in the case of a rejection decision there can easily be a considerable gain. Again consider, the same double sampling plan, and suppose that $d_1=2$ which means that we go into the second sample as favorably as possible. But even so, three defectives in the second sample could occur on anywhere from three to 50 pieces. Thus the possible truncated second sample sizes run from $Re_2-d_1=5-2=3$ to $n_2=50$, or the decrease in amount of inspection anywhere from 47 on down to zero. Hence, it is often well to truncate inspection as soon as rejection becomes certain (on the second sample).

Curtailing the second sample only when the decision is for rejection is called "semi-truncation."

Let us now consider the effect of "semi-truncation" on the four characteristics of a double sampling plan. Quite obviously the probability of acceptance Pa is unaffected, because we only truncate when the decision is already assured. Moreover, for the average outgoing quality, AOQ, using interpretation 1 of Section 9.5, we again have very closely AOQ = p'Pa, as in (9.42). The other interpretations, if worked exactly, give complicated formulas, but the results are very nearly the same as for 1.

The main difference occurs for the average sample number, ASN. For the calculation of the ASN we need to use a formula available in [10]. Suppose that in a sample of n pieces, k defectives calls for rejection, but k-1 or fewer yields acceptance. (For single sampling k would be Re, while for double sampling on the second sample, k would be Re_2-d_1). We are assuming the binomial distribution (Type B calculations). Then the average sample number to a decision, whether acceptance or rejection for given p' is

$$\text{ASN} = (k/p')\cdot P(k+1 \text{ or more defectives in } n+1) + $$
$$+ nP(k-1 \text{ or fewer defectives in } n). \qquad (9.46)$$

The first term on the right is the contribution under rejection, and the second term is obviously the contribution under acceptance, the full sample being completed.

We thus have for our example, the calculation

$$\text{ASN}_{p'} = 50+P(2|50,p')[(3/p')P(4 \text{ or more}|51,p') + 50P(2 \text{ or less}|50,p')] +$$
$$+ P(3|50,p')[(2/p')P(3 \text{ or more}|51, p') + 50P(0 \text{ or } 1|50,p')]$$

where the first 50 is for the first sample, the inspection of which is assumed to be always completed. Then the other two terms are the respective relative additional contributions from going into the second sample with 2 and with 3 defectives, so that the k's are 3 and 2. Such calculations are something of a chore, but tables can be useful. The calculations can be readily programmed on a computer.

A little reflection will make it clear to the reader that semi-truncation has little effect on the ASN curve if p' is relatively low, but for higher p', when the probability of a second sample is substantial it can have a rather marked effect.

As for the average total inspection ATI, semi-truncation produces no effect, because when the eventual decision is acceptance either n_1 or n_1+n_2 pieces are inspected just as for the non-truncation case; and whenever the decision is rejection we must look at all N pieces whether under truncation or not, the probability of such being 1-Pa.

9.9. Further Notations on OC Curves. In this Section we introduce a few additional notations, related to the OC curve. First we may consider the inverse job of finding p' for a given Pa. Consider Table 9.4 for the single sampling plan n=100, Ac=5. To find Pa, one begins at the first column, proceeds to the next and then to the third column. To solve the inverse problem, finding p' for given Pa, we start at the third column with our Pa, then look in Table VII, in the c=5 column for the given Pa=.500, say. We find $P(5 \text{ or less}|c'=5.6) = .512$ and $P(5 \text{ or less}|c'=5.8) = .478$. Interpolation gives $P(5 \text{ or less}|c'=5.67) = .500$. Then we divide this np' value by n to find p'. Such a p' is designated p'_{50}. Two other such special points of interest, p'_{95} and p'_{10}, are also given in the table.

Table 9.4. OC Curve, Type B calculations: binomial (exact), Poisson (approximate) for n=100, Ac=5. Additional points p'_{95}, p'_{50}, p'_{10} also calculated.

Process Fraction p'	Expectation np'	Poisson $Pa = P(d \leq 5)$	Binomial $Pa = P(d \leq 5)$
.01	1	.999	.999
.02	2	.983	.985
.03	3	.916	.919
.04	4	.785	.788
.05	5	.616	.616
.06	6	.446	.441
.07	7	.301	.291
.08	8	.191	.180
.09	9	.116	.105
.10	10	.067	.058
$.0261 = p'_{95}$	2.61	.950	
$.0567 = p'_{50}$	5.67	.500	
$.0930 = p'_{10}$	9.30	.100	

The fraction defective p'_{50}, is commonly called "indifference quality." Quite a few writers and practitioners call p'_{95} acceptable quality level, and p'_{10} lot tolerance proportion defective. But this practice does not follow other common practice, for example the military standard, MIL-STD-105D [11]. There the acceptable quality level AQL is a nominal process per cent defective, such that, if a producer runs at this level, the great majority of his lots will be accepted. This type of definition is given in [12]:

"Acceptable Quality Level [AQL] The maximum percent defective (or the maximum number of defects per hundred units) that, for purposes of acceptance sampling, can be considered satisfactory as a process average.

"Note: When a consumer designates some specific value of AQL for a certain characteristic or group of characteristics, he indicates to the supplier that his (the consumer's) acceptance sampling plan will accept the great majority of the lots that the supplier submits, provided that the process average level of percent defective in these lots be no greater than the designated value of AQL. Thus the AQL is a designated value of percent defective (or of defects per hundred units) that the consumer indicates will be accepted a great majority of the time by the acceptance sampling procedure to be used."

Therefore, although often an AQL does equal p'_{95}, this is by no means always the case. When it is true then the producer's risk, $PR = \alpha = .05$.

Similarly we find the following definition in [12]:

"Lot Tolerance Percent Defective [LTPD] Expressed in percent defective, the poorest quality in an individual lot that should be accepted."

Here again LTPD is not tied to a definite risk value. However, many think of an LTPD in terms of a consumer's risk of .10. If this is the case then we have in various segments of the literature the following notations:

$$\text{LTPD} = p_t = p'_{10} = LQ_{10}.$$

The last is "limiting quality" in the Military Standard MIL-STD 105D [11].

We have shown in Figure 9.6 some of these notations. The regions along the p' axis are not specifically bounded by p'_{95} and p'_{10}, but often would be so thought of in practice. In Table 9.4 are also given the exact probabilities (to the best three places) from the binomial. Note that these exact Pa's lie above the Pa's from the Poisson approximation, for p' up to .04, and below from $p' \geq .06$. Thus, as is typical, the binomial gives somewhat more square-shouldered or discriminating OC curves than does the Poisson approximation. If we were to use a Type A OC curve calculation with the hypergeometric distribution, the OC "curve" (actually a series of discrete points for the possible D values) would be even more square-shouldered, depending upon N. Thus usage of approximations tends to overstate the risks, both PR and CR. The OC curves cross at about $p' = Ac/n$.

9.10. Characteristics for Defects Plans. As has been mentioned earlier acceptance sampling plans might also be given in terms of defects per hundred units rather than the per cent defective. Defects plans can be adapted to

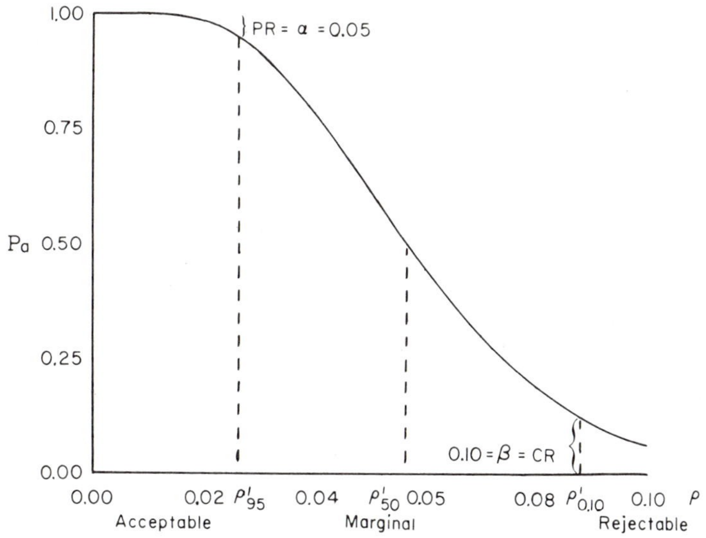

FIG. 9.6. An OC curve for n=100, Ac=5, showing various notations. Probabilities from Table 9.4.

opportunities for defects on bulk product, for example areas of paper or tinplate. But, commonly, defects plans are used for counts of defects on samples of n, or n_1 and n_2 discrete pieces.

When the average number of defects per hundred pieces is small, such as 1.0 or 1.5, it makes little difference which basis is used. For example, if we have a lot of 10,000 pieces, and the acceptable quality level AQL is 1 defect per 100 pieces, then we might think of randomly "throwing" 100 defects at the lot of 10,000. They could well all "hit" different pieces, thus giving 100 defective pieces. But more than one defect could conceivably reach the same piece, thus giving slightly fewer than 100 defective pieces in the 10,000. But we would only expect about one piece to contain two defects. So C' = 1.0 defects per hundred pieces is about equivalent to .99 per cent defective.

On the other hand if the AQL = 10 defects per hundred pieces then there is likely to be sufficient duplication to make this equivalent to 9 defectives per hundred units. Above AQL's of 10, the usual thing is to use defects per hundred units rather than per cent defective, P', for AQL. In fact we can sometimes regard as many as 1000 defects per hundred units as satisfactory quality. It all depends on the type of product and the "defects" in question.

As to calculations for an OC curve, the Type B calculations, where we are concerned with a process average of C' defects per hundred units, the Poisson distribution is not approximate but is exact, if the usual Poisson assumptions are met. When Type A calculations are to be made, we have a finite lot to draw from, instead of a process. Defects on the pieces in the finite lot cannot be exactly Poisson, but for most all practical purposes we can assume a Poisson distribution.

The OC and ASN calculations are unchanged essentially. For AOQ and ATI, in which 100% sorting of the lot occurs, such sorting must be possible, which is not always the case, for example with some kinds of bulk product.

9.11. Devising a Sampling Plan to Match Two Points on an OC Curve. There is considerable intuitive appeal to the following approach: Decide that we approve of a fraction defective of, say, $p'_1 = .01$, or less, but frown upon one of $p'_2 = .040$, or more. Then we may decide upon risks respectively of .05 and .10. Thus

$$AQL = p'_1 = .010 \qquad PR = \alpha = .05$$

$$RQL = p'_2 = .040 \qquad CR = \beta = .10.$$

Now suppose we seek a single sampling plan to satisfy these two conditions. We can only choose whole numbers for Ac: 0, 1, 2, etc. Therefore we cannot expect to hit exactly both points desired. We may choose to match either point as well as possible, using the Poisson approximation, and then for the plan so determined, the risk at the other point will be somewhat decreased, if anything. It leads to smaller sample sizes if we choose to match the (p'_2, β) point on the OC curve as closely as the Poisson approximation permits, instead of seeking to match the $(p'_1, 1-\alpha)$ point. But if we are anxious to cut the β risk rather than the α risk we can match the $(p'_1, 1-\alpha)$ point, at a cost of a somewhat larger n.

Table 9.5 considerably facilitates the finding of a desired plan. For it we have assumed $\alpha = .05$, $\beta = .10$ as is widely used. Then we find the "operating ratio" [13]:

$$R_o = p'_2/p'_1. \tag{9.47}$$

This ratio measures the relative strength of discriminating power, smaller ratios being related to greater power.

To use the table, we need to agree upon $\alpha = .05$, $\beta = .10$, then we divide our specified p'_2 by p'_1 and have the desired R_0. This will likely not be

Table 9.5. Operating Ratios Ro for Single Sampling Plans, Using Poisson Approximation, For Fitting Two Points on OC Curve: p'_1, PR=α= .05; p'_2, CR=β= .10. Acceptance Numbers Ac, and Expected numbers of Defectives per Sample also Given.

R_0*	$c = A_c$	np'_2
44.7	0	2.30
10.9	1	3.89
6.51	2	5.32
4.89	3	6.68
4.06	4	7.99
3.55	5	9.27
3.21	6	10.53
2.96	7	11.77
2.77	8	12.99
2.62	9	14.21
2.50	10	15.41
2.40	11	16.60
2.31	12	17.78
2.24	13	18.96
2.18	14	20.13
2.12	15	21.29
2.07	16	22.45
2.03	17	23.61
1.99	18	24.76

*Since sums of terms of the Poisson may be obtained from the chi-square distribution, $R_0 = \frac{\chi^2(2Ac + 2, \beta \text{ above})}{\chi^2(2Ac + 2, 1-\alpha \text{ above})}$. Also $np'_2 = .5\ \chi^2(2Ac + 2, \beta \text{ above})$.

an entry in the table. If not, we take the next smaller entry. In the row so determined is the acceptance number Ac. Also in this row we find the Poisson parameter np'_2. Given such an entry and p'_2 specified we find n by division. This gives us n and Ac, thus determining the plan. The α risk will be less than .05, and as a matter of fact, if we make the probability calculation using the binomial distribution for p'_2 and n, the Pa at p'_2, that is,

$$P(d \leq Ac|n,p'_2)$$

will be less than β also, because the binomial gives a slightly more discriminating OC curve than does the Poisson approximation.

Let us consider the example at the beginning of the section: $p'_1 = .010$, $PR = \alpha = .05$; $p'_2 = .040$, $CR = \beta = .10$. Then the desired $R_0 = 4.00$. This says use Ac = 5. Also $np'_2 = 9.27$. Therefore $n(.040) = 9.27$ or $n = 232$. So the plan is n = 232, Ac = 5. Now how does this plan do on $p'_1 = .01$? $np'_1 = 2.32$, and

$$Pa = P(d \leq 5|np'_1 = 2.32) = .968$$

and hence $\alpha = .032$.

A set of plans of this nature is given in [14]. If one wishes a <u>double</u> sampling plan to match two given points on the OC curve, the problem is not at all easy. But one can seek a single sampling plan in [11] with the desired OC curve, and then this standard provides a double sampling plan with similar characteristics.

9.12. Use of Control Chart Record with Acceptance Sampling. It is good practice to keep a control chart record of each producer on each part number which is sample inspected. Each lot is represented by the one single sample or by the first sample in double sampling. Since the sample sizes are equal, in general, an np or a c chart can be used very easily. In this way a history of the <u>level</u> and <u>stability</u> of performance of the producer can be obtained. This is of much help in dealing with the producer.

In one case the chief inspector of a manufacturer of hot water heaters announced that they wanted to use a sampling plan with an AOQL of .0025 or .25% for the temperature control which, while inexpensive, costs considerably to be removed and replaced. The manufacturer, was immediately descended upon by the producer's president, production manager and chief inspector, all of whom said in various words "We cannot live with this." Then the manufacturer's chief inspector said "What do you mean you cannot? Look at these records. For months you have been running at .001 or .10%." The visitors were gratified that they were shown to be so good, but a bit chagrined that they did not know how good they were.

If a producer's record on a part is sufficiently good, the consumer may be able to safely reduce the sample size used. This is the philosophy behind reduced sampling in the Military Standard [11]. It is simply that if the probability of a lot coming in at say 2% or worse is extremely low, then we may well use a sampling plan with a β risk at $p' = .02$, of as much as .50. This is because past history strongly indicates that such a "bad" lot just is not going to be coming along!

On the other hand if $\bar{p}$ is relatively high and/or performance is erratic, then we may wish to use a tighter sampling plan, usually in the form of lower acceptance numbers, but it could be with a larger sample size. Thus we might change from a plan with an acceptable quality level AQL of 1% to one with an average outgoing quality limit AOQL of 1%. The former is lenient toward lots at 1% and fairly lenient with lots somewhat worse, while the AOQL plan will be tighter and will make sure the outgoing quality averages no worse than 1%.

If the control chart record indicates poor enough quality we may even go off from sampling inspection altogether.

9.13. Summary - General Principles of Sampling Acceptance and Some Misconceptions. We now wish to emphasize certain basics of acceptance sampling.

9.13.1. Prerequisites to Sound Decision Making. We again emphasize four prerequisites. We first need a clear, unequivocal definition for each defect or characteristic to be inspected for. Secondly, we need an objective way to inspect. In the case of visual defects this may take the form of limit samples, which provide the dividing line between a defect and a non-defect. For a measurable characteristic we need a good measuring technique, or at least a good go-not-go gage. Thirdly, we need to obtain by proper training and supervision, careful, accurate inspection. Finally, we need good representative samples from each lot. The pieces for the sample can be drawn purely at random from small lots. This can be done by use of random sampling numbers if each piece can be assigned a number. At the very least the sampler should draw pieces from all around the lot. If a lot is large, then such pure random sampling may be infeasible. A stratified approach can be used by drawing the same size small sample from each container or carton. Or, if there are too many such cartons, we might choose a random sampling of cartons by random numbers, then from each one chosen, a small sample is drawn to make up the required total sample. In double sampling, many people draw enough pieces for the two samples, should both be needed. The whole aim is to obtain as representative and unbiased a sample as possible.

In one actual industrial case a sample of 100 piston ring castings was drawn from a box containing 3000. The sample contained 25 defective ones or 25% defective. They thought that the uninspected 2900 might well have over 700 defectives. So the 2900 were sorted 100% to salvage the good ones. But in the 2900 only 4 more defectives were found! Was the sample of 100 randomly chosen?

9.13.2. Conditional Character of Probabilities of Acceptance. We have earlier mentioned that the probability of acceptance $Pa = P(acc|p')$, always depends upon the incoming quality, for example p' or c'. Thus if, say, we let $p' = .06$ be p'_{10}, then $Pa = .10$. This leads some unwary individuals to think that the probability of having a lot at 6% to work on is .10, that is, one lot in ten of those passed will be at 6%. But such a situation can only arise if all incoming lots are at 6%. Meanwhile nine lots in ten will have been rejected, and the sensible consumer would have quickly tightened up inspection or have begun sorting each lot 100%.

In actuality, if there can somehow be ascertained an a priori distribution of incoming p' qualities for the producer, it is then possible to use this distribution and the OC curve to find the a posteriori distribution of outgoing quality levels by use of

$$P(p')P(acc|p')$$

for a set of frequency classes of p'. Occasionally it is possible from past records to obtain such an a priori distribution of lot parameters p', but probably not very commonly.

It is therefore best to think of the OC curve as giving the probability of acceptance Pa of a lot at p', should such a one be offered. We may then perhaps from control chart records, have some subjective idea of the likelihood of a bad quality lot coming along for acceptance. From this we may choose an appropriate OC curve. Or probably better, we may use a sampling acceptance system such as [11], as described in the next chapter.

9.13.3. What Is a Defective? This question has become more and more important especially as greater protection is being provided for the consuming public. Judges, attorneys and juries may well take an unfriendly attitude toward the presence of a "defective" in any product being used by the public. For this reason many organizations have begun speaking of "nonconforming" pieces rather than "defectives." We have in this book been using the terms "defective" and "defect" frequently and consistently. But it must be emphasized that there is an enormous range in the seriousness of a "defect" from something quite dangerous, to the mildest sort of blemish. For example, we may have categories of defects: critical, major A, major B, minor and incidental. A critical defect may be so important or dangerous that we must inspect each piece in the lot not once, but perhaps many times. Only after a long history of perfect production may we permit some sampling. It is for major and minor defects that we can well use sampling acceptance plans suitably chosen. For incidental defects, if sampling, then only a very small sample size need be used. Or perhaps a spot check only need be given.

Of all the characteristics of industrial parts, some deliberate choice should be made between (a) inspecting and testing each piece many times, (b) inspecting 100%, (c) sampling inspection, (d) spot checking, and (e) no inspection at all for some characteristic. Things influencing the choice are (a) the seriousness of the defect, (b) the difficulty and cost of inspecting for it, (c) the likelihood of its occurrence, and (d) whether or not the test destroys the part.*

The author sometimes wonders what proportion of characteristics can actually be carefully patrolled. For example, of 20 pages of specifications for insulated wire, how many characteristics are actually checked regularly?

9.13.4. The Proportion of Sample Size to Lot Size is Quite Unimportant. We have been giving major emphasis to Type B calculations, namely where we consider lots to have been taken from a process in control at p' or c', or where we assume infinite lot size, so that the binomial and Poisson apply exactly. Now if we are concerned with defectives, how much difference does it make to take into account the lot size, and thus to use Type A calculations, which call for the use of the hypergeometric distribution? As previously mentioned, if $n/N \leq .1$, it makes very little difference. We may even go further. Consider Figure 9.7, where all the OC curves are for the one single plan: n=10, Ac=0. However, the lot sizes vary: 20, 40, 100, 300 and in-

---

* Sometimes the test may weaken the part or increase its likelihood of failure. We may well treat such cases like destructive testing.

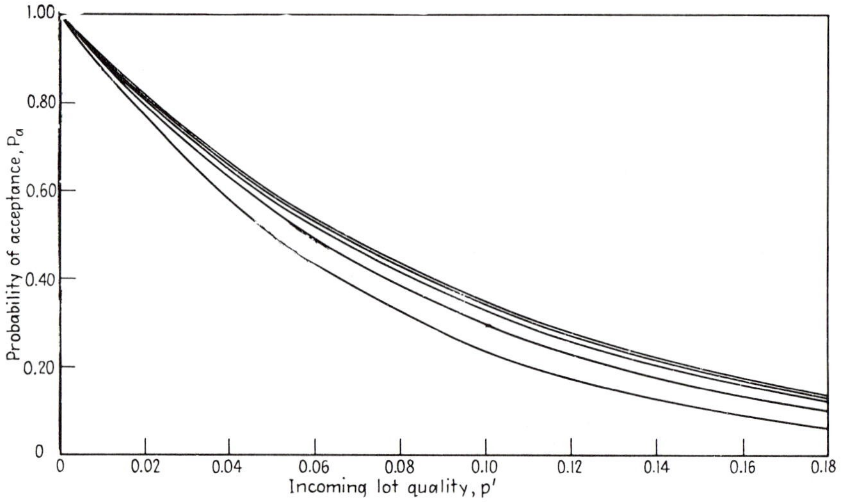

FIG. 9.7. OC curves for the plan n=10, Ac=0. The lot sizes are 20 for the bottom curve, 40 for the next, 100, 300, finally infinity for the top curve. Last is a Type B curve, the others are Type A curves. For most practical purposes the top four curves may be considered alike. Reproduced with permission from I. W. Burr, "Engineering Statistics and Quality Control", McGraw-Hill, New York, 1953, p. 316.

finity. The last is a Type B OC curve, the others Type A. The lowest curve is for N=20, in which we are sampling 50% of the lot. This OC curve is not greatly below that for N=40, in which we are sampling 25% of the lot. All five of the curves are really quite similar and scarcely differ from a practical viewpoint.

Actually the bottom four curves were drawn to guide the eye, because they really consist only of a series of discrete points. For example, if N=20, then the only possible p' values are .00, .05, .10, .15 etc., for D=0, 1, 2, etc.

From such studies we can say that a sampling plan of say n=10, Ac=0 will do just about the same job on N=1,000,000 as it will on N=100, or even N=40.

Now there are two practical considerations which suggest that somewhat larger sample sizes be used with larger N's than smaller N's. First, it is more difficult to sample randomly a very large lot. Secondly, it is a more serious error, perhaps, to erroneously reject a very large lot and thus call for 100% sorting or salvaging. But the amount of increase in n is not proportional to N. This may be seen in MIL-STD 105D [11].

It is not the proportion that the sample is of the lot that determines the discriminating power of the sample, but is the gross size of the sample and the acceptance-rejection criteria.

9.13.5. Misconception 1: "We are being scientific and fair by sampling the same percentage of each lot." As we have just seen in Section 9.13.4, the percentage of the lot sampled (n/N)100% is relatively unimportant, certainly so in comparison with n itself. But we still find people taking, say, 10% of each lot for the sample. This malady is called "percentitis."

To further consolidate this viewpoint consider Figure 9.8, in which are shown a number of OC curves, in each of which the sample is 10% of the lot.

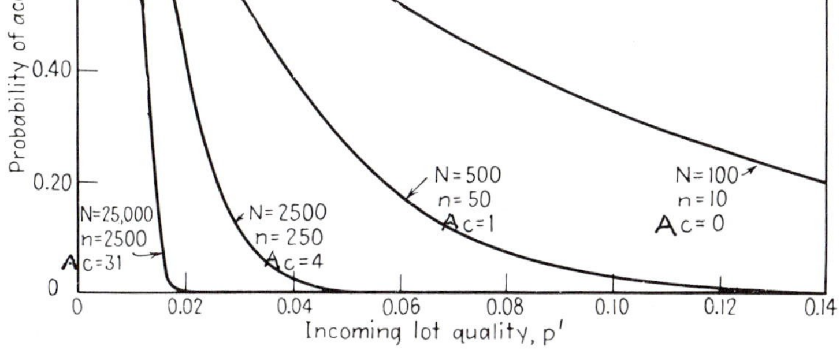

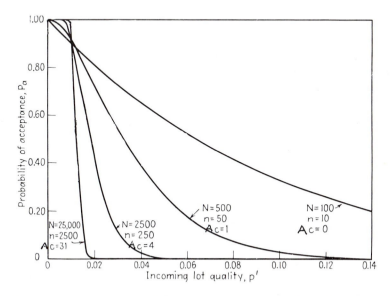

FIG. 9.8. OC curves for four 10% sampling plans, each with Pa = .90 at p' = .01. Reproduced with permission from I. W. Burr, "Engineering Statistics and Quality Control", McGraw-Hill, New York, 1953, p. 317.

Each curve has about the same Pa of .90 at p'=.01. But there the resemblence ends. For the N=100, n=10, Ac=0 plan, Pa decreases very slowly. This is quite an undiscriminating plan. As N increases we use appropriate Ac numbers to go along with n=.1N, to give $P(a|p'=.01)=.90$, and the curves become more and more sharply discriminating. The plan for N=100 is very likely much too undiscriminating, suggesting a greater sampling than 10%, whereas that for N=25,000 is likely to be far more sharply discriminating than is needed. Thus we need much less sampling than 10%, when N is very large.

Moral: pay attention to the OC curve much more than to the proportion n/N sampled.

9.13.6. Even 100% Inspection Does Not Give Complete Assurance. The author has seen many instances of the inefficiency of 100% sorting. In one case an inspector was using a dial gage with a pointer which was supposed to stop in a green sector, if the piece was between limits. The assistant plant manager noted that the inspector seemed to be just going through the motions of passing the pieces. So he covered up the dial with his hand, and the inspector passed the next 40 pieces, without being able to see the dial at all! In another case, a lot of washing machine motors, supposedly checked 100% electrically, all were packed up and shipped without any rotors (the rotating shaft and core).

Of course not all 100% inspection is this bad. But studies have shown that seldom is 100% inspection more than 80% efficient. Thus only 80% of the defectives present <u>or</u> <u>less</u>, will actually be found. Meanwhile many good pieces may well be rejected!

Efficiency of such sorting tends to be highest where there is but one objective defect possible per piece and the inspection is not too hurried. But efficiency can be quite low if there are many <u>visual</u> defects to be looked for and the pieces move along rather rapidly on a conveyor, say. Finally efficiency tends to be lower when defects are quite rare.

9.13.7. Misconception 2: "We can only use perfect lots." As has just been mentioned, 100% inspection does not give complete assurance of perfect lots. In fact even 300% or 400% inspection may not weed out all of the defectives present. How then can one hope to have perfect lots of articles to work with? The biggest hope of success is to concentrate on improving the process to make virtually all product right in the first place. It is still true, as has often been said "Quality is made at the machine, not at the inspection station."

But still there simply is no way to provide absolute guarantees of perfect lots. Commonly we must settle for something slightly less than perfection. In fact we are often forced into trade-offs between extremely high quality at very high cost and less high quality at lower cost. Fortunately a great many types of defects will show up in assembly or at final inspection and be removed or fixed.

9.13.8. Misconception 3: "We have a strong sampling plan because we do not tolerate any defectives in the sample." Many make such a boast. But the true picture shows up in the OC curve. Using Ac=0, we of course find a lazy OC curve if n is small, say, 10 or 20. There is a good chance for inferior lots to get by, if offered. See Figure 9.7 again. On the other hand if n=200, say, then for p'=.01, Pa=.1334 from a binomial table, which is very low. But if p'= .01 would be satisfactory, such a Pa would be most unfortunate. Or if p'=.001, Pa is still only up to .8186. If this is what you like and need, fine. But if so you must spell out your needs carefully to the producer, otherwise you will have trouble with him, for you are forcing him to make virtually perfect lots to have a good chance of their passing. However, perfection always comes only at a premium price.

The principle here is that OC curves for Ac=0 always rise relatively steeply to Pa=1 at p'=0, being concave upwards throughout. The OC curves for Ac > 0 all have points of inflection having the concave side down for low p' and up for high p', which gives an OC curve commonly making more sense.

9.13.9. Misconception 4: "If a sampling plan is any good it ought to give the same decision every time on a given lot." This bit of nonsense only comes from those who have not thought through the implications of the OC curve. Only if p'=0 are we sure of the same decision (acceptance) every-time. When $p' > 0$, there is always some chance for each of the two decisions to occur. Only when Pa is very close to 1, or to 0, will the decision be the same nearly all the time.

9.13.10. Misconception 5: "Acceptance sampling can only be used if the process is in control." Such an idea comes about perhaps in our study of OC curve calculations for the Type B case. But that is only to find out how the plan operates on lots from a process at p'. There is nothing that says p' remains fixed. It can vary wildly, and Type A OC calculations tell us what Pa is for a lot of any given fraction defective p'=D/N.

9.13.11. Summary. We have the following characteristics in the form of four curves:

A. Protection

1. OC curve, Pa vs. p' or c'
2. AOQ curve, AOQ vs. p' or c'

B. Cost of inspection

1. ASN curve, ASN vs. p' or c' for a decision
2. ATI curve, ATI vs. p' or c' for total inspection.

These may be based on p' or c' for the underlying process at these levels: Type B; or for an individual lot at these levels: Type A. Quality is often defined in terms of defectives per hundred units P', or defects per hundred units C'.

PROBLEMS

For the following single sampling plans find an appropriate four or five points and sketch the following curves, labeling axes: (a) OC, (b) AOQ, (c) ASN, (d) ATI. (e) What is the AOQL for the plan?

9.1. N = 4000, n = 200, Ac = 1, Re = 2 defectives [11]

9.2. N = 4000, n = 200, Ac = 3, Re = 4 defectives [11]

9.3. N = 4000, n = 200, Ac = 5, Re = 6 defectives [11]

9.4. N = 1000, n = 80, Ac = 2, Re = 3 defectives [11]

9.5. N = 1000, n = 80, Ac = 5, Re = 6 defectives [11]

9.6. N = 500, n = 50, Ac = 7, Re = 8 defects [11]

9.7. N = 500, n = 50, Ac = 10, Re = 11 defects [11]

9.8. N = 4000 n = 95, Ac = 3, Re = 4 defectives [1]

9.9. N = 1000 n = 65, Ac = 2, Re = 3 defectives [1]

9.10. N = 500 n = 39, Ac = 1, Re = 2 defectives [1]

The following double sampling plans correspond respectively to those above, 9.1 to 9.10, within the cited references. Follow the same instructions for parts (a) through (e). Hint: Use p''s for the OC curve which give simple n_1p''s to avoid interpolation in Table VII.

9.11. N=4000, $n_1=n_2=125$, $Ac_1=0$, $Re_1=2$, $Ac_2=1$, $Re_2=2$

9.12. N=4000, $n_1=n_2=125$, $Ac_1=1$, $Re_1=4$, $Ac_2=4$, $Re_2=5$

9.13. N=4000, $n_1=n_2=125$, $Ac_1=2$, $Re_1=5$, $Ac_2=6$, $Re_2=7$

9.14. N=1000, $n_1=n_2=50$, $Ac_1=0$, $Re_1=3$, $Ac_2=3$, $Re_2=4$

9.15. N=1000, $n_1=n_2=50$, $Ac_1=2$, $Re_1=5$, $Ac_2=6$, $Re_2=7$

9.16. N=500, $n_1=n_2=32$, $Ac_1=3$, $Re_1=7$, $Ac_2=8$, $Re_2=9$

9.17. N=500, $n_1=n_2=32$, $Ac_1=5$, $Re_1=9$, $Ac_2=12$, $Re_2=13$

9.18. N=4000, $n_1=80$, $n_2=140$, $Ac_1=1$, $Re_1=8$, $Ac_2=7$, $Re_2=8$

9.19. N=1000, $n_1=38$, $n_2=87$, $Ac_1=0$, $Re_1=5$, $Ac_2=4$, $Re_2=5$

9.20. N=500, $n_1=34$, $n_2=56$, $Ac_1=0$, $Re_1=4$, $Ac_2=3$, $Re_2=4$

9.21. For the plan N=5000, n=300, Ac=4, Re=5, find $p'_{95}, p'_{50}, p'_{10}$.

9.22. For the plan N=4000, n=200, Ac=2, Re=3, find $p'_{95}, p'_{50}, p'_{10}$.

Devise a single sampling plan to meet (approximately) as well as possible the following pairs of points on the OC curve:

9.23. $p_1'=.01$, $\alpha=.05$; $p_2'=.04$, $\beta=.10$

9.24. $p_1'=.005$, $\alpha=.05$; $p_2'=.06$, $\beta=.10$

9.25. $p_1'=.02$, $\alpha=.05$; $p_2'=.07$, $\beta=.10$

9.26. $p_1'=.015$, $\alpha=.05$; $p_2'=.08$, $\beta=.10$

9.27. How might you go about finding p_{10}' for a double sampling plan?

9.28. Why is it in the nature of a double sampling plan that rejection is never made on the occurrence of but one "defective" piece in the one or two samples drawn?

9.29. The following plan was devised for a part that was extremely expensive to test, lot size being 100: $n_1=2$, $Ac_1=0$, $Re_1=2$, $n_2=1$, $Ac_2=1$, $Re_2=2$. Sketch the OC curve and comment. (Hint: It is easy to derive an equation for Pa).

9.30. What form would a triple sampling plan take? Give an example of one to illustrate. What advantage might it have?

9.31. In Problems 9.12 to 9.17, usually $Re_1 < Re_2$, whereas in the older 9.18 to 9.20 $Re_1=Re_2$. What advantage is there to setting $Re_1 < Re_2$? (Consider 9.18, suppose $d_1=7$, how good is the conditional probability of acceptance now?)

9.32. Compare single and double sampling as to the relative practical advantages of each.

9.33. When is the Poisson distribution a reasonable approximation to the binomial distribution? The binomial for the hypergeometric? The Poisson for the hypergeometric?

9.34. What might be wrong with using two sampling plans, single and double with similar OC curves as follows: Start out on the double plan, then if one takes a second sample, only continue on the second sample till we reach a total sample equivalent to n of the single plan and now use Ac of the single pla[n]. Does this give the same OC curve?

9.35. Verify the other three probabilities for the example in Section 9.4.2.

Find for the hypergeometric distribution the following terms, either using direct calculation and Table IX, or [8].

9.36. $P(3|50,10,8)$

9.37. $P(1|40,6,4)$

9.38. $P(4|60,12,6)$

9.39. $P(2|30,5,4)$

9.40. One of the helpful symmetries used in [8] is that n and D may be interchanged without changing the value. Thus $P(2|30,5,4) = P(2|30,4,5)$ in (9.29). Prove this is true.

9.41. Note the recursion relation (9.35) for the hypergeometric distribution Use the same approach to derive simpler recursion relations for the binomial (2.30) and the Poisson (2.35).

9.42. Find the ASN without and with truncation of second sample (under rejection), for the double sampling plan: N=2000, $n_1=n_2=40$, $Ac_1=1$, $Re_1=4$, $Ac_2=4$, $Re_2=5$ at p'=.05. Use binomial calculation, and appropriate tables if available.

References

1. H. F. Dodge and H. G. Romig, "Sampling Inspection Tables." Wiley, New York, 1959.

2. U. S. Dept. of the Army, "Tables of the Binomial Distribution." Natl. Bur. Standards, Appl. Math. Ser. 6 U. S. Govt. Printing Office, Washington, DC, 1950.

3. Harvard Univ., Comput. Lab., "Tables of the Cumulative Binomial Probability Distribution." Harvard Univ. Press, Cambridge, Massachusetts, 1955.

4. W. H. Robertson, "Tables of the Binomial Distribution Function for Small Values of p." Tech. Services, Dept. of Commerce, Washington, DC, 1960.

5. H. G. Romig, "Fifty to 100 Binomial Tables." Wiley, New York, 1947.

6. S. Weintraub, "Cumulative Binomial Probability Distribution for Small Values of p." Free Press of Glencoe, Collier-Macmillan, London, 1963.

7. H. Working, "A Guide to Utilization of the Binomial and Poisson Distributions." Stanford Univ. Press, Stanford, California, 1943.

8. G. J. Lieberman and D. B. Owen, "Tables of the Hypergeometric Probability Distribution." Stanford Univ. Press, Stanford, California, 1961.

9. I. W. Burr, Some approximate relations between terms of the hypergeometric, binomial and Poisson distributions. Communications in Stat. 1, 297-301 (1973).

10. Statistical Research Group, Columbia Univ., "Sampling Inspection." McGraw-Hill, New York, 1948.

11. Military Standard MIL-105D (ABC), "Sampling Procedures and Tables for Inspection by Attributes." Dept. of Defense, 1963. Also Amer. Nat. Standards Inst. Z1.4, and available from Amer. Soc. for Quality Control, Milwaukee, Wisconsin.

12. ASQC Standard A2-1971, "Definitions and Symbols for Acceptance Sampling by Attributes," Amer. Soc. for Quality Control, Milwaukee, Wisconsin.

13. P. Peach and S. B. Littauer, A note on sampling inspection by attributes. Ann. Math. Statist. 17, 81-84 (1946).

14. R. L. Kirkpatrick, Binomial sampling plans indexed by AQL and LTPD. Indust. Quality Control 22 (No. 6), 290-292 (1965).

CHAPTER 10

SOME STANDARD PLANS FOR ATTRIBUTES

10.1. Introduction. Chapter 9 discussed fundamental characteristics of acceptance sampling plans for attributes. In particular two of the characteristics were related to protection, namely the OC and AOQ curves, while the other two were related to cost, that is, the ASN and ATI curves. Calculations of these various curves for single and double sampling plans were discussed and illustrated. In Section 9.11 there was presented a method for finding a plan matching two points on the OC curve, a table being given for the case when p'_{95} and p'_{10} are both specified.

In the foregoing we have been concerned with individual sampling plans, and their analysis. In the present chapter we consider collections of sampling plans or systems, integrated to satisfy appropriate goals. Different aims give rise to differing sets of plans. It is a great convenience to have a set of plans all calculated out and available. But to make sound use of them one should be very familiar with the basic elements as discussed in Chapter 9.

Also we take up multiple sampling, a natural extension of single and double sampling.

The early work in acceptance sampling by attributes was largely concentrated in the Bell Telephone Laboratories, beginning about 1923. Among the concepts originated by engineers and mathematicians in that organization were acceptable quality level (AQL), lot tolerance per cent defective (LTPD), consumer's risk (CR or β), producer's risk (PR or α), average outgoing quality (AOQ), average outgoing-quality limit (AOQL), average total inspection (ATI), reduced inspection, tightened inspection, and double and multiple sampling. Among the men responsible for the developments were (alphabetically) H. F. Dodge, G. D. Edwards, T. C. Fry, E. C. Molina, P. S. Olmstead, H. G. Romig, and W. A. Shewhart. W. Bartky working with Western Electric Co., did important work in multiple sampling. The famous Dodge-Romig tables were one outgrowth of this Bell Laboratories work.

In order to meet the enormous problems of acceptance of material during World War II, the Army Ordnance Department carried on further work in acceptance sampling procedures. Several key people from the Bell Laboratories worked on the project with Army Ordnance personnel. Among the latter were H. R. Bellinson, G. R. Gause, L. W. Shaw, L. E. Simon, and A. Stein. The ultimate result was the development of a set of 12 integrated tables of sampling plans. These were commonly called the Army Ordnance Tables.

Early in 1945, the Statistical Research Group of Columbia University, which had been working on statistical problems arising in national defense, was asked by the United States Navy to prepare a manual of tables, procedures, and principles for sampling inspection. The Navy manual, or Navy Tables, for sampling inspection by attributes was thus produced. Similar work for variables was discontinued in August, 1945. Subsequently the attribute tables were adopted for use by both the United States Army and United States Navy, and they became known as the Joint Army-Navy Service Forces Tables, or JAN Tables. The men chiefly responsible for the preparation of the Navy Tables were H. A. Freeman, Milton Friedman, Frederick Mosteller, L. J. Savage, D. H. Schwartz, and W. A. Wallis. See [1].

The history of acceptance sampling by attributes is of much interest. An excellent and authoritative presentation of this history is given in [2], which covers the evolution through the currently most widely used system ABC-STD-105, which is also called MIL-STD-105D. It is also an American National Standards

Institute ANSI-Z1.4. We will discuss this system in Section 10.3. It is presently in world-wide use.

10.2. The Dodge-Romig Sampling Inspection Tables. These tables were originally published in 1929, coming out in a book in 1944. An enlarged and extended second edition came out in 1959 [3]. The tables aim at providing consumer protection on a minimum inspection cost. Basically and specifically

Aim of Dodge-Romig Sampling Inspection Tables.

1. To provide specified consumer protection, either a specified LTPD (p'_{10}), or a specified AOQL value.
2. Among the many sampling plans which could supply the specified kind and amount of consumer protection, the tables give that one plan which gives this protection on a minimum average total inspection ATI for given lot size N and quality level $\bar{p}$.

Let us illustrate this aim in reference to one of the Dodge-Romig tables here reproduced as Table 10.1. This table has six column sections corresponding to classes of process average $\bar{p}$. Then too the 19 rows are for classes of lot size N. If one wants a double sampling plan providing an AOQL protection of 1%, he uses this table and enters it via his N and his estimated p', that is $\bar{p}$. (If no $\bar{p}$ is yet available the rightmost section for $\bar{p}$ = .81 to 1.00% can well be used, till some history is built up.) Thus N and $\bar{p}$ lead to a block in the table. There one finds sample sizes n_1 and n_2, the acceptance number c_1 for the first sample (Ac_1 in Chapter 9), the acceptance number c_2 for the combined sample of $n_1 + n_2$ (Ac_2 in Chapter 9). In this system $c_2 + 1$ is the rejection number for both the first and for the combined samples (that is, both Re_1 and Re_2).

Now suppose that we have N = 1500. In the appropriate rows we find six plans

TABLE 10.1. A double sampling, average quality Dodge-Romig table.

Double Sampling Table for
Average Outgoing Quality Limit (AOQL) = 1.0%

Lot Size	Process Average 0 to 0.02%						Process Average 0.03 to 0.20%						Process Average 0.21 to 0.40%					
	Trial 1		Trial 2			p_t %	Trial 1		Trial 2			p_t %	Trial 1		Trial 2			p_t %
	n_1	c_1	n_2	n_1+n_2	c_2		n_1	c_1	n_2	n_1+n_2	c_2		n_1	c_1	n_2	n_1+n_2	c_2	
1–25	All	0	–	–	–	–	All	0	–	–	–	–	All	0	–	–	–	–
26–50	22	0	–	–	–	7.7	22	0	–	–	–	7.7	22	0	–	–	–	7.7
51–100	33	0	17	50	1	6.9	33	0	17	50	1	6.9	33	0	17	50	1	6.9
101–200	43	0	22	65	1	5.8	43	0	22	65	1	5.8	43	0	22	65	1	5.8
201–300	47	0	28	75	1	5.5	47	0	28	75	1	5.5	47	0	28	75	1	5.5
301–400	49	0	31	80	1	5.4	49	0	31	80	1	5.4	55	0	60	115	2	4.8
401–500	50	0	30	80	1	5.4	50	0	30	80	1	5.4	55	0	65	120	2	4.7
501–600	50	0	30	80	1	5.4	50	0	30	80	1	5.4	60	0	65	125	2	4.6
601–800	50	0	35	85	1	5.3	60	0	70	130	2	4.5	60	0	70	130	2	4.5
801–1000	55	0	30	85	1	5.2	60	0	75	135	2	4.4	60	0	75	135	2	4.4
1001–2000	55	0	35	90	1	5.1	65	0	75	140	2	4.3	75	0	120	195	3	3.8
2001–3000	65	0	80	145	2	4.2	65	0	80	145	2	4.2	75	0	125	200	3	3.7
3001–4000	70	0	80	150	2	4.1	70	0	80	150	2	4.1	80	0	175	255	4	3.5
4001–5000	70	0	80	150	2	4.1	70	0	80	150	2	4.1	80	0	180	260	4	3.4
5001–7000	70	0	80	150	2	4.1	75	0	125	200	3	3.7	80	0	180	260	4	3.4
7001–10,000	70	0	80	150	2	4.1	80	0	125	205	3	3.6	85	0	180	265	4	3.3
10,001–20,000	70	0	80	150	2	4.1	80	0	130	210	3	3.6	90	0	230	320	5	3.2
20,001–50,000	75	0	80	155	2	4.0	80	0	135	215	3	3.6	95	0	300	395	6	2.9
50,001–100,000	75	0	80	155	2	4.0	85	0	180	265	4	3.3	170	1	380	550	8	2.6

Lot Size	Process Average 0.41 to 0.60%						Process Average 0.61 to 0.80%						Process Average 0.81 to 1.00%					
	Trial 1		Trial 2			p_t %	Trial 1		Trial 2			p_t %	Trial 1		Trial 2			p_t %
	n_1	c_1	n_2	n_1+n_2	c_2		n_1	c_1	n_2	n_1+n_2	c_2		n_1	c_1	n_2	n_1+n_2	c_2	
1–25	All	0	–	–	–	–	All	0	–	–	–	–	All	0	–	–	–	–
26–50	22	0	–	–	–	7.7	22	0	–	–	–	7.7	22	0	–	–	–	7.7
51–100	33	0	17	50	1	6.9	33	0	17	50	1	6.9	33	0	17	50	1	6.9
101–200	43	0	22	65	1	5.8	43	0	22	65	1	5.8	47	0	43	90	2	5.4
201–300	55	0	50	105	2	4.9	55	0	50	105	2	4.9	55	0	50	105	2	4.9
301–400	55	0	60	115	2	4.8	55	0	60	115	2	4.8	60	0	80	140	3	4.5
401–500	55	0	65	120	2	4.7	60	0	95	155	3	4.3	60	0	95	155	3	4.3
501–600	60	0	65	125	2	4.6	65	0	100	165	3	4.2	65	0	100	165	3	4.2
601–800	65	0	105	170	3	4.1	65	0	105	170	3	4.1	70	0	140	210	4	3.9
801–1000	65	0	110	175	3	4.0	70	0	150	220	4	3.8	125	1	180	305	6	3.5
1001–2000	80	0	165	245	4	3.7	135	1	200	335	6	3.3	140	1	245	385	7	3.2
2001–3000	80	0	170	250	4	3.6	150	1	265	415	7	3.0	215	2	355	570	10	2.8
3001–4000	85	0	220	305	5	3.3	160	1	330	490	8	2.8	225	2	455	680	12	2.7
4001–5000	145	1	225	370	6	3.1	225	2	375	600	10	2.7	240	2	595	835	14	2.5
5001–7000	155	1	285	440	7	2.9	235	2	440	675	11	2.6	310	3	665	975	16	2.4
7001–10,000	165	1	355	520	8	2.7	250	2	585	835	13	2.4	385	4	785	1170	19	2.3
10,001–20,000	175	1	415	590	9	2.6	325	3	655	980	15	2.3	520	6	980	1500	24	2.2
20,001–50,000	250	2	490	740	11	2.4	340	3	910	1250	19	2.2	610	7	1410	2020	32	2.1
50,001–100,000	275	2	700	975	14	2.2	420	4	1050	1470	22	2.1	770	9	1850	2620	41	2.0

Trial 1: n_1 = first sample size; c_1 = acceptance number for first sample
"All" indicates that each piece in the lot is to be inspected
Trial 2: n_2 = second sample size; c_2 = acceptance number for first and second samples combined
p_t = lot tolerance per cent defective with a Consumer's Risk (P_C) of 0.10

Reproduced with permission from H. F. Dodge and H. G. Romig, "Sampling Inspection Tables," Wiley, New York, 1959, p. 210.

n_1	c_1	n_2	n_1+n_2	c_2
55	0	35	90	1
65	0	75	140	2
75	0	120	195	3
80	0	165	245	4
135	1	200	335	6
140	1	245	385	7

Each one of these six plans has approximately the same AOQL, namely 1%, but they have differing ATI's at each incoming quality. Now if p has been running at $\bar{p}$ = .5%, then we should use the 80, 165 plan to obtain the lowest of the six ATI's when $\bar{p}$ = .5%. But if $\bar{p}$ is say .9% we obtain the lowest ATI at .9% by using the 140, 245 plan. Thus, knowing N and how $\bar{p}$ has been running we can find a plan providing protection at an AOQL of 1%, and which does so on a minimum ATI, for this $\bar{p}$.

We note also that in each block there is listed the lot tolerance per cent defective p_t, for which the consumer's risk is .10, that is, p'_{10}. These run from 5.1% down to 3.2%. Therefore we obtain this degree of LTPD protection, along with our 1% AOQL protection. So in reality we have two kinds of protection for the consumer.

The Dodge-Romig tables necessarily involve some compromises or approximations. It is essential that classes be used for N's and $\bar{p}$'s to avoid having infinitely many cases to handle. Thus with only six sections (classes of $\bar{p}$) we cannot absolutely guarantee that we have minimized the ATI for each and every combination of $\bar{p}$ and N leading to that block. But the ATI will be very nearly a minimum. With whole numbers for c_1 and c_2 we also may not hit the desired AOQL exactly. Finally, n_1+n_2 is always made divisible by 5, and the Poisson distribution is used as an approximation to the hypergeometric, whenever feasible.

We here give also as Table 10.2 an LTPD table for double sampling plans, with LTPD = p_t = 4.0%. In this table all plans have p_t = 4.0% closely, and the AOQL's vary, increasing in general as $\bar{p}$ and n's increase.

Further, we reproduce in Table 10.3 a single sample, average quality table, analogous to Table 10.1.

The available LTPD's in the older tables for p_t protection are .5, 1.0, 2.0, 3.0, 4.0, 5.0, 7.0 and 10.0%, while for the newer ones for average quality protection, the AOQL's are .1, .25, .50, .75, 1.0, 1.5, 2.0, 2.5, 3.0, 4.0, 5.0, 7.0 and 10.0%.

In the book there is provided a great collection of Type B (binomial) OC curves. All OC curves for the AOQL plans, both single and double are given. Then in another appendix there are single sample OC curves for c = 0, 1, 2, 3 and $n \leq 500$. These can be very useful.

The introductory chapters are also of much interest. The tables are still widely useful.

We also mention that by replacing _per cents defective_ by _defects per hundred units_ the tables may be used for defects instead of defectives. Thus for acceptance one would count defects in samples rather than defectives to compare with acceptance and rejection numbers.

In summary we give the decisions and requirements needed to find a desired Dodge-Romig sampling plan:

1. Decision on whether to use single or double sampling.
2. Decision on whether to specify AOQL or LTPD protection and the desired percent for that type chosen. Decisions in 1 and 2 lead to the appropriate table.
3. Decide whether to use defectives or defects.
4. Specify the lot size N.

TABLE 10.2. A double sampling, lot quality Dodge-Romig Table.

Double Sampling Table for Lot Tolerance Per Cent Defective (LTPD) = 4.0%

Lot Size	Process Average 0 to 0.04% — Trial 1		Trial 2			AOQL in %	Process Average 0.05 to 0.40% — Trial 1		Trial 2			AOQL in %	Process Average 0.41 to 0.80% — Trial 1		Trial 2			AOQL in %
	n_1	c_1	n_2	n_1+n_2	c_2		n_1	c_1	n_2	n_1+n_2	c_2		n_1	c_1	n_2	n_1+n_2	c_2	
1–35	All	0	–	–	–	0	All	0	–	–	–	0	All	0	–	–	–	0
36–50	34	0	–	–	–	0.35	34	0	–	–	–	0.35	34	0	–	–	–	0.35
51–75	40	0	–	–	–	0.43	40	0	–	–	–	0.43	40	0	–	–	–	0.43
76–100	50	0	25	75	1	0.46	50	0	25	75	1	0.46	50	0	25	75	1	0.46
101–150	55	0	30	85	1	0.55	55	0	30	85	1	0.55	55	0	30	85	1	0.55
151–200	60	0	30	90	1	0.64	60	0	30	90	1	0.64	60	0	30	90	1	0.64
201–300	60	0	35	95	1	0.70	60	0	35	95	1	0.70	60	0	65	125	2	0.75
301–400	65	0	35	100	1	0.71	65	0	35	100	1	0.71	65	0	65	130	2	0.80
401–500	65	0	40	105	1	0.73	65	0	70	135	2	0.83	65	0	70	135	2	0.83
501–600	65	0	40	105	1	0.74	65	0	75	140	2	0.85	65	0	100	165	3	0.93
601–800	65	0	40	105	1	0.75	65	0	75	140	2	0.87	65	0	110	175	3	0.97
801–1000	70	0	40	110	1	0.76	70	0	75	145	2	0.90	70	0	105	175	3	0.98
1001–2000	70	0	40	110	1	0.78	70	0	80	150	2	0.94	70	0	145	215	4	1.1
2001–3000	70	0	80	150	2	0.95	70	0	115	185	3	1.1	70	0	180	250	5	1.2
3001–4000	70	0	80	150	2	0.96	70	0	115	185	3	1.1	110	1	175	285	6	1.3
4001–5000	70	0	80	150	2	0.97	70	0	115	185	3	1.1	115	1	170	285	6	1.3
5001–7000	70	0	80	150	2	0.98	70	0	115	185	3	1.1	115	1	205	320	7	1.4
7001–10,000	70	0	80	150	2	0.98	70	0	150	220	4	1.2	115	1	205	320	7	1.4
10,001–20,000	70	0	80	150	2	0.98	70	0	150	220	4	1.2	115	1	235	350	8	1.5
20,001–50,000	70	0	80	150	2	0.99	70	0	150	220	4	1.2	115	1	270	385	9	1.6
50,001–100,000	70	0	80	150	2	0.99	70	0	185	255	5	1.3	115	1	300	415	10	1.7

Lot Size	Process Average 0.81 to 1.20% — Trial 1		Trial 2			AOQL in %	Process Average 1.21 to 1.60% — Trial 1		Trial 2			AOQL in %	Process Average 1.61 to 2.00% — Trial 1		Trial 2			AOQL in %
	n_1	c_1	n_2	n_1+n_2	c_2		n_1	c_1	n_2	n_1+n_2	c_2		n_1	c_1	n_2	n_1+n_2	c_2	
1–35	All	0	–	–	–	0	All	0	–	–	–	0	All	0	–	–	–	0
36–50	34	0	–	–	–	0.35	34	0	–	–	–	0.35	34	0	–	–	–	0.35
51–75	40	0	–	–	–	0.43	40	0	–	–	–	0.43	40	0	–	–	–	0.43
76–100	50	0	25	75	1	0.46	50	0	25	75	1	0.46	50	0	25	75	1	0.46
101–150	55	0	30	85	1	0.55	55	0	30	85	1	0.55	55	0	30	85	1	0.55
151–200	60	0	55	115	2	0.68	60	0	55	115	2	0.68	60	0	55	115	2	0.68
201–300	60	0	65	125	2	0.75	60	0	90	150	3	0.84	60	0	90	150	3	0.84
301–400	65	0	95	160	3	0.86	65	0	95	160	3	0.86	65	0	120	185	4	0.92
401–500	65	0	100	165	3	0.92	65	0	130	195	4	0.96	105	1	140	245	6	1.0
501–600	65	0	135	200	4	1.0	105	1	145	250	6	1.1	105	1	175	280	7	1.1
601–800	65	0	140	205	4	1.0	105	1	185	290	7	1.2	105	1	210	315	8	1.2
801–1000	110	1	155	265	6	1.2	110	1	210	320	8	1.2	145	2	230	375	10	1.3
1001–2000	110	1	195	305	7	1.3	150	2	240	390	10	1.5	180	3	295	475	13	1.6
2001–3000	110	1	260	370	9	1.4	185	3	305	490	13	1.6	220	4	410	630	18	1.7
3001–4000	150	2	255	405	10	1.5	185	3	340	525	14	1.6	285	6	465	750	22	1.8
4001–5000	150	2	285	435	11	1.6	185	3	395	580	16	1.7	285	6	520	805	24	1.9
5001–7000	150	2	320	470	12	1.6	185	3	435	620	17	1.7	320	7	585	905	27	2.0
7001–10,000	150	2	325	475	12	1.7	220	4	460	680	19	1.9	320	7	645	965	29	2.1
10,001–20,000	150	2	355	505	13	1.7	220	4	495	715	20	1.9	350	8	790	1140	35	2.2
20,001–50,000	150	2	420	570	15	1.7	255	5	575	830	24	2.0	385	9	895	1280	40	2.3
50,001–100,000	150	2	450	600	16	1.8	255	5	665	920	27	2.1	415	10	985	1400	44	2.4

Trial 1: n_1 = first sample size; c_1 = acceptance number for first sample
"All" indicates that each piece in the lot is to be inspected
Trial 2: n_2 = second sample size; c_2 = acceptance number for first and second samples combined
AOQL = Average Outgoing Quality Limit

TABLE 10.3. Single sampling, average quality Dodge-Romig tables.

Single Sampling Table for
Average Outgoing Quality Limit (AOQL) = 1.0%

Lot Size	Process Average 0 to 0.02%			Process Average 0.03 to 0.20%			Process Average 0.21 to 0.40%			Process Average 0.41 to 0.60%			Process Average 0.61 to 0.80%			Process Average 0.81 to 1.00%.		
	n	c	p_t %	n	c	p_t %	n	c	p_t %	n	c	p_t %	n	c	p_t %	n	c	p_t %
1–25	All	0	–	All	0	–	All	0	–	All	0	–	All	0	–	All	0	–
26–50	22	0	7.7	22	0	7.7	22	0	7.7	22	0	7.7	22	0	7.7	22	0	7.7
51–100	27	0	7.1	27	0	7.1	27	0	7.1	27	0	7.1	27	0	7.1	27	0	7.1
101–200	32	0	6.4	32	0	6.4	32	0	6.4	32	0	6.4	32	0	6.4	32	0	6.4
201–300	33	0	6.3	33	0	6.3	33	0	6.3	33	0	6.3	33	0	6.3	65	1	5.0
301–400	34	0	6.1	34	0	6.1	34	0	6.1	70	1	4.6	70	1	4.6	70	1	4.6
401–500	35	0	6.1	35	0	6.1	35	0	6.1	70	1	4.7	70	1	4.7	70	1	4.7
501–600	35	0	6.1	35	0	6.1	75	1	4.4	75	1	4.4	75	1	4.4	75	1	4.4
601–800	35	0	6.2	35	0	6.2	75	1	4.4	75	1	4.4	75	1	4.4	120	2	4.2
801–1000	35	0	6.3	35	0	6.3	80	1	4.4	80	1	4.4	120	2	4.3	120	2	4.3
1001–2000	36	0	6.2	80	1	4.5	80	1	4.5	130	2	4.0	130	2	4.0	180	3	3.7
2001–3000	36	0	6.2	80	1	4.6	80	1	4.6	130	2	4.0	185	3	3.6	235	4	3.3
3001–4000	36	0	6.2	80	1	4.7	135	2	3.9	135	2	3.9	185	3	3.6	295	5	3.1
4001–5000	36	0	6.2	85	1	4.6	135	2	3.9	190	3	3.5	245	4	3.2	300	5	3.1
5001–7000	37	0	6.1	85	1	4.6	135	2	3.9	190	3	3.5	305	5	3.0	420	7	2.8
7001–10,000	37	0	6.2	85	1	4.6	135	2	3.9	245	4	3.2	310	5	3.0	430	7	2.7
10,001–20,000	85	1	4.6	135	2	3.9	195	3	3.4	250	4	3.2	435	7	2.7	635	10	2.4
20,001–50,000	85	1	4.6	135	2	3.9	255	4	3.1	380	6	2.8	575	9	2.5	990	15	2.1
50,001–100,000	85	1	4.6	135	2	3.9	255	4	3.1	445	7	2.6	790	12	2.3	1520	22	1.9

Single Sampling Table for
Average Outgoing Quality Limit (AOQL) = 1.5%

Lot Size	Process Average 0 to 0.03%			Process Average 0.04 to 0.30%			Process Average 0.31 to 0.60%			Process Average 0.61 to 0.90%			Process Average 0.91 to 1.20%			Process Average 1.21 to 1.50%		
	n	c	p_t %	n	c	p_t %	n	c	p_t %	n	c	p_t %	n	c	p_t %	n	c	p_t %
1–15	All	0	–	All	0	–	All	0	–	All	0	–	All	0	–	All	0	–
16–50	16	0	11.6	16	0	11.6	16	0	11.6	16	0	11.6	16	0	11.6	16	0	11.6
51–100	20	0	9.8	20	0	9.8	20	0	9.8	20	0	9.8	20	0	9.8	20	0	9.8
101–200	22	0	9.5	22	0	9.5	22	0	9.5	22	0	9.5	22	0	9.5	44	1	8.2
201–300	23	0	9.2	23	0	9.2	23	0	9.2	47	1	7.9	47	1	7.9	47	1	7.9
301–400	23	0	9.3	23	0	9.3	49	1	7.8	49	1	7.8	49	1	7.8	49	1	7.8
401–500	23	0	9.4	23	0	9.4	50	1	7.7	50	1	7.7	50	1	7.7	50	1	7.7
501–600	24	0	9.0	24	0	9.0	50	1	7.7	50	1	7.7	50	1	7.7	50	1	7.7
601–800	24	0	9.1	24	0	9.1	50	1	7.8	50	1	7.8	80	2	6.4	80	2	6.4
801–1000	24	0	9.1	55	1	7.0	55	1	7.0	85	2	6.2	85	2	6.2	85	2	6.2
1001–2000	24	0	9.1	55	1	7.0	55	1	7.0	85	2	6.2	120	3	5.4	155	4	5.0
2001–3000	24	0	9.2	55	1	7.1	90	2	5.9	125	3	5.3	160	4	4.9	200	5	4.6
3001–4000	24	0	9.2	55	1	7.1	90	2	5.9	125	3	5.3	165	4	4.8	240	6	4.4
4001–5000	24	0	9.2	55	1	7.1	90	2	5.9	125	3	5.3	205	5	4.6	280	7	4.2
5001–7000	24	0	9.2	55	1	7.1	90	2	5.9	165	4	4.8	205	5	4.6	325	8	4.0
7001–10,000	24	0	9.2	55	1	7.1	130	3	5.2	165	4	4.8	250	6	4.2	375	9	3.8
10,001–20,000	55	1	7.1	90	2	5.9	130	3	5.2	210	5	4.4	340	8	3.8	515	12	3.4
20,001–50,000	55	1	7.1	90	2	5.9	170	4	4.7	295	7	4.0	480	11	3.5	860	19	3.0
50,001–100,000	55	1	7.1	130	3	5.2	210	5	4.4	340	8	3.8	625	14	3.3	1120	24	2.8

n = sample size; c = acceptance number
"All" indicates that each piece in the lot is to be inspected
p_t = lot tolerance per cent defective with a Consumer's Risk (P_C) of 0.10

Reproduced with permission from H. F. Dodge and H. G. Romig, "Sampling Inspection Tables," Wiley, New York, 1959, p. 200.

5. Use $\bar{p}$ in percent. If $\bar{p}$ is unknown the conservative approach is to use the section corresponding to the largest $\bar{p}$'s. For defects basis, given $\bar{c}$ as the average number of defects per individual piece or unit, then $\bar{c} \cdot 100$, the average defects per 100 units, corresponds to $\bar{p}$ in %.
6. N and $\bar{p}$ in 4 and 5 lead to the specific plan in the appropriate table. The amount of the opposite consumer protection is shown.
7. As information on $\bar{p}$ accumulates from single samples or first samples in double sampling, different sections of the table may be used.

10.3. The ABC or Military Standard 105D Plan. This sampling system is the outgrowth of the Army Service Forces and the Joint Army-Navy tables. The present system is the third revision of the Military Standard MIL-STD 105A. It was developed by an international committee from the United States, Great Britain and Canada, thus the current designation "ABC." This evolution is detailed in Part III of [2]. It was the work of many people!

We are here reproducing the introductory material, most of the summary tables, special descriptive tables, and an example table for one of the "code letters." We shall describe various aspects of the plans in relation to this material.

10.3.1. Basic Aim. The ABC Standard 105D aims to insure that, on the average, the consumer will be using material at the specified acceptable quality level AQL, or better. It is designed so that if a producer runs his process at the AQL exactly, the great majority of his lots will be accepted. This probability of acceptance at the AQL runs from about .88 for small lots and relatively tight AQL's up to about .995 for large lots and/or relatively loose

AQL's. [4]. Of course if his process is better than the AQL, then the Pa will be correspondingly higher.

But if the producer runs at a worse p' than the AQL, he may well have some lots accepted, but sooner or later the sampling plan will go onto "tightened inspection," and then a smaller proportion of lots will be accepted. Moreover if the producer does not soon improve his process, so as to justify a return to "normal inspection," then 100% inspection will be instituted and stern action taken. But if he improves his process sufficiently the plan will reinstitute "normal inspection."

Thus the basic aim of the ABC Standard is to enable the outgoing quality to be maintained at the specified AQL or better.

10.3.2. General Description. The sampling plans are contained in nine summary tables: Single, Table II; Double, Table III, and Multiple, Table IV. As conditions warrant, one may specify any one of the three kinds of sampling. See below for "multiple sampling." Then each of the three tables is divided into three parts for "normal," "tightened" and "reduced inspection," the last being for small sample sizes when quality is consistently excellent relative to the AQL. Institution of tightened inspection is essential when called for, but use of reduced inspection, when available is optional.

The AQL's are to be specified in terms of <u>defectives</u> per hundred units, that is, percent defective, or in terms of <u>defects</u> per hundred units. The latter may be used for all AQL's whereas the former is only used for AQL's of 10.0 or lower. See Sections 4.2 to 4.6 in the tables.

There are various "inspection levels," providing differing sample sizes to give relatively lower or higher discriminating power to the OC curve. The standard levels are I, II and III, in which II is in general assumed, or used if none is specified. Level I gives less discrimination, III greater.

SAMPLING PROCEDURES AND TABLES FOR INSPECTION BY ATTRIBUTES

1. SCOPE

1.1 PURPOSE. This publication establishes sampling plans and procedures for inspection by attributes. When specified by the responsible authority, this publication shall be referenced in the specification, contract, inspection instructions, or other documents and the provisions set forth herein shall govern. The "responsible authority" shall be designated in one of the above documents.

1.2 APPLICATION. Sampling plans designated in this publication are applicable, but not limited, to inspection of the following:

a. End items.

b. Components and raw materials.

c. Operations.

d. Materials in process.

e. Supplies in storage.

f. Maintenance operations.

g. Data or records.

h. Administrative procedures.

These plans are intended primarily to be used for a continuing series of lots or batches. The plans may also be used for the inspection of isolated lots or batches, but, in this latter case, the user is cautioned to consult the operating characteristic curves to find a plan which will yield the desired protection (see 11.6).

1.3 INSPECTION. Inspection is the process of measuring, examining, testing, or otherwise comparing the unit of product (see 1.5) with the requirements.

1.4 INSPECTION BY ATTRIBUTES. Inspection by attributes is inspection whereby either the unit of product is classified simply as defective or nondefective, or the number of defects in the unit of product is counted, with respect to a given requirement or set of requirements.

1.5 UNIT OF PRODUCT. The unit of product is the thing inspected in order to determine its classification as defective or nondefective or to count the number of defects. It may be a single article, a pair, a set, a length, an area, an operation, a volume, a component of an end product, or the end product itself. The unit of product may or may not be the same as the unit of purchase, supply, production, or shipment.

2. CLASSIFICATION OF DEFECTS AND DEFECTIVES

2.1 METHOD OF CLASSIFYING DEFECTS. A classification of defects is the enumeration of possible defects of the unit of product classified according to their seriousness. A defect is any nonconformance of the unit of product with specified requirements. Defects will normally be grouped into one or more of the following classes; however, defects may be grouped into other classes, or into subclasses within these classes.

2.1.1 CRITICAL DEFECT. A critical defect is a defect that judgment and experience indicate is likely to result in hazardous or unsafe conditions for individuals using, maintaining, or depending upon the product; or a defect that judgment and experience indicate is likely to prevent performance of the tactical function of a major end item such as a ship, aircraft, tank, missile or space vehicle. NOTE: For a special provision relating to critical defects, see 6.3.

2.1.2 MAJOR DEFECT. A major defect is a defect, other than critical, that is likely to result in failure, or to reduce materially the usability of the unit of product for its intended purpose.

2.1.3 MINOR DEFECT. A minor defect is a defect that is not likely to reduce materially the usability of the unit of product for its intended purpose, or is a departure from established standards having little bearing on the effective use or operation of the unit.

2.2 METHOD OF CLASSIFYING DEFECTIVES. A defective is a unit of product which contains one or more defects. Defectives will usually be classified as follows:

2.2.1 CRITICAL DEFECTIVE. A critical defective contains one or more critical defects and may also contain major and or minor defects. NOTE: For a special provision relating to critical defectives, see 6.3.

2.2.2 MAJOR DEFECTIVE. A major defective contains one or more major defects, and may also contain minor defects but contains no critical defect.

2.2.3 MINOR DEFECTIVE. A minor defective contains one or more minor defects but contains no critical or major defect.

3. PERCENT DEFECTIVE AND DEFECTS PER HUNDRED UNITS

3.1 EXPRESSION OF NONCONFORMANCE. The extent of nonconformance of product shall be expressed either in terms of percent defective or in terms of defects per hundred units.

3.2 PERCENT DEFECTIVE. The percent defective of any given quantity of units of product is one hunderd times the number of defective units of product contained therein divided by the total number of units of product, i.e.:

$$\text{Percent defective} = \frac{\text{Number of defectives}}{\text{Number of units inspected}} \times 100$$

3.3 DEFECTS PER HUNDRED UNITS. The number of defects per hundred units of any given quantity of units of product is one hundred times the number of defects contained therein (one or more defects being possible in any unit of product) divided by the total number of units of product, i.e.:

$$\text{Defects per hundred units} = \frac{\text{Number of defects}}{\text{Number of units inspected}} \times 100$$

4. ACCEPTABLE QUALITY LEVEL (AQL)

4.1 USE. The AQL, together with the Sample Size Code Letter, is used for indexing the sampling plans provided herein.

4.2 DEFINITION. The AQL is the maximum percent defective (or the maximum number of defects per hundred units) that, for purposes of sampling inspection, can be considered satisfactory as a process average (see 11.2).

4.3 NOTE ON THE MEANING OF AQL. When a consumer designates some specific value of AQL for a certain defect or group of defects, he indicates to the supplier that his (the consumer's) acceptance sampling plan will accept the great majority of the lots or batches that the supplier submits, provided the process average level of percent defective (or defects per hundred units) in these lots or batches be no greater than the designated value of AQL. Thus, the AQL is a designated value of percent defective (or defects per hundred units) that the consumer indicates will be accepted most of the time by the acceptance sampling procedure to be used. The sampling plans provided herein are so arranged that the probability of acceptance at the designated AQL value depends upon the sample size, being generally higher for large samples than for small ones, for a given AQL. The AQL alone does not describe the protection to the consumer for individual lots or batches but more directly relates to what might be expected from a series of lots or batches, provided the steps indicated in this publication are taken. It is necessary to refer to the operating characteristic curve of the plan, to determine what protection the consumer will have.

4.4 LIMITATION. The designation of an AQL shall not imply that the supplier has the right to supply knowingly any defective unit of product.

4.5 SPECIFYING AQLs. The AQL to be used will be designated in the contract or by the responsible authority. Different AQLs may be designated for groups of defects considered collectively, or for individual defects. An AQL for a group of defects may be designated in addition to AQLs for individual defects, or subgroups, within that group. AQL values of 10.0 or less may be expressed either in percent defective or in defects per hundred units; those over 10.0 shall be expressed in defects per hundred units only.

4.6 PREFERRED AQLs. The values of AQLs given in these tables are known as preferred AQLs. If, for any product, an AQL be designated other than a preferred AQL, these tables are not applicable.

5. SUBMISSION OF PRODUCT

5.1 LOT OR BATCH. The term lot or batch shall mean "inspection lot" or "inspection batch," i.e., a collection of units of product from which a sample is to be drawn and inspected to determine conformance with the acceptability criteria, and may differ from a collection of units designated as a lot or batch for other purposes (e.g., production, shipment, etc.).

5.2 FORMATION OF LOTS OR BATCHES. The product shall be assembled into identifiable lots, sublots, batches, or in such other manner as may be prescribed (see 5.4). Each lot or batch shall, as far as is practicable,

5. SUBMISSION OF PRODUCT (Continued)

consist of units of product of a single type, grade, class, size, and composition, manufactured under essentially the same conditions, and at essentially the same time.

5.3 LOT OR BATCH SIZE. The lot or batch size is the number of units of product in a lot or batch.

5.4 PRESENTATION OF LOTS OR BATCHES. The formation of the lots or batches, lot or batch size, and the manner in which each lot or batch is to be presented and identified by the supplier shall be designated or approved by the responsible authority. As necessary, the supplier shall provide adequate and suitable storage space for each lot or batch, equipment needed for proper identification and presentation, and personnel for all handling of product required for drawing of samples.

6. ACCEPTANCE AND REJECTION

6.1 ACCEPTABILITY OF LOTS OR BATCHES. Acceptability of a lot or batch will be determined by the use of a sampling plan or plans associated with the designated AQL or AQLs.

6.2 DEFECTIVE UNITS. The right is reserved to reject any unit of product found defective during inspection whether that unit of product forms part of a sample or not, and whether the lot or batch as a whole is accepted or rejected. Rejected units may be repaired or corrected and resubmitted for inspection with the approval of, and in the manner specified by, the responsible authority.

6.3 SPECIAL RESERVATION FOR CRITICAL DEFECTS. The supplier may be required at the discretion of the responsible authority to inspect every unit of the lot or batch for critical defects. The right is reserved to inspect every unit submitted by the supplier for critical defects, and to reject the lot or batch immediately, when a critical defect is found. The right is reserved also to sample, for critical defects, every lot or batch submitted by the supplier and to reject any lot or batch if a sample drawn therefrom is found to contain one or more critical defects.

6.4 RESUBMITTED LOTS OR BATCHES. Lots or batches found unacceptable shall be resubmitted for reinspection only after all units are re-examined or retested and all defective units are removed or defects corrected. The responsible authority shall determine whether normal or tightened inspection shall be used, and whether reinspection shall include all types or classes of defects or for the particular types or classes of defects which caused initial rejection.

7. DRAWING OF SAMPLES

7.1 SAMPLE. A sample consists of one or more units of product drawn from a lot or batch, the units of the sample being selected at random without regard to their quality. The number of units of product in the sample is the sample size.

7.2 REPRESENTATIVE SAMPLING. When appropriate, the number of units in the sample shall be selected in proportion to the size of sublots or subbatches, or parts of the lot or batch, identified by some rational criterion.

7. DRAWING OF SAMPLES (Continued)

When representative sampling is used, the units from each part of the lot or batch shall be selected at random.

7.3 TIME OF SAMPLING. Samples may be drawn after all the units comprising the lot or batch have been assembled, or samples may be drawn during assembly of the lot or batch.

7.4 DOUBLE OR MULTIPLE SAMPLING. When double or multiple sampling is to be used, each sample shall be selected over the entire lot or batch.

8. NORMAL, TIGHTENED AND REDUCED INSPECTION

8.1 INITIATION OF INSPECTION. Normal inspection will be used at the start of inspection unless otherwise directed by the responsible authority.

8.2 CONTINUATION OF INSPECTION. Normal, tightened or reduced inspection shall continue unchanged for each class of defects or defectives on successive lots or batchs except where the switching procedures given below require change. The switching procedures given below require a change. The switching procedures shall be applied to each class of defects or defectives independently.

8.3 SWITCHING PROCEDURES.

8.3.1 NORMAL TO TIGHTENED. When normal inspection is in effect, tightened inspection shall be instituted when 2 out of 5 consecutive lots or batches have been rejected on original inspection (i.e., ignoring resubmitted lots or batches for this procedure).

8.3.2 TIGHTENED TO NORMAL. When tightened inspection is in effect, normal inspection shall be instituted when 5 consecutive lots or batches have been considered acceptable on original inspection.

8.3.3 NORMAL TO REDUCED. When normal inspection is in effect, reduced inspection shall be instituted providing that all of the following conditions are satisfied:

a. The preceding 10 lots or batches (or more, as indicated by the note to Table VIII) have been on normal inspection and none has been rejected on original inspection; and

b. The total number of defectives (or defects) in the samples from the preceding 10 lots or batches (or such other number as was used for condition "a" above) is equal to or less than the applicable number given in Table VIII. If double or multiple sampling is in use, all samples inspected should be included, not "first" samples only; and

c. Production is at a steady rate; and

d. Reduced inspection is considered desirable by the responsible authority.

8.3.4 REDUCED TO NORMAL. When reduced inspection is in effect, normal inspection shall be instituted if any of the following occur on original inspection:

a. A lot or batch is rejected; or

b. A lot or batch is considered acceptable under the procedures of 10.1.4; or

c. Production becomes irregular or delayed; or

d Other conditions warrant that normal inspection shall be instituted.

8.4 DISCONTINUATION OF INSPECTION. In the event that 10 consecutive lots or batches remain on tightened inspection (or such other number as may be designated by the responsible authority), inspection under the provisions of this document should be discontinued pending action to improve the quality of submitted material.

9. SAMPLING PLANS

9.1 SAMPLING PLAN. A sampling plan indicates the number of units of product from each lot or batch which are to be inspected (sample size or series of sample sizes) and the criteria for determining the acceptability of the lot or batch (acceptance and rejection numbers).

9.2 INSPECTION LEVEL. The inspection level determines the relationship between the lot or batch size and the sample size. The inspection level to be used for any particular requirement will be prescribed by the responsible authority. Three inspection levels: I, II, and III, are given in Table I for general use. Unless otherwise specified, Inspection Level II will be used. However, Inspection Level I may be specified when less discrimination is needed, or Level III may be specified for greater discrimination. Four additional special levels: S–1, S–2, S–3 and S–4, are given in the same table and may be used where relatively small sample sizes are necessary and large sampling risks can or must be tolerated.

NOTE: In the designation of inspection levels S–1 to S–4, care must be exercised to avoid AQLs inconsistent with these inspection levels.

9.3 CODE LETTERS. Sample sizes are designated by code letters. Table I shall be used to find the applicable code letter for the particular lot or batch size and the prescribed inspection level.

9.4 OBTAINING SAMPLING PLAN. The AQL and the code letter shall be used to obtain the sampling plan from Tables II, III or IV. When no sampling plan is available for a given combination of AQL and code letter, the tables direct the user to a different letter. The sample size to be used is given by the new code letter not by the original letter. If this procedure leads to different sample sizes for different classes of defects, the code letter corresponding to the largest sample size derived may be used for all classes of defects when designated or approved by the responsible authority. As an alternative to a single sampling plan with an acceptance number of 0, the plan with an acceptance number of 1 with its correspondingly larger sample size for a designated AQL (where available), may be used when designated or approved by the responsible authority.

9.5 TYPES OF SAMPLING PLANS. Three types of sampling plans: Single, Double and Multiple, are given in Tables II, III and IV, respectively. When several types of plans are available for a given AQL and code letter, any one may be used. A decision as to type of plan, either single, double, or multiple, when available for a given AQL and code letter, will usually be based upon the comparison between the administrative difficulty and the average sample sizes of the available plans. The average sample size of multiple plans is less than for double (except in the case corresponding to single acceptance number 1) and both of these are always less than a single sample size. Usually the administrative difficulty for single sampling and the cost per unit of the sample are less than for double or multiple.

10. DETERMINATION OF ACCEPTABILITY

10.1 PERCENT DEFECTIVE INSPECTION. To determine acceptability of a lot or batch under percent defective inspection, the applicable sampling plan shall be used in accordance with 10.1.1, 10.1.2, 10.1.3, 10.1.4, and 10.1.5.

10.1.1 SINGLE SAMPLING PLAN. The number of sample units inspected shall be equal to the sample size given by the plan. If the number of defectives found in the sample is equal to or less than the acceptance number, the lot or batch shall be considered acceptable. If the number of defectives is equal to or greater than the rejection number, the lot or batch shall be rejected.

10.1.2 DOUBLE SAMPLING PLAN. The number of sample units inspected shall be equal to the first sample size given by the plan. If the number of defectives found in the first sample is equal to or less than the first acceptance number, the lot or batch shall be considered acceptable. If the number of defectives found in the first sample is equal to or greater than the first rejection number, the lot or batch shall be rejected. If the number of defectives found in the first sample is between the first acceptance and rejection numbers, a second sample of the size given by the plan shall be inspected. The number of defectives found in the first and second samples shall be accumulated. If the cumulative number of defectives is equal to or less than the second acceptance number, the lot or batch shall be considered acceptable. If the cumulative number of defectives is equal to or greater than the second rejection number, the lot or batch shall be rejected.

10.1.3 MULTIPLE SAMPLE PLAN. Under multiple sampling, the procedure shall be similar to that specified in 10.1.2, except that the number of successive samples required to reach a decision may be more than two.

10.1.4 SPECIAL PROCEDURE FOR REDUCED INSPECTION. Under reduced inspection, the sampling procedure may terminate without either acceptance or rejection criteria having been met. In these circumstances, the lot or batch will be considered acceptable, but normal inspection will be reinstated starting with the next lot or batch (see 8.3.4 (b)).

10.2 DEFECTS PER HUNDRED UNITS INSPECTION. To determine the acceptability of a lot or batch under Defects per Hundred Units inspection, the procedure specified for Percent Defective inspection above shall be used, except that the word "defects" shall be substituted for "defectives."

11. SUPPLEMENTARY INFORMATION

11.1 OPERATING CHARACTERISTIC CURVES. The operating characteristic curves for normal inspection, shown in Table X (pages 30–62), indicate the percentage of lots or batches which may be expected to be accepted under the various sampling plans for a given process quality. The curves shown are for single sampling; curves for double and multiple sampling are matched as closely as practicable. The O. C. curves shown for AQLs greater than 10.0 are based on the Poisson distribution and are applicable for defects per hundred units inspection; those for AQLs of 10.0 or less and sample sizes of 80 or less are based on the binomial distribution and are applicable for percent defec-

11. SUPPLEMENTARY INFORMATION (Continued)

tive inspection; those for AQLs of 10.0 or less and sample sizes larger then 80 are based on the Poisson distribution and are applicable either for defects per hundred units inspection, or for percent defective inspection (the Poisson distribution being an adequate approximation to the binomial distribution under these conditions). Tabulated values, corresponding to selected values of probabilities of acceptance (P_a, in percent) are given for each of the curves shown, and, in addition, for tightened inspection, and for defects per hundred units for AQLs of 10.0 or less and sample sizes of 80 or less.

11.2 PROCESS AVERAGE. The process average is the average percent defective or average number of defects per hundred units (whichever is applicable) of product submitted by the supplier for original inspection. Original inspection is the first inspection of a particular quantity of product as distinguished from the inspection of product which has been resubmitted after prior rejection.

11.3 AVERAGE OUTGOING QUALITY (AOQ). The AOQ is the average quality of outgoing product including all accepted lots or batches, plus all rejected lots or batches after the rejected lots or batches have been effectively 100 percent inspected and all defectives replaced by nondefectives.

11.4 AVERAGE OUTGOING QUALITY LIMIT (AOQL). The AOQL is the maximum of the AOQs for all possible incoming qualities for a given acceptance sampling plan. AOQL values are given in Table V–A for each of the single sampling plans for normal inspection and in Table V–B for each of the single sampling plans for tightened inspection.

11.5 AVERAGE SAMPLE SIZE CURVES. Average sample size curves for double and multiple sampling are in Table IX. These show the average sample sizes which may be expected to occur under the various sampling plans for a given process quality. The curves assume no curtailment of inspection and are approximate to the extent that they are based upon the Poisson distribution, and that the sample sizes for double and multiple sampling are assumed to be 0.631n and 0.25n respectively, where n is the equivalent single sample size.

11.6 LIMITING QUALITY PROTECTION. The sampling plans and associated procedures given in this publication were designed for use where the units of product are produced in a continuing series of lots or batches over a period of time. However, if the lot or batch is of an isolated nature, it is desirable to limit the selection of sampling plans to those, associated with a designated AQL value, that provide not less than a specified limiting quality protection. Sampling plans for this purpose can be selected by choosing a Limiting Quality (LQ) and a consumer's risk to be associated with it. Tables VI and VII give values of LQ for the commonly used consumer's risks of 10 percent and 5 percent respectively. If a different value of consumer's risk is required, the O.C. curves and their tabulated values may be used. The concept of LQ may also be useful in specifying the AQL and Inspection Levels for a series of lots or batches, thus fixing minimum sample size where there is some reason for avoiding (with more than a given consumer's risk) more than a limiting proportion of defectives (or defects) in any single lot or batch.

TABLE I—*Sample size code letters*

Lot or batch size	Special inspection levels				General inspection levels		
	S-1	S-2	S-3	S-4	I	II	III
2 to 8	A	A	A	A	A	A	B
9 to 15	A	A	A	A	A	B	C
16 to 25	A	A	B	B	B	C	D
26 to 50	A	B	B	C	C	D	E
51 to 90	B	B	C	C	C	E	F
91 to 150	B	B	C	D	D	F	G
151 to 280	B	C	D	E	E	G	H
281 to 500	B	C	D	E	F	H	J
501 to 1200	C	C	E	F	G	J	K
1201 to 3200	C	D	E	G	H	K	L
3201 to 10000	C	D	F	G	J	L	M
10001 to 35000	C	D	F	H	K	M	N
35001 to 150000	D	E	G	J	L	N	P
150001 to 500000	D	E	G	J	M	P	Q
500001 and over	D	E	H	K	N	Q	R

TABLE II-A—Single sampling plans for normal inspection (Master table)

Sample size code letter	Sample size	Acceptable Quality Levels (normal inspection) 0.010	0.015	0.025	0.040	0.065	0.10	0.15	0.25	0.40	0.65	1.0	1.5	2.5	4.0	6.5	10	15	25	40	65	100	150	250	400	650	1000
		Ac Re	Ac Re	Ac Re	Ac Re	Ac Re	Ac Re	Ac Re	Ac Re	Ac Re	Ac Re	Ac Re	Ac Re	Ac Re	Ac Re	Ac Re	Ac Re	Ac Re	Ac Re	Ac Re	Ac Re	Ac Re	Ac Re	Ac Re	Ac Re	Ac Re	Ac Re
A	2	↓	↓	↓	↓	↓	↓	↓	↓	↓	↓	↓	↓	↓	↓	0 1	↓	↓	1 2	2 3	3 4	5 6	7 8	10 11	14 15	21 22	30 31
B	3													↓	0 1	↑	↓	1 2	2 3	3 4	5 6	7 8	10 11	14 15	21 22	30 31	44 45
C	5												↓	0 1	↑	↓	1 2	2 3	3 4	5 6	7 8	10 11	14 15	21 22	30 31	44 45	↑
D	8											↓	0 1	↑	↓	1 2	2 3	3 4	5 6	7 8	10 11	14 15	21 22	30 31	44 45	↑	
E	13										↓	0 1	↑	↓	1 2	2 3	3 4	5 6	7 8	10 11	14 15	21 22	30 31	44 45	↑		
F	20									↓	0 1	↑	↓	1 2	2 3	3 4	5 6	7 8	10 11	14 15	21 22	↑	↑	↑			
G	32								↓	0 1	↑	↓	1 2	2 3	3 4	5 6	7 8	10 11	14 15	21 22	↑						
H	50							↓	0 1	↑	↓	1 2	2 3	3 4	5 6	7 8	10 11	14 15	21 22	↑							
J	80						↓	0 1	↑	↓	1 2	2 3	3 4	5 6	7 8	10 11	14 15	21 22	↑								
K	125					↓	0 1	↑	↓	1 2	2 3	3 4	5 6	7 8	10 11	14 15	21 22	↑									
L	200				↓	0 1	↑	↓	1 2	2 3	3 4	5 6	7 8	10 11	14 15	21 22	↑										
M	315			↓	0 1	↑	↓	1 2	2 3	3 4	5 6	7 8	10 11	14 15	21 22	↑											
N	500		↓	0 1	↑	↓	1 2	2 3	3 4	5 6	7 8	10 11	14 15	21 22	↑												
P	800	↓	0 1	↑	↓	1 2	2 3	3 4	5 6	7 8	10 11	14 15	21 22	↑													
Q	1250	0 1	↑	↓	1 2	2 3	3 4	5 6	7 8	10 11	14 15	21 22	↑														
R	2000	↑	↑	1 2	2 3	3 4	5 6	7 8	10 11	14 15	21 22	↑															

↓ = Use first sampling plan below arrow. If sample size equals, or exceeds, lot or batch size, do 100 percent inspection.
↑ = Use first sampling plan above arrow.
Ac = Acceptance number.
Re = Rejection number.

TABLE II-B — Single sampling plans for tightened inspection (Master table)

Sample size code letter	Sample size	Acceptable Quality Levels (tightened inspection)																									
		0.010	0.015	0.025	0.040	0.065	0.10	0.15	0.25	0.40	0.65	1.0	1.5	2.5	4.0	6.5	10	15	25	40	65	100	150	250	400	650	1000
		Ac Re	Ac Re	Ac Re	Ac Re	Ac Re	Ac Re	Ac Re	Ac Re	Ac Re	Ac Re	Ac Re	Ac Re	Ac Re	Ac Re	Ac Re	Ac Re	Ac Re	Ac Re	Ac Re	Ac Re	Ac Re	Ac Re	Ac Re	Ac Re	Ac Re	Ac Re
A	2	↓	↓	↓	↓	↓	↓	↓	↓	↓	↓	↓	↓	↓	↓	↓	↓	↓	↓	1 2	2 3	3 4	5 6	8 9	12 13	18 19	27 28
B	3	↓	↓	↓	↓	↓	↓	↓	↓	↓	↓	↓	↓	↓	↓	0 1	↓	↓	1 2	2 3	3 4	5 6	8 9	12 13	18 19	27 28	41 42
C	5	↓	↓	↓	↓	↓	↓	↓	↓	↓	↓	↓	↓	↓	0 1	↓	↓	1 2	2 3	3 4	5 6	8 9	12 13	18 19	27 28	41 42	↑
D	8	↓	↓	↓	↓	↓	↓	↓	↓	↓	↓	↓	↓	0 1	↓	↓	1 2	2 3	3 4	5 6	8 9	12 13	18 19	27 28	41 42	↑	↑
E	13	↓	↓	↓	↓	↓	↓	↓	↓	↓	↓	↓	0 1	↓	↓	1 2	2 3	3 4	5 6	8 9	12 13	18 19	27 28	41 42	↑	↑	↑
F	20	↓	↓	↓	↓	↓	↓	↓	↓	↓	↓	0 1	↓	↓	1 2	2 3	3 4	5 6	8 9	12 13	18 19	↑	↑	↑	↑	↑	↑
G	32	↓	↓	↓	↓	↓	↓	↓	↓	↓	0 1	↓	↓	1 2	2 3	3 4	5 6	8 9	12 13	18 19	↑	↑	↑	↑	↑	↑	↑
H	50	↓	↓	↓	↓	↓	↓	↓	↓	0 1	↓	↓	1 2	2 3	3 4	5 6	8 9	12 13	18 19	↑	↑	↑	↑	↑	↑	↑	↑
J	80	↓	↓	↓	↓	↓	↓	↓	0 1	↓	↓	1 2	2 3	3 4	5 6	8 9	12 13	18 19	↑	↑	↑	↑	↑	↑	↑	↑	↑
K	125	↓	↓	↓	↓	↓	↓	0 1	↓	↓	1 2	2 3	3 4	5 6	8 9	12 13	18 19	↑	↑	↑	↑	↑	↑	↑	↑	↑	↑
L	200	↓	↓	↓	↓	↓	0 1	↓	↓	1 2	2 3	3 4	5 6	8 9	12 13	18 19	↑	↑	↑	↑	↑	↑	↑	↑	↑	↑	↑
M	315	↓	↓	↓	↓	0 1	↓	↓	1 2	2 3	3 4	5 6	8 9	12 13	18 19	↑	↑	↑	↑	↑	↑	↑	↑	↑	↑	↑	↑
N	500	↓	↓	↓	0 1	↓	↓	1 2	2 3	3 4	5 6	8 9	12 13	18 19	↑	↑	↑	↑	↑	↑	↑	↑	↑	↑	↑	↑	↑
P	800	↓	↓	0 1	↓	↓	1 2	2 3	3 4	5 6	8 9	12 13	18 19	↑	↑	↑	↑	↑	↑	↑	↑	↑	↑	↑	↑	↑	↑
Q	1250	↓	0 1	↓	↓	1 2	2 3	3 4	5 6	8 9	12 13	18 19	↑	↑	↑	↑	↑	↑	↑	↑	↑	↑	↑	↑	↑	↑	↑
R	2000	0 1	↑	↓	1 2	2 3	3 4	5 6	8 9	12 13	18 19	↑	↑	↑	↑	↑	↑	↑	↑	↑	↑	↑	↑	↑	↑	↑	↑
S	3150			1 2																							

↓ = Use first sampling plan below arrow. If sample size equals or exceeds lot or batch size, do 100 percent inspection.
↑ = Use first sampling plan above arrow.
Ac = Acceptance number.
Re = Rejection number.

TABLE II-C—Single sampling plans for reduced inspection (Master table)

Sample size code letter	Sample size	Acceptable Quality Levels (reduced inspection)†																									
		0.010	0.015	0.025	0.040	0.065	0.10	0.15	0.25	0.40	0.65	1.0	1.5	2.5	4.0	6.5	10	15	25	40	65	100	150	250	400	650	1000
		Ac Re	Ac Re	Ac Re	Ac Re	Ac Re	Ac Re	Ac Re	Ac Re	Ac Re	Ac Re	Ac Re	Ac Re	Ac Re	Ac Re	Ac Re	Ac Re	Ac Re	Ac Re	Ac Re	Ac Re	Ac Re	Ac Re	Ac Re	Ac Re	Ac Re	Ac Re
A	2	↓	↓	↓	↓	↓	↓	↓	↓	↓	↓	↓	↓	↓	↓	0 1	↓	↓	1 2	2 3	3 4	5 6	7 8	10 11	14 15	21 22	30 31
B	2	↓	↓	↓	↓	↓	↓	↓	↓	↓	↓	↓	↓	↓	0 1	↑	↓	0 2	1 3	2 4	3 5	5 6	7 8	10 11	14 15	21 22	30 31
C	2	↓	↓	↓	↓	↓	↓	↓	↓	↓	↓	↓	↓	0 1	↑	↓	0 2	1 3	1 4	2 5	3 6	5 8	7 10	10 13	14 17	21 24	↑
D	3	↓	↓	↓	↓	↓	↓	↓	↓	↓	↓	↓	0 1	↑	↓	0 2	1 3	1 4	2 5	3 6	5 8	7 10	10 13	14 17	21 24	↑	↑
E	5	↓	↓	↓	↓	↓	↓	↓	↓	↓	↓	0 1	↑	↓	0 2	1 3	1 4	2 5	3 6	5 8	7 10	10 13	14 17	21 24	↑	↑	↑
F	8	↓	↓	↓	↓	↓	↓	↓	↓	↓	0 1	↑	↓	0 2	1 3	1 4	2 5	3 6	5 8	7 10	10 13	↑	↑	↑	↑	↑	↑
G	13	↓	↓	↓	↓	↓	↓	↓	↓	0 1	↑	↓	0 2	1 3	1 4	2 5	3 6	5 8	7 10	10 13	↑	↑	↑	↑	↑	↑	↑
H	20	↓	↓	↓	↓	↓	↓	↓	0 1	↑	↓	0 2	1 3	1 4	2 5	3 6	5 8	7 10	10 13	↑	↑	↑	↑	↑	↑	↑	↑
J	32	↓	↓	↓	↓	↓	↓	0 1	↑	↓	0 2	1 3	1 4	2 5	3 6	5 8	7 10	10 13	↑	↑	↑	↑	↑	↑	↑	↑	↑
K	50	↓	↓	↓	↓	↓	0 1	↑	↓	0 2	1 3	1 4	2 5	3 6	5 8	7 10	10 13	↑	↑	↑	↑	↑	↑	↑	↑	↑	↑
L	80	↓	↓	↓	↓	0 1	↑	↓	0 2	1 3	1 4	2 5	3 6	5 8	7 10	10 13	↑	↑	↑	↑	↑	↑	↑	↑	↑	↑	↑
M	125	↓	↓	↓	0 1	↑	↓	0 2	1 3	1 4	2 5	3 6	5 8	7 10	10 13	↑	↑	↑	↑	↑	↑	↑	↑	↑	↑	↑	↑
N	200	↓	↓	0 1	↑	↓	0 2	1 3	1 4	2 5	3 6	5 8	7 10	10 13	↑	↑	↑	↑	↑	↑	↑	↑	↑	↑	↑	↑	↑
P	315	↓	0 1	↑	↓	0 2	1 3	1 4	2 5	3 6	5 8	7 10	10 13	↑	↑	↑	↑	↑	↑	↑	↑	↑	↑	↑	↑	↑	↑
Q	500	0 1	↑	↓	0 2	1 3	1 4	2 5	3 6	5 8	7 10	10 13	↑	↑	↑	↑	↑	↑	↑	↑	↑	↑	↑	↑	↑	↑	↑
R	800	↑	↑	0 2	1 3	1 4	2 5	3 6	5 8	7 10	10 13	↑	↑	↑	↑	↑	↑	↑	↑	↑	↑	↑	↑	↑	↑	↑	↑

↓ = Use first sampling plan below arrow. If sample size equals or exceeds lot or batch size, do 100 percent inspection.
↑ = Use first sampling plan above arrow.
Ac = Acceptance number.
Re = Rejection number.
† = If the acceptance number has been exceeded, but the rejection number has not been reached, accept the lot, but reinstate normal inspection (see 10.1.4).

TABLE III-A—Double sampling plans for normal inspection (Master table)

Sample size code letter	Sample	Sample size	Cumulative sample size	Acceptable Quality Levels (normal inspection)																									
				0.010	0.015	0.025	0.040	0.065	0.10	0.15	0.25	0.40	0.65	1.0	1.5	2.5	4.0	6.5	10	15	25	40	65	100	150	250	400	650	1000
				Ac Re	Ac Re	Ac Re	Ac Re	Ac Re	Ac Re	Ac Re	Ac Re	Ac Re	Ac Re	Ac Re	Ac Re	Ac Re	Ac Re	Ac Re	Ac Re	Ac Re	Ac Re	Ac Re	Ac Re	Ac Re	Ac Re	Ac Re	Ac Re	Ac Re	Ac Re
A				↓	↓	↓	↓	↓	↓	↓	↓	↓	↓	↓	↓	↓	↓	•	↓	↓	•	•	•	•	•	•	•	•	•
B	First	2	2	↓	↓	↓	↓	↓	↓	↓	↓	↓	↓	↓	↓	↓	•	↑	↓	0 2	0 3	1 4	2 5	3 7	5 9	7 11	11 16	17 22	25 31
	Second	2	4																	1 2	3 4	4 5	6 7	8 9	12 13	18 19	26 27	37 38	56 57
C	First	3	3	↓	↓	↓	↓	↓	↓	↓	↓	↓	↓	↓	↓	•	↑	↓	0 2	0 3	1 4	2 5	3 7	5 9	7 11	11 16	17 22	25 31	↑
	Second	3	6																1 2	3 4	4 5	6 7	8 9	12 13	18 19	26 27	37 38	56 57	
D	First	5	5	↓	↓	↓	↓	↓	↓	↓	↓	↓	↓	↓	•	↑	↓	0 2	0 3	1 4	2 5	3 7	5 9	7 11	11 16	17 22	25 31	↑	↑
	Second	5	10															1 2	3 4	4 5	6 7	8 9	12 13	18 19	26 27	37 38	56 57		
E	First	8	8	↓	↓	↓	↓	↓	↓	↓	↓	↓	↓	•	↑	↓	0 2	0 3	1 4	2 5	3 7	5 9	7 11	11 16	17 22	25 31	↑	↑	↑
	Second	8	16														1 2	3 4	4 5	6 7	8 9	12 13	18 19	26 27	37 38	56 57			
F	First	13	13	↓	↓	↓	↓	↓	↓	↓	↓	↓	•	↑	↓	0 2	0 3	1 4	2 5	3 7	5 9	7 11	11 16	↑	↑	↑	↑	↑	↑
	Second	13	26													1 2	3 4	4 5	6 7	8 9	12 13	18 19	26 27						
G	First	20	20	↓	↓	↓	↓	↓	↓	↓	↓	•	↑	↓	0 2	0 3	i 4	2 5	3 7	5 9	7 11	11 16	↑	↑	↑	↑	↑	↑	↑
	Second	20	40												1 2	3 4	4 5	6 7	8 9	12 13	18 19	26 27							
H	First	32	32	↓	↓	↓	↓	↓	↓	↓	•	↑	↓	0 2	0 3	1 4	2 5	3 7	5 9	7 11	11 16	↑	↑	↑	↑	↑	↑	↑	↑
	Second	32	64											1 2	3 4	4 5	6 7	8 9	12 13	18 19	26 27								
J	First	50	50	↓	↓	↓	↓	↓	↓	•	↑	↓	0 2	0 3	1 4	2 5	3 7	5 9	7 11	11 16	↑	↑	↑	↑	↑	↑	↑	↑	↑
	Second	50	100										1 2	3 4	4 5	6 7	8 9	12 13	18 19	26 27									
K	First	80	80	↓	↓	↓	↓	↓	•	↑	↓	0 2	0 3	1 4	2 5	3 7	5 9	7 11	11 16	↑	↑	↑	↑	↑	↑	↑	↑	↑	↑
	Second	80	160									1 2	3 4	4 5	6 7	8 9	12 13	18 19	26 27										
L	First	125	125	↓	↓	↓	↓	•	↑	↓	0 2	0 3	1 4	2 5	3 7	5 9	7 11	11 16	↑	↑	↑	↑	↑	↑	↑	↑	↑	↑	↑
	Second	125	250								1 2	3 4	4 5	6 7	8 9	12 13	18 19	26 27											
M	First	200	200	↓	↓	↓	•	↑	↓	0 2	0 3	1 4	2 5	3 7	5 9	7 11	11 16	↑	↑	↑	↑	↑	↑	↑	↑	↑	↑	↑	↑
	Second	200	400							1 2	3 4	4 5	6 7	8 9	12 13	18 19	26 27												
N	First	315	315	↓	↓	•	↑	↓	0 2	0 3	1 4	2 5	3 7	5 9	7 11	11 16	↑	↑	↑	↑	↑	↑	↑	↑	↑	↑	↑	↑	↑
	Second	315	630						1 2	3 4	4 5	6 7	8 9	12 13	18 19	26 27													
P	First	500	500	↓	•	↑	↓	0 2	0 3	1 4	2 5	3 7	5 9	7 11	11 16	↑	↑	↑	↑	↑	↑	↑	↑	↑	↑	↑	↑	↑	↑
	Second	500	1000					1 2	3 4	4 5	6 7	8 9	12 13	18 19	26 27														
Q	First	800	800	•	↑	↓	0 2	0 3	1 4	2 5	3 7	5 9	7 11	11 16	↑	↑	↑	↑	↑	↑	↑	↑	↑	↑	↑	↑	↑	↑	↑
	Second	800	1600				1 2	3 4	4 5	6 7	8 9	12 13	18 19	26 27															
R	First	1250	1250	↑	↑	0 2	0 3	1 4	2 5	3 7	5 9	7 11	11 16	↑	↑	↑	↑	↑	↑	↑	↑	↑	↑	↑	↑	↑	↑	↑	↑
	Second	1250	2500			1 2	3 4	4 5	6 7	8 9	12 13	18 19	26 27																

↓ = Use first sampling plan below arrow. If sample size equals or exceeds lot or batch size, do 100 percent inspection.

↑ = Use first sampling plan above arrow.

Ac = Acceptance number

Re = Rejection number

• = Use corresponding single sampling plan (or alternatively, use double sampling plan below, where available).

TABLE III-B — Double sampling plans for tightened inspection (Master table)

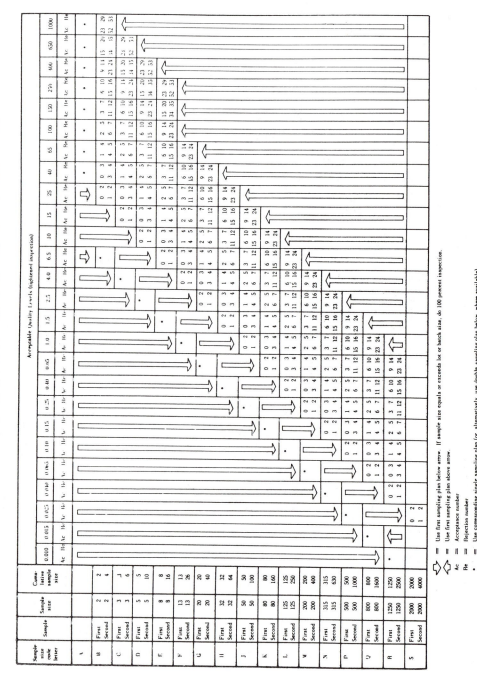

Sample size code letter	Sample	Sample size	Cumulative sample size	Acceptable Quality Levels (tightened inspection) 0.010	0.015	0.025	0.040	0.065	0.10	0.15	0.25	0.40	0.65	1.0	1.5	2.5	4.0	6.5	10	15	25	40	65	100	150	250	400	650	1000
				Ac Re	Ac Re	Ac Re	Ac Re	Ac Re	Ac Re	Ac Re	Ac Re	Ac Re	Ac Re	Ac Re	Ac Re	Ac Re	Ac Re	Ac Re	Ac Re	Ac Re	Ac Re	Ac Re	Ac Re	Ac Re	Ac Re	Ac Re	Ac Re	Ac Re	Ac Re
A				↓	↓	↓	↓	↓	↓	↓	↓	↓	↓	↓	↓	↓	↓	↓	↓	↓	↓	•	•	•	•	•	•	•	•
B	First	2	2	↓	↓	↓	↓	↓	↓	↓	↓	↓	↓	↓	↓	↓	↓	•	↓	↓	0 2	0 3	1 4	2 5	3 7	6 10	9 14	15 20	23 29
	Second	2	4																		1 2	3 4	4 5	6 7	11 12	15 16	23 24	34 35	52 53
C	First	3	3	↓	↓	↓	↓	↓	↓	↓	↓	↓	↓	↓	↓	↓	•	↓	↓	0 2	0 3	1 4	2 5	3 7	6 10	9 14	15 20	23 29	↑
	Second	3	6																	1 2	3 4	4 5	6 7	11 12	15 16	23 24	34 35	52 53	
D	First	5	5	↓	↓	↓	↓	↓	↓	↓	↓	↓	↓	↓	↓	•	↓	↓	0 2	0 3	1 4	2 5	3 7	6 10	9 14	15 20	23 29	↑	↑
	Second	5	10															1 2	3 4	4 5	6 7	11 12	15 16	23 24	34 35	52 53			
E	First	8	8	↓	↓	↓	↓	↓	↓	↓	↓	↓	↓	↓	•	↓	↓	0 2	0 3	1 4	2 5	3 7	6 10	9 14	15 20	23 29	↑	↑	↑
	Second	8	16														1 2	3 4	4 5	6 7	11 12	15 16	23 24	34 35	52 53				
F	First	13	13	↓	↓	↓	↓	↓	↓	↓	↓	↓	↓	•	↓	↓	0 2	0 3	1 4	2 5	3 7	6 10	9 14	↑	↑	↑	↑	↑	↑
	Second	13	26													1 2	3 4	4 5	6 7	11 12	15 16	23 24							
G	First	20	20	↓	↓	↓	↓	↓	↓	↓	↓	↓	•	↓	↓	0 2	0 3	1 4	2 5	3 7	6 10	9 14	↑	↑	↑	↑	↑	↑	↑
	Second	20	40												1 2	3 4	4 5	6 7	11 12	15 16	23 24								
H	First	32	32	↓	↓	↓	↓	↓	↓	↓	↓	•	↓	↓	0 2	0 3	1 4	2 5	3 7	6 10	9 14	↑	↑	↑	↑	↑	↑	↑	↑
	Second	32	64											1 2	3 4	4 5	6 7	11 12	15 16	23 24									
J	First	50	50	↓	↓	↓	↓	↓	↓	↓	•	↓	↓	0 2	0 3	1 4	2 5	3 7	6 10	9 14	↑	↑	↑	↑	↑	↑	↑	↑	↑
	Second	50	100										1 2	3 4	4 5	6 7	11 12	15 16	23 24										
K	First	80	80	↓	↓	↓	↓	↓	↓	•	↓	↓	0 2	0 3	1 4	2 5	3 7	6 10	9 14	↑	↑	↑	↑	↑	↑	↑	↑	↑	↑
	Second	80	160									1 2	3 4	4 5	6 7	11 12	15 16	23 24											
L	First	125	125	↓	↓	↓	↓	↓	•	↓	↓	0 2	0 3	1 4	2 5	3 7	6 10	9 14	↑	↑	↑	↑	↑	↑	↑	↑	↑	↑	↑
	Second	125	250								1 2	3 4	4 5	6 7	11 12	15 16	23 24												
M	First	200	200	↓	↓	↓	↓	•	↓	↓	0 2	0 3	1 4	2 5	3 7	6 10	9 14	↑	↑	↑	↑	↑	↑	↑	↑	↑	↑	↑	↑
	Second	200	400							1 2	3 4	4 5	6 7	11 12	15 16	23 24													
N	First	315	315	↓	↓	↓	•	↓	↓	0 2	0 3	1 4	2 5	3 7	6 10	9 14	↑	↑	↑	↑	↑	↑	↑	↑	↑	↑	↑	↑	↑
	Second	315	630							1 2	3 4	4 5	6 7	11 12	15 16	23 24													
P	First	500	500	↓	↓	•	↓	↓	0 2	0 3	1 4	2 5	3 7	6 10	9 14	↑	↑	↑	↑	↑	↑	↑	↑	↑	↑	↑	↑	↑	↑
	Second	500	1000						1 2	3 4	4 5	6 7	11 12	15 16	23 24														
Q	First	800	800	↓	•	↓	↓	0 2	0 3	1 4	2 5	3 7	6 10	9 14	↑	↑	↑	↑	↑	↑	↑	↑	↑	↑	↑	↑	↑	↑	↑
	Second	800	1600					1 2	3 4	4 5	6 7	11 12	15 16	23 24															
R	First	1250	1250	•	↑	↓	0 2	0 3	1 4	2 5	3 7	6 10	9 14	↑	↑	↑	↑	↑	↑	↑	↑	↑	↑	↑	↑	↑	↑	↑	↑
	Second	1250	2500				1 2	3 4	4 5	6 7	11 12	15 16	23 24																
S	First	2000	2000			0 2																							
	Second	2000	4000			1 2																							

↓ = Use first sampling plan below arrow. If sample size equals or exceeds lot or batch size, do 100 percent inspection.

↑ = Use first sampling plan above arrow.

Ac = Acceptance number

Re = Rejection number

• = Use corresponding single sampling plan (or, alternatively, use double sampling plan below, where available).

TABLE III-C—Double sampling plans for reduced inspection (Master table)

Sample size code letter	Sample	Sample size	Cumulative sample size	Acceptable Quality Levels (reduced inspection)†																									
				0.010	0.015	0.025	0.040	0.065	0.10	0.15	0.25	0.40	0.65	1.0	1.5	2.5	4.0	6.5	10	15	25	40	65	100	150	250	400	650	1000
				Ac Re	Ac Re	Ac Re	Ac Re	Ac Re	Ac Re	Ac Re	Ac Re	Ac Re	Ac Re	Ac Re	Ac Re	Ac Re	Ac Re	Ac Re	Ac Re	Ac Re	Ac Re	Ac Re	Ac Re	Ac Re	Ac Re	Ac Re	Ac Re	Ac Re	Ac Re
A				↓	↓	↓	↓	↓	↓	↓	↓	↓	↓	↓	↓	↓	↓	•	↓	↓	•	•	•	•	•	•	•	•	•
B				↓	↓	↓	↓	↓	↓	↓	↓	↓	↓	↓	↓	↓	•	↑	↓	•	•	•	•	•	•	•	•	•	•
C				↓	↓	↓	↓	↓	↓	↓	↓	↓	↓	↓	↓	•	↑	↓	•	•	•	•	•	•	•	•	•	•	↑
D	First	2	2	↓	↓	↓	↓	↓	↓	↓	↓	↓	↓	↓	•	↑	↓	0 2	0 3	0 4	0 4	1 5	2 7	3 8	5 10	7 12	11 17	↑	↑
	Second	2	4	↓	↓	↓	↓	↓	↓	↓	↓	↓	↓	↓	•	↑	↓	0 2	0 4	1 5	3 6	4 7	6 9	8 12	12 16	18 22	26 30	↑	↑
E	First	3	3	↓	↓	↓	↓	↓	↓	↓	↓	↓	↓	•	↑	↓	0 2	0 3	0 4	0 4	1 5	2 7	3 8	5 10	7 12	11 17	↑	↑	↑
	Second	3	6	↓	↓	↓	↓	↓	↓	↓	↓	↓	↓	•	↑	↓	0 2	0 4	1 5	3 6	4 7	6 9	8 12	12 16	18 22	26 30	↑	↑	↑
F	First	5	5	↓	↓	↓	↓	↓	↓	↓	↓	↓	•	↑	↓	0 2	0 3	0 4	0 4	1 5	2 7	3 8	5 10	↑	↑	↑	↑	↑	↑
	Second	5	10	↓	↓	↓	↓	↓	↓	↓	↓	↓	•	↑	↓	0 2	0 4	1 5	3 6	4 7	6 9	8 12	12 16	↑	↑	↑	↑	↑	↑
G	First	8	8	↓	↓	↓	↓	↓	↓	↓	↓	•	↑	↓	0 2	0 3	0 4	0 4	1 5	2 7	3 8	5 10	↑	↑	↑	↑	↑	↑	↑
	Second	8	16	↓	↓	↓	↓	↓	↓	↓	↓	•	↑	↓	0 2	0 4	1 5	3 6	4 7	6 9	8 12	12 16	↑	↑	↑	↑	↑	↑	↑
H	First	13	13	↓	↓	↓	↓	↓	↓	↓	•	↑	↓	0 2	0 3	0 4	0 4	1 5	2 7	3 8	5 10	↑	↑	↑	↑	↑	↑	↑	↑
	Second	13	26	↓	↓	↓	↓	↓	↓	↓	•	↑	↓	0 2	0 4	1 5	3 6	4 7	6 9	8 12	12 16	↑	↑	↑	↑	↑	↑	↑	↑
J	First	20	20	↓	↓	↓	↓	↓	↓	•	↑	↓	0 2	0 3	0 4	0 4	1 5	2 7	3 8	5 10	↑	↑	↑	↑	↑	↑	↑	↑	↑
	Second	20	40	↓	↓	↓	↓	↓	↓	•	↑	↓	0 2	0 4	1 5	3 6	4 7	6 9	8 12	12 16	↑	↑	↑	↑	↑	↑	↑	↑	↑
K	First	32	32	↓	↓	↓	↓	↓	•	↑	↓	0 2	0 3	0 4	0 4	1 5	2 7	3 8	5 10	↑	↑	↑	↑	↑	↑	↑	↑	↑	↑
	Second	32	64	↓	↓	↓	↓	↓	•	↑	↓	0 2	0 4	1 5	3 6	4 7	6 9	8 12	12 16	↑	↑	↑	↑	↑	↑	↑	↑	↑	↑
L	First	50	50	↓	↓	↓	↓	•	↑	↓	0 2	0 3	0 4	0 4	1 5	2 7	3 8	5 10	↑	↑	↑	↑	↑	↑	↑	↑	↑	↑	↑
	Second	50	100	↓	↓	↓	↓	•	↑	↓	0 2	0 4	1 5	3 6	4 7	6 9	8 12	12 16	↑	↑	↑	↑	↑	↑	↑	↑	↑	↑	↑
M	First	80	80	↓	↓	↓	•	↑	↓	0 2	0 3	0 4	0 4	1 5	2 7	3 8	5 10	↑	↑	↑	↑	↑	↑	↑	↑	↑	↑	↑	↑
	Second	80	160	↓	↓	↓	•	↑	↓	0 2	0 4	1 5	3 6	4 7	6 9	8 12	12 16	↑	↑	↑	↑	↑	↑	↑	↑	↑	↑	↑	↑
N	First	125	125	↓	↓	•	↑	↓	0 2	0 3	0 4	0 4	1 5	2 7	3 8	5 10	↑	↑	↑	↑	↑	↑	↑	↑	↑	↑	↑	↑	↑
	Second	125	250	↓	↓	•	↑	↓	0 2	0 4	1 5	3 6	4 7	6 9	8 12	12 16	↑	↑	↑	↑	↑	↑	↑	↑	↑	↑	↑	↑	↑
P	First	200	200	↓	•	↑	↓	0 2	0 3	0 4	0 4	1 5	2 7	3 8	5 10	↑	↑	↑	↑	↑	↑	↑	↑	↑	↑	↑	↑	↑	↑
	Second	200	400	↓	•	↑	↓	0 2	0 4	1 5	3 6	4 7	6 9	8 12	12 16	↑	↑	↑	↑	↑	↑	↑	↑	↑	↑	↑	↑	↑	↑
Q	First	315	315	•	↑	↓	0 2	0 3	0 4	0 4	1 5	2 7	3 8	5 10	↑	↑	↑	↑	↑	↑	↑	↑	↑	↑	↑	↑	↑	↑	↑
	Second	315	630	•	↑	↓	0 2	0 4	1 5	3 6	4 7	6 9	8 12	12 16	↑	↑	↑	↑	↑	↑	↑	↑	↑	↑	↑	↑	↑	↑	↑
R	First	500	500	↑	↑	0 2	0 3	0 4	0 4	1 5	2 7	3 8	5 10	↑	↑	↑	↑	↑	↑	↑	↑	↑	↑	↑	↑	↑	↑	↑	↑
	Second	500	1000	↑	↑	0 2	0 4	1 5	3 6	4 7	6 9	8 12	12 16	↑	↑	↑	↑	↑	↑	↑	↑	↑	↑	↑	↑	↑	↑	↑	↑

↓ = Use first sampling plan below arrow. If sample size equals or exceeds lot or batch size, do 100 percent inspection.
↑ = Use first sampling plan above arrow.
Ac = Acceptance number.
Re = Rejection number.
• = Use corresponding single sampling plan (or alternatively, use double sampling plan below, when available.)
† = If, after the second sample, the acceptance number has been exceeded, but the rejection number has not been reached, accept the lot, but reinstate normal inspection (see 10.14).

TABLE IV-A—Multiple sampling plans for normal inspection (Master table)

Sample size code letter	Sample	Sample size	Cumulative sample size	Acceptable Quality Levels (normal inspection)																									
				0.010	0.015	0.025	0.040	0.065	0.10	0.15	0.25	0.40	0.65	1.0	1.5	2.5	4.0	6.5	10	15	25	40	65	100	150	250	400	650	1000
				Ac Re	Ac Re	Ac Re	Ac Re	Ac Re	Ac Re	Ac Re	Ac Re	Ac Re	Ac Re	Ac Re	Ac Re	Ac Re	Ac Re	Ac Re	Ac Re	Ac Re	Ac Re	Ac Re	Ac Re	Ac Re	Ac Re	Ac Re	Ac Re	Ac Re	Ac Re
A				↓	↓	↓	↓	↓	↓	↓	↓	↓	↓	↓	↓	↓	↓	•	↓	↓	•	•	•	•	•	•	•	•	•
B				↓	↓	↓	↓	↓	↓	↓	↓	↓	↓	↓	↓	↓	•	↑	↓	++	++	++	++	++	++	++	++	++	++
C				↓	↓	↓	↓	↓	↓	↓	↓	↓	↓	↓	↓	•	↑	↓	++	++	++	++	++	++	++	++	++	++	↑
D	First	2	2	↓	↓	↓	↓	↓	↓	↓	↓	↓	↓	↓	•	↑	↓	# 2	# 2	# 3	# 4	0 4	0 5	1 7	2 9	4 12	6 16	↑	↑
	Second	2	4															# 2	0 3	0 3	1 5	1 6	3 8	4 10	7 14	11 19	17 27		
	Third	2	6															0 2	0 3	1 4	2 6	3 8	6 10	8 13	13 19	19 27	29 39		
	Fourth	2	8															0 3	1 4	2 5	3 7	5 10	8 13	12 17	19 25	27 34	40 49		
	Fifth	2	10															1 3	2 4	3 6	5 8	7 11	11 15	17 20	25 29	36 40	53 58		
	Sixth	2	12															1 3	3 5	4 6	7 9	10 12	14 17	21 23	31 33	45 47	65 68		
	Seventh	2	14															2 3	4 5	6 7	9 10	13 14	18 19	25 26	37 38	53 54	77 78		
E	First	3	3	↓	↓	↓	↓	↓	↓	↓	↓	↓	↓	•	↑	↓	# 2	# 2	# 3	# 4	0 4	0 5	1 7	2 9	4 12	6 16	↑	↑	↑
	Second	3	6														# 2	0 3	0 3	1 5	1 6	3 8	4 10	7 14	11 19	17 27			
	Third	3	9														0 2	0 3	1 4	2 6	3 8	6 10	8 13	13 19	19 27	29 39			
	Fourth	3	12														0 3	1 4	2 5	3 7	5 10	8 13	12 17	19 25	27 34	40 49			
	Fifth	3	15														1 3	2 4	3 6	5 8	7 11	11 15	17 20	25 29	36 40	53 58			
	Sixth	3	18														1 3	3 5	4 6	7 9	10 12	14 17	21 23	31 33	45 47	65 68			
	Seventh	3	21														2 3	4 5	6 7	9 10	13 14	18 19	25 26	37 38	53 54	77 78			
F	First	5	5	↓	↓	↓	↓	↓	↓	↓	↓	↓	•	↑	↓	# 2	# 2	# 3	# 4	0 4	0 5	1 7	2 9	↑	↑	↑	↑	↑	↑
	Second	5	10													# 2	0 3	0 3	1 5	1 6	3 8	4 10	7 14						
	Third	5	15													0 2	0 3	1 4	2 6	3 8	6 10	8 13	13 19						
	Fourth	5	20													0 3	1 4	2 5	3 7	5 10	8 13	12 17	19 25						
	Fifth	5	25													1 3	2 4	3 6	5 8	7 11	11 15	17 20	25 29						
	Sixth	5	30													1 3	3 5	4 6	7 9	10 12	14 17	21 23	31 33						
	Seventh	5	35													2 3	4 5	6 7	9 10	13 14	18 19	25 26	37 38						
G	First	8	8	↓	↓	↓	↓	↓	↓	↓	↓	•	↑	↓	# 2	# 2	# 3	# 4	0 4	0 5	1 7	2 9	↑	↑	↑	↑	↑	↑	↑
	Second	8	16												# 2	0 3	0 3	1 5	1 6	3 8	4 10	7 14							
	Third	8	24												0 2	0 3	1 4	2 6	3 8	6 10	8 13	13 19							
	Fourth	8	32												0 3	1 4	2 5	3 7	5 10	8 13	12 17	19 25							
	Fifth	8	40												1 3	2 4	3 6	5 8	7 11	11 15	17 20	25 29							
	Sixth	8	48												1 3	3 5	4 6	7 9	10 12	14 17	21 23	31 33							
	Seventh	8	56												2 3	4 5	6 7	9 10	13 14	18 19	25 26	37 38							
H	First	13	13	↓	↓	↓	↓	↓	↓	↓	•	↑	↓	# 2	# 2	# 3	# 4	0 4	0 5	1 7	2 9	↑	↑	↑	↑	↑	↑	↑	↑
	Second	13	26											# 2	0 3	0 3	1 5	1 6	3 8	4 10	7 14								
	Third	13	39											0 2	0 3	1 4	2 6	3 8	6 10	8 13	13 19								
	Fourth	13	52											0 3	1 4	2 5	3 7	5 10	8 13	12 17	19 25								
	Fifth	13	65											1 3	2 4	3 6	5 8	7 11	11 15	17 20	25 29								
	Sixth	13	78											1 3	3 5	4 6	7 9	10 12	14 17	21 23	31 33								
	Seventh	13	91											2 3	4 5	6 7	9 10	13 14	18 19	25 26	37 38								
J	First	20	20	↓	↓	↓	↓	↓	↓	•	↑	↓	# 2	# 2	# 3	# 4	0 4	0 5	1 7	2 9	↑	↑	↑	↑	↑	↑	↑	↑	↑
	Second	20	40										# 2	0 3	0 3	1 5	1 6	3 8	4 10	7 14									
	Third	20	60										0 2	0 3	1 4	2 6	3 8	6 10	8 13	13 19									
	Fourth	20	80										0 3	1 4	2 5	3 7	5 10	8 13	12 17	19 25									
	Fifth	20	100										1 3	2 4	3 6	5 8	7 11	11 15	17 20	25 29									
	Sixth	20	120										1 3	3 5	4 6	7 9	10 12	14 17	21 23	31 33									
	Seventh	20	140										2 3	4 5	6 7	9 10	13 14	18 19	25 26	37 38									

↓ = Use first sampling plan below arrow (refer to continuation of table on following page, when necessary). If sample size equals or exceeds lot or batch size, do 100 percent inspection.
↑ = Use first sampling plan above arrow.
Ac = Acceptance number.
Re = Rejection number.
• = Use corresponding single sampling plan (or alternatively, use multiple sampling plan below, where available).
++ = Use corresponding double sampling plan (or alternatively, use multiple sampling plan below, where available).
# = Acceptance not permitted at this sample size.

TABLE IV-A — Multiple sampling plans for normal inspection (Master table) (Continued)

Sample size code letter	Sample	Sample size	Cumulative sample size	Acceptable Quality Levels (normal inspection)																									
				0.010	0.015	0.025	0.040	0.065	0.10	0.15	0.25	0.40	0.65	1.0	1.5	2.5	4.0	6.5	10	15	25	40	65	100	150	250	400	650	1000
				Ac Re	Ac Re	Ac Re	Ac Re	Ac Re	Ac Re	Ac Re	Ac Re	Ac Re	Ac Re	Ac Re	Ac Re	Ac Re	Ac Re	Ac Re	Ac Re	Ac Re	Ac Re	Ac Re	Ac Re	Ac Re	Ac Re	Ac Re	Ac Re	Ac Re	Ac Re
K	First	32	32	↓	↓	↓	↓	↓	•	↑	↓	* 2	* 2	* 3	* 4	0 4	0 5	1 7	2 9	↑	↑	↑	↑	↑	↑	↑	↑	↑	↑
	Second	32	64									* 2	0 3	0 3	1 5	1 6	3 8	4 10	7 14										
	Third	32	96									0 2	0 3	1 4	2 6	3 8	6 10	8 13	13 19										
	Fourth	32	128									0 3	1 4	2 5	3 7	5 10	8 13	12 17	19 25										
	Fifth	32	160									1 3	2 4	3 6	5 8	7 11	11 15	17 20	25 29										
	Sixth	32	192									1 3	3 5	4 6	7 9	10 12	14 17	21 23	31 33										
	Seventh	32	224									2 3	4 5	6 7	9 10	13 14	18 19	25 26	37 38										
L	First	50	50	↓	↓	↓	↓	•	↑	↓	* 2	* 2	* 3	* 4	0 4	0 5	1 7	2 9	↑	↑	↑	↑	↑	↑	↑	↑	↑	↑	↑
	Second	50	100								* 2	0 3	0 3	1 5	1 6	3 8	4 10	7 14											
	Third	50	150								0 2	0 3	1 4	2 6	3 8	6 10	8 13	13 19											
	Fourth	50	200								0 3	1 4	2 5	3 7	5 10	8 13	12 17	19 25											
	Fifth	50	250								1 3	2 4	3 6	5 8	7 11	11 15	17 20	25 29											
	Sixth	50	300								1 3	3 5	4 6	7 9	10 12	14 17	21 23	31 33											
	Seventh	50	350								2 3	4 5	6 7	9 10	13 14	18 19	25 26	37 38											
M	First	80	80	↓	↓	↓	•	↑	↓	* 2	* 2	* 3	* 4	0 4	0 5	1 7	2 9	↑	↑	↑	↑	↑	↑	↑	↑	↑	↑	↑	↑
	Second	80	160							* 2	0 3	0 3	1 5	1 6	3 8	4 10	7 14												
	Third	80	240							0 2	0 3	1 4	2 6	3 8	6 10	8 13	13 19												
	Fourth	80	320							0 3	1 4	2 5	3 7	5 10	8 13	12 17	19 25												
	Fifth	80	400							1 3	2 4	3 6	5 8	7 11	11 15	17 20	25 29												
	Sixth	80	480							1 3	3 5	4 6	7 9	10 12	14 17	21 23	31 33												
	Seventh	80	560							2 3	4 5	6 7	9 10	13 14	18 19	25 26	37 38												
N	First	125	125	↓	↓	•	↑	↓	* 2	* 2	* 3	* 4	0 4	0 5	1 7	2 9	↑	↑	↑	↑	↑	↑	↑	↑	↑	↑	↑	↑	↑
	Second	125	250						* 2	0 3	0 3	1 5	1 6	3 8	4 10	7 14													
	Third	125	375						0 2	0 3	1 4	2 6	3 8	6 10	8 13	13 19													
	Fourth	125	500						0 3	1 4	2 5	3 7	5 10	8 13	12 17	19 25													
	Fifth	125	625						1 3	2 4	3 6	5 8	7 11	11 15	17 20	25 29													
	Sixth	125	750						1 3	3 5	4 6	7 9	10 12	14 17	21 23	31 33													
	Seventh	125	875						2 3	4 5	6 7	9 10	13 14	18 19	25 26	37 38													
P	First	200	200	↓	•	↑	↓	* 2	* 2	* 3	* 4	0 4	0 5	1 7	2 9	↑	↑	↑	↑	↑	↑	↑	↑	↑	↑	↑	↑	↑	↑
	Second	200	400					* 2	0 3	0 3	1 5	1 6	3 8	4 10	7 14														
	Third	200	600					0 2	0 3	1 4	2 6	3 8	6 10	8 13	13 19														
	Fourth	200	800					0 3	1 4	2 5	3 7	5 10	8 13	12 17	19 25														
	Fifth	200	1000					1 3	2 4	3 6	5 8	7 11	11 15	17 20	25 29														
	Sixth	200	1200					1 3	3 5	4 6	7 9	10 12	14 17	21 23	31 33														
	Seventh	200	1400					2 3	4 5	6 7	9 10	13 14	18 19	25 26	37 38														
Q	First	315	315	•	↑	↓	* 2	* 2	* 3	* 4	0 4	0 5	1 7	2 9	↑	↑	↑	↑	↑	↑	↑	↑	↑	↑	↑	↑	↑	↑	↑
	Second	315	630				* 2	0 3	0 3	1 5	1 6	3 8	4 10	7 14															
	Third	315	945				0 2	0 3	1 4	2 6	3 8	6 10	8 13	13 19															
	Fourth	315	1260				0 3	1 4	2 5	3 7	5 10	8 13	12 17	19 25															
	Fifth	315	1575				1 3	2 4	3 6	5 8	7 11	11 15	17 20	25 29															
	Sixth	315	1890				1 3	3 5	4 6	7 9	10 12	14 17	21 23	31 33															
	Seventh	315	2205				2 3	4 5	6 7	9 10	13 14	18 19	25 26	37 38															
R	First	500	500	↑	↑	* 2	* 2	* 3	* 4	0 4	0 5	1 7	2 9	↑	↑	↑	↑	↑	↑	↑	↑	↑	↑	↑	↑	↑	↑	↑	↑
	Second	500	1000			* 2	0 3	0 3	1 5	1 6	3 8	4 10	7 14																
	Third	500	1500			0 2	0 3	1 4	2 6	3 8	6 10	8 13	13 19																
	Fourth	500	2000			0 3	1 4	2 5	3 7	5 10	8 13	12 17	19 25																
	Fifth	500	2500			1 3	2 4	3 6	5 8	7 11	11 15	17 20	25 29																
	Sixth	500	3000			1 3	3 5	4 6	7 9	10 12	14 17	21 23	31 33																
	Seventh	500	3500			2 3	4 5	6 7	9 10	13 14	18 19	25 26	37 38																

↓ = Use first sampling plan below arrow. If sample size equals or exceeds lot or batch size, do 100 percent inspection.
↑ = Use first sampling plan above arrow (refer to preceding page, when necessary).
Ac = Acceptance number.
Re = Rejection number.
• = Use corresponding single sampling plan (or alternatively, use multiple plan below, where available).
* = Acceptance not permitted at this sample size.

TABLE IV-B—Multiple sampling plans for tightened inspection (Master table)

Sample size code letter	Sample	Sample size	Cumulative sample size	Acceptable Quality Levels (tightened inspection)																									
				0.010	0.015	0.025	0.040	0.065	0.10	0.15	0.25	0.40	0.65	1.0	1.5	2.5	4.0	6.5	10	15	25	40	65	100	150	250	400	650	1000
				Ac Re	Ac Re	Ac Re	Ac Re	Ac Re	Ac Re	Ac Re	Ac Re	Ac Re	Ac Re	Ac Re	Ac Re	Ac Re	Ac Re	Ac Re	Ac Re	Ac Re	Ac Re	Ac Re	Ac Re	Ac Re	Ac Re	Ac Re	Ac Re	Ac Re	Ac Re
A				↓	↓	↓	↓	↓	↓	↓	↓	↓	↓	↓	↓	↓	↓	↓	↓	↓	↓	•	•	•	•	•	•	•	•
B				↓	↓	↓	↓	↓	↓	↓	↓	↓	↓	↓	↓	↓	↓	•	↓	↓	↔	↔	↔	↔	↔	↔	↔	↔	↔
C				↓	↓	↓	↓	↓	↓	↓	↓	↓	↓	↓	↓	↓	•	↓	↓	↔	↔	↔	↔	↔	↔	↔	↔	↔	↑
D	First	2	2	↓	↓	↓	↓	↓	↓	↓	↓	↓	↓	↓	↓	•	↓	↓	# 2	# 2	# 3	# 4	0 4	0 6	1 8	3 10	6 15	↑	↑
	Second	2	4																# 2	0 3	0 3	1 5	2 7	3 9	6 12	10 17	16 25		
	Third	2	6																0 2	0 3	1 4	2 6	4 9	7 12	11 17	17 24	26 36		
	Fourth	2	8																0 3	1 4	2 5	3 7	6 11	10 15	16 22	24 31	37 46		
	Fifth	2	10																1 3	2 4	3 6	5 8	9 12	14 17	22 25	32 37	49 55		
	Sixth	2	12																1 3	3 5	4 6	7 9	12 14	18 20	27 29	40 43	61 64		
	Seventh	2	14																2 3	4 5	6 7	9 10	14 15	21 22	32 33	48 49	72 73		
E	First	3	3	↓	↓	↓	↓	↓	↓	↓	↓	↓	↓	↓	•	↓	↓	# 2	# 2	# 3	# 4	0 4	0 6	1 8	3 10	6 15	↑	↑	↑
	Second	3	6															# 2	0 3	0 3	1 5	2 7	3 9	6 12	10 17	16 25			
	Third	3	9															0 2	0 3	1 4	2 6	4 9	7 12	11 17	17 24	26 36			
	Fourth	3	12															0 3	1 4	2 5	3 7	6 11	10 15	16 22	24 31	37 46			
	Fifth	3	15															1 3	2 4	3 6	5 8	9 12	14 17	22 25	32 37	49 55			
	Sixth	3	18															1 3	3 5	4 6	7 9	12 14	18 20	27 29	40 43	61 64			
	Seventh	3	21															2 3	4 5	6 7	9 10	14 15	21 22	32 33	48 49	72 73			
F	First	5	5	↓	↓	↓	↓	↓	↓	↓	↓	↓	↓	•	↓	↓	# 2	# 2	# 3	# 4	0 4	0 6	1 8	↑	↑	↑	↑	↑	↑
	Second	5	10														# 2	0 3	0 3	1 5	2 7	3 9	6 12						
	Third	5	15														0 2	0 3	1 4	2 6	4 9	7 12	11 17						
	Fourth	5	20														0 3	1 4	2 5	3 7	6 11	10 15	16 22						
	Fifth	5	25														1 3	2 4	3 6	5 8	9 12	14 17	22 25						
	Sixth	5	30														1 3	3 5	4 6	7 9	12 14	18 20	27 29						
	Seventh	5	35														2 3	4 5	6 7	9 10	14 15	21 22	32 33						
G	First	8	8	↓	↓	↓	↓	↓	↓	↓	↓	↓	•	↓	↓	# 2	# 2	# 3	# 4	0 4	0 6	1 8	↑	↑	↑	↑	↑	↑	↑
	Second	8	16													# 2	0 3	0 3	1 5	2 7	3 9	6 12							
	Third	8	24													0 2	0 3	1 4	2 6	4 9	7 12	11 17							
	Fourth	8	32													0 3	1 4	2 5	3 7	6 11	10 15	16 22							
	Fifth	8	40													1 3	2 4	3 6	5 8	9 12	14 17	22 25							
	Sixth	8	48													1 3	3 5	4 6	7 9	12 14	18 20	27 29							
	Seventh	8	56													2 3	4 5	6 7	9 10	14 15	21 22	32 33							
H	First	13	13	↓	↓	↓	↓	↓	↓	↓	↓	•	↓	↓	# 2	# 2	# 3	# 4	0 4	0 6	1 8	↑	↑	↑	↑	↑	↑	↑	↑
	Second	13	26												# 2	0 3	0 3	1 5	2 7	3 9	6 12								
	Third	13	39												0 2	0 3	1 4	2 6	4 9	7 12	11 17								
	Fourth	13	52												0 3	1 4	2 5	3 7	6 11	10 15	16 22								
	Fifth	13	65												1 3	2 4	3 6	5 8	9 12	14 17	22 25								
	Sixth	13	78												1 3	3 5	4 6	7 9	12 14	18 20	27 29								
	Seventh	13	91												2 3	4 5	6 7	9 10	14 15	21 22	32 33								
J	First	20	20	↓	↓	↓	↓	↓	↓	↓	•	↓	↓	# 2	# 2	# 3	# 4	0 4	0 6	1 8	↑	↑	↑	↑	↑	↑	↑	↑	↑
	Second	20	40										# 2	0 3	0 3	1 5	2 7	3 9	6 12									
	Third	20	60										0 2	0 3	1 4	2 6	4 9	7 12	11 17									
	Fourth	20	80										0 3	1 4	2 5	3 7	6 11	10 15	16 22									
	Fifth	20	100										1 3	2 4	3 6	5 8	9 12	14 17	22 25									
	Sixth	20	120										1 3	3 5	4 6	7 9	12 14	18 20	27 29									
	Seventh	20	140										2 3	4 5	6 7	9 10	14 15	21 22	32 33									

↓ = Use first sampling plan below arrow (refer to continuation of table on following page, when necessary). If sample size equals or exceeds lot or batch size, do 100 percent inspection.
↑ = Use first sampling plan above arrow.
Ac = Acceptance number
Re = Rejection number
• = Use corresponding single sampling plan (or alternatively, use multiple sampling plan below, where available).
↔ = Use corresponding double sampling plan (or alternatively, use multiple sampling plan below, where available).
= Acceptance not permitted at this sample size.

TABLE IV-B — Multiple sampling plans for tightened inspection (Master table) (Continued)

Sample size code letter	Sample	Sample size	Cumulative sample size	0.010 Ac Re	0.015 Ac Re	0.025 Ac Re	0.040 Ac Re	0.065 Ac Re	0.10 Ac Re	0.15 Ac Re	0.25 Ac Re	0.40 Ac Re	0.65 Ac Re	1.0 Ac Re	1.5 Ac Re	2.5 Ac Re	4.0 Ac Re	6.5 Ac Re	10 Ac Re	15 Ac Re	25 Ac Re	40 Ac Re	65 Ac Re	100 Ac Re	150 Ac Re	250 Ac Re	400 Ac Re	650 Ac Re	1000 Ac Re
K	First	32	32	↓	↓	↓	↓	↓	↓	•	↓	↓	* 2	* 2	* 3	* 4	0 4	0 6	1 8	↑	↑	↑	↑	↑	↑	↑	↑	↑	↑
	Second	32	64										* 2	0 3	0 3	1 5	2 7	3 9	6 12										
	Third	32	96										0 2	0 3	1 4	2 6	4 9	7 12	11 17										
	Fourth	32	128										0 3	1 4	2 5	3 7	6 11	10 15	16 22										
	Fifth	32	160										1 3	2 4	3 6	5 8	9 12	14 17	22 25										
	Sixth	32	192										1 3	3 5	4 6	7 9	12 14	18 20	27 29										
	Seventh	32	224										2 3	4 5	6 7	9 10	14 15	21 22	32 33										
L	First	50	50	↓	↓	↓	↓	↓	•	↓	↓	* 2	* 2	* 3	* 4	0 4	0 6	1 8	↑	↑	↑	↑	↑	↑	↑	↑	↑	↑	↑
	Second	50	100									* 2	0 3	0 3	1 5	2 7	3 9	6 12											
	Third	50	150									0 2	0 3	1 4	2 6	4 9	7 12	11 17											
	Fourth	50	200									0 3	1 4	2 5	3 7	6 11	10 15	16 22											
	Fifth	50	250									1 3	2 4	3 6	5 8	9 12	14 17	22 25											
	Sixth	50	300									1 3	3 5	4 6	7 9	12 14	18 20	27 29											
	Seventh	50	350									2 3	4 5	6 7	9 10	14 15	21 22	32 33											
M	First	80	80	↓	↓	↓	↓	•	↓	↓	* 2	* 2	* 3	* 4	0 4	0 6	1 8	↑	↑	↑	↑	↑	↑	↑	↑	↑	↑	↑	↑
	Second	80	160								* 2	0 3	0 3	1 5	2 7	3 9	6 12												
	Third	80	240								0 2	0 3	1 4	2 6	4 9	7 12	11 17												
	Fourth	80	320								0 3	1 4	2 5	3 7	6 11	10 15	16 22												
	Fifth	80	400								1 3	2 4	3 6	5 8	9 12	14 17	22 25												
	Sixth	80	480								1 3	3 5	4 6	7 9	12 14	18 20	27 29												
	Seventh	80	560								2 3	4 5	6 7	9 10	14 15	21 22	32 33												
N	First	125	125	↓	↓	↓	•	↓	↓	* 2	* 2	* 3	* 4	0 4	0 6	1 8	↑	↑	↑	↑	↑	↑	↑	↑	↑	↑	↑	↑	↑
	Second	125	250							* 2	0 3	0 3	1 5	2 7	3 9	6 12													
	Third	125	375							0 2	0 3	1 4	2 6	4 9	7 12	11 17													
	Fourth	125	500							0 3	1 4	2 5	3 7	6 11	10 15	16 22													
	Fifth	125	625							1 3	2 4	3 6	5 8	9 12	14 17	22 25													
	Sixth	125	750							1 3	3 5	4 6	7 9	12 14	18 20	27 29													
	Seventh	125	875							2 3	4 5	6 7	9 10	14 15	21 22	32 33													
P	First	200	200	↓	↓	•	↓	↓	* 2	* 2	* 3	* 4	0 4	0 6	1 8	↑	↑	↑	↑	↑	↑	↑	↑	↑	↑	↑	↑	↑	↑
	Second	200	400						* 2	0 3	0 3	1 5	2 7	3 9	6 12														
	Third	200	600						0 2	0 3	1 4	2 6	4 9	7 12	11 17														
	Fourth	200	800						0 3	1 4	2 5	3 7	6 11	10 15	16 22														
	Fifth	200	1000						1 3	2 4	3 6	5 8	9 12	14 17	22 25														
	Sixth	200	1200						1 3	3 5	4 6	7 9	12 14	18 20	27 29														
	Seventh	200	1400						2 3	4 5	6 7	9 10	14 15	21 22	32 33														
Q	First	315	315	↓	•	↓	↓	* 2	* 2	* 3	* 4	0 4	0 6	1 8	↑	↑	↑	↑	↑	↑	↑	↑	↑	↑	↑	↑	↑	↑	↑
	Second	315	630					* 2	0 3	0 3	1 5	2 7	3 9	6 12															
	Third	315	945					0 2	0 3	1 4	2 6	4 9	7 12	11 17															
	Fourth	315	1260					0 3	1 4	2 5	3 7	6 11	10 15	16 22															
	Fifth	315	1575					1 3	2 4	3 6	5 8	9 12	14 17	22 25															
	Sixth	315	1890					1 3	3 5	4 6	7 9	12 14	18 20	27 29															
	Seventh	315	2205					2 3	4 5	6 7	9 10	14 15	21 22	32 33															
R	First	500	500	•	↑	↓	* 2	* 2	* 3	* 4	0 4	0 6	1 8	↑	↑	↑	↑	↑	↑	↑	↑	↑	↑	↑	↑	↑	↑	↑	↑
	Second	500	1000				* 2	0 3	0 3	1 5	2 7	3 9	6 12																
	Third	500	1500				0 2	0 3	1 4	2 6	4 9	7 12	11 17																
	Fourth	500	2000				0 3	1 4	2 5	3 7	6 11	10 15	16 22																
	Fifth	500	2500				1 3	2 4	3 6	5 8	9 12	14 17	22 25																
	Sixth	500	3000				1 3	3 5	4 6	7 9	12 14	18 20	27 29																
	Seventh	500	3500				2 3	4 5	6 7	9 10	14 15	21 22	32 33																
S	First	800	800			* 2																							
	Second	800	1600			* 2																							
	Third	800	2400			0 2																							
	Fourth	800	3200			0 3																							
	Fifth	800	4000			1 3																							
	Sixth	800	4800			1 3																							
	Seventh	800	5600			2 3																							

↓ = Use first sampling plan below arrow. If sample size equals or exceeds lot or batch size, do 100 percent inspection.
↑ = Use first sampling plan above arrow (refer to preceding page, when necessary).
Ac = Acceptance number
Re = Rejection number
• = Use corresponding single sampling plan (or alternatively, use multiple sampling plan below, where available).
* = **Acceptance not permitted at this sample size.**

TABLE V-B — *Average Outgoing Quality Limit Factors for Tightened Inspection (Single sampling)*

Code letter	Sample size	Acceptable Quality Level																							
		0.010	0.015	0.025	0.040	0.065	0.10	0.15	0.25	0.40	0.65	1.0	1.5	2.5	4.0	6.5	10	15	25	40	65	100	150	250	400	650	1000
A	2																			42	69	97	160	260	400	620	970
B	3															12			28	46	65	110	170	270	410	650	1100
C	5														7.4			17	27	39	63	100	160	250	390	610	
D	8													4.6			11	17	24	40	64	99	160	240	380		
E	13												2.8			6.5	11	15	24	40	61	95	150	240			
F	20											1.8			4.2	6.9	9.7	16	26	40	62						
G	32										1.2			2.6	4.3	6.1	9.9	16	25	39							
H	50									0.74			1.7	2.7	3.9	6.3	10	16	25								
J	80								0.46			1.1	1.7	2.4	4.0	6.4	9.9	16									
K	125							0.29			0.67	1.1	1.6	2.5	4.1	6.4	9.9										
L	200						0.18			0.42	0.69	0.97	1.6	2.6	4.0	6.2											
M	315					0.12			0.27	0.44	0.62	1.0	1.6	2.5	3.9												
N	500				0.074			0.17	0.27	0.39	0.63	1.0	1.6	2.5													
P	800			0.046			0.11	0.17	0.24	0.40	0.64	0.99	1.6														
Q	1250		0.029			0.067	0.11	0.16	0.25	0.41	0.64	0.99															
R	2000	0.018			0.042	0.069	0.097	0.16	0.26	0.40	0.62																
S	3150			0.027																							

Note: For the exact AOQL, the above values must be multiplied by $\left(1 - \frac{\text{Sample size}}{\text{Lot or Batch size}}\right)$ (see 11.4)

TABLE VI-A—Limiting Quality (in percent defective) for which P_a = 10 Percent (for Normal Inspection, Single sampling)

Code letter	Sample size	Acceptable Quality Level															
		0.010	0.015	0.025	0.040	0.065	0.10	0.15	0.25	0.40	0.65	1.0	1.5	2.5	4.0	6.5	10
A	2															68	
B	3														54		
C	5													37			58
D	8												25			41	54
E	13											16			27	36	44
F	20										11			18	25	30	42
G	32									6.9			12	16	20	27	34
H	50								4.5			7.6	10	13	18	22	29
J	80							2.8			4.8	6.5	8.2	11	14	19	24
K	125						1.8			3.1	4.3	5.4	7.4	9.4	12	16	23
L	200					1.2			2.0	2.7	3.3	4.6	5.9	7.7	10	14	
M	315				0.73			1.2	1.7	2.1	2.9	3.7	4.9	6.4	9.0		
N	500			0.46			0.78	1.1	1.3	1.9	2.4	3.1	4.0	5.6			
P	800		0.29			0.49	0.67	0.84	1.2	1.5	1.9	2.5	3.5				
Q	1250	0.18			0.31	0.43	0.53	0.74	0.94	1.2	1.6	2.3					
R	2000			0.20	0.27	0.33	0.46	0.59	0.77	1.0	1.4						

TABLE VI-B—Limiting Quality (in defects per hundred units) for which P_a = 10 Percent (for Normal Inspection, Single sampling)

Code letter	Sample size	Acceptable Quality Level																									
		0.010	0.015	0.025	0.040	0.065	0.10	0.15	0.25	0.40	0.65	1.0	1.5	2.5	4.0	6.5	10	15	25	40	65	100	150	250	400	650	1000
A	2															120			200	270	330	460	590	770	1000	1400	1900
B	3														77			130	180	220	310	390	510	670	940	1300	1800
C	5													46			78	110	130	190	240	310	400	560	770	1100	
D	8												29			49	67	84	120	150	190	250	350	480	670		
E	13											18			30	41	51	71	91	120	160	220	300	410			
F	20										12			20	27	33	46	59	77	100	140						
G	32									7.2			12	17	21	29	37	48	63	88							
H	50								4.6			7.8	11	13	19	24	31	40	56								
J	80							2.9			4.9	6.7	8.4	12	15	19	25	35									
K	125						1.8			3.1	4.3	5.4	7.4	9.4	12	16	23										
L	200					1.2			2.0	2.7	3.3	4.6	5.9	7.7	10	14											
M	315				0.73			1.2	1.7	2.1	2.9	3.7	4.9	6.4	9.0												
N	500			0.46			0.78	1.1	1.3	1.9	2.4	3.1	4.0	5.6													
P	800		0.29			0.49	0.67	0.84	1.2	1.5	1.9	2.5	3.5														
Q	1250	0.18			0.31	0.43	0.53	0.74	0.94	1.2	1.6	2.3															
R	2000			0.20	0.27	0.33	0.46	0.59	0.77	1.0	1.4																

TABLE VIII — *Limit Numbers for Reduced Inspection*

Number of sample units from last 10 lots or batches	Acceptable Quality Level																									
	0.010	0.015	0.025	0.040	0.065	0.10	0.15	0.25	0.40	0.65	1.0	1.5	2.5	4.0	6.5	10	15	25	40	65	100	150	250	400	650	1000
20 - 29	*	*	*	*	*	*	*	*	*	*	*	*	*	*	*	0	0	2	4	8	14	22	40	68	115	181
30 - 49	*	*	*	*	*	*	*	*	*	*	*	*	*	*	0	0	1	3	7	13	22	36	63	105	178	277
50 - 79	*	*	*	*	*	*	*	*	*	*	*	*	*	0	0	2	3	7	14	25	40	63	110	181	301	
80 - 129	*	*	*	*	*	*	*	*	*	*	*	*	0	0	2	4	7	14	24	42	68	105	181	297		
130 - 199	*	*	*	*	*	*	*	*	*	*	*	0	0	2	4	7	13	25	42	72	115	177	301	490		
200 - 319	*	*	*	*	*	*	*	*	*	*	0	0	2	4	8	14	22	40	68	115	181	277	471			
320 - 499	*	*	*	*	*	*	*	*	*	0	0	1	4	8	14	24	39	68	113	189						
500 - 799	*	*	*	*	*	*	*	*	0	0	2	3	7	14	25	40	63	110	181							
800 - 1249	*	*	*	*	*	*	*	0	0	2	4	7	14	24	42	68	105	181								
1250 - 1999	*	*	*	*	*	*	0	0	2	4	7	13	24	40	69	110	169									
2000 - 3149	*	*	*	*	*	0	0	2	4	8	14	22	40	68	115	181										
3150 - 4999	*	*	*	*	0	0	1	4	8	14	24	38	67	111	186											
5000 - 7999	*	*	*	0	0	2	3	7	14	25	40	63	110	181												
8000 - 12499	*	*	0	0	2	4	7	14	24	42	68	105	181													
12500 - 19999	*	0	0	2	4	7	13	24	40	69	110	169														
20000 - 31499	0	0	2	4	8	14	22	40	68	115	181															
31500 - 49999	0	1	4	8	14	24	38	67	111	186																
50000 & Over	2	3	7	14	25	40	63	110	181	301																

Denotes that the number of sample units from the last ten lots or batches is not sufficient for reduced inspection for this AQL. In this instance more than ten lots or batches may be used for the calculation, provided that the lots or batches used are the most recent ones in sequence, that they have all been on normal inspection, and that none has been rejected while on original inspection.

TABLE IX—*Average sample size curves for double and multiple sampling (normal and tightened inspection)*

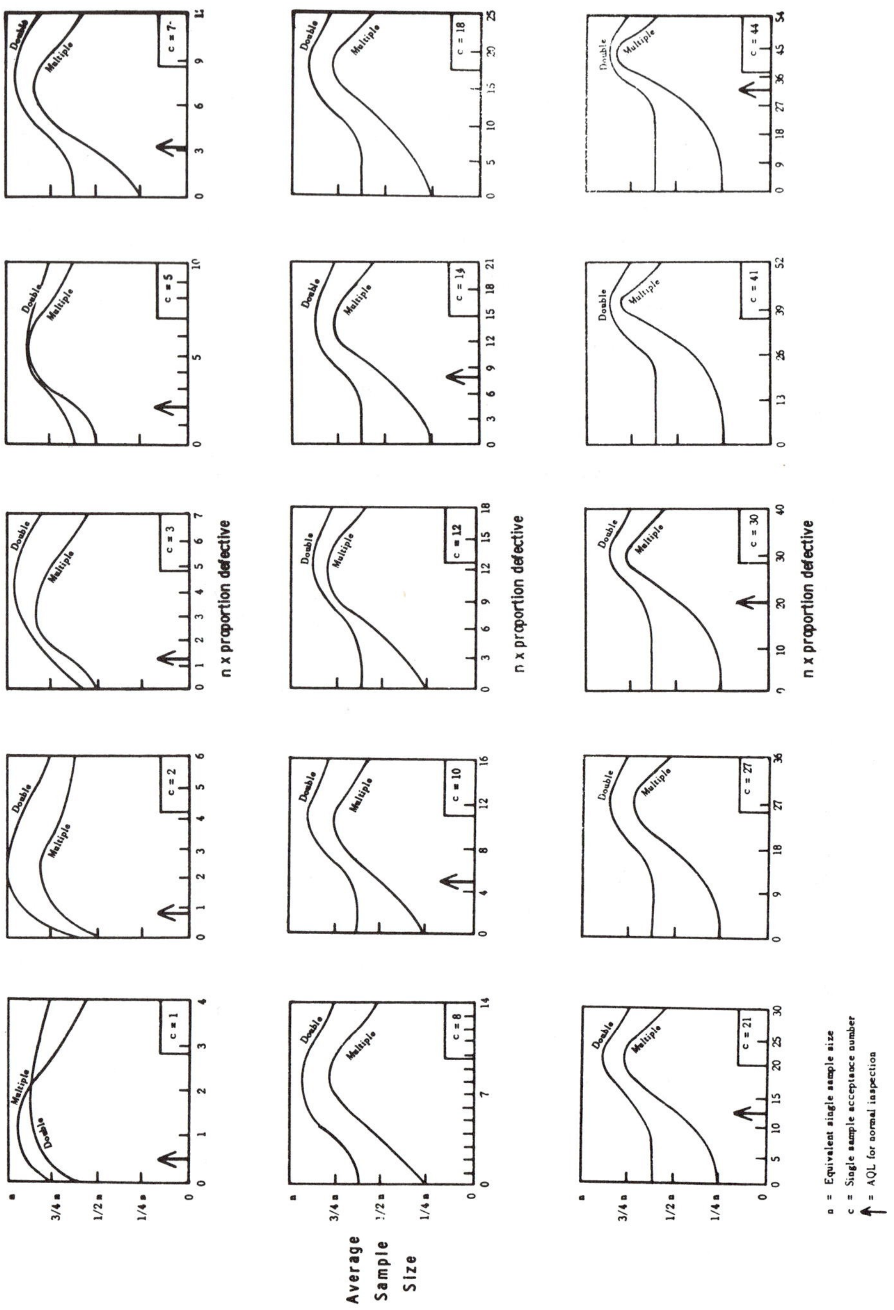

TABLE X-J — *Tables for sample size code letter: J*

CHART J - OPERATING CHARACTERISTIC CURVES FOR SINGLE SAMPLING PLANS

(Curves for double and multiple sampling are matched as closely as practicable)

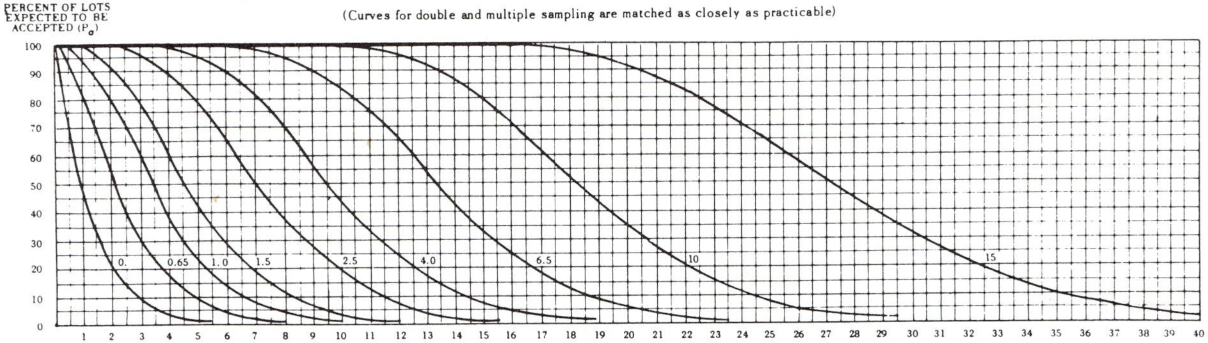

QUALITY OF SUBMITTED LOTS (p, in percent defective for AQL's ≤10; in defects per hundred units for AQL's >10)

Note: Figures on curves are Acceptable Quality Levels (AQL's) for normal inspection.

TABLE X-J-1 - TABULATED VALUES FOR OPERATING CHARACTERISTIC CURVES FOR SINGLE SAMPLING PLANS

P_a	Acceptable Quality Levels (normal inspection)																					
	0.15	0.65	1.0	1.5	2.5	4.0	X	6.5	X	10	0.15	0.65	1.0	1.5	2.5	4.0	X	6.5	X	10	X	15
	p (in percent defective)										p (in defects per hundred units)											
99.0	0.013	0.188	0.550	1.05	2.30	3.72	4.50	6.13	7.88	9.75	0.013	0.186	0.545	1.03	2.23	3.63	4.38	5.96	7.62	9.35	12.9	15.7
95.0	0.064	0.444	1.03	1.73	3.32	5.06	5.98	7.91	9.89	11.9	0.064	0.444	1.02	1.71	3.27	4.98	5.87	7.71	9.61	11.6	15.6	18.6
90.0	0.132	0.666	1.38	2.20	3.98	5.91	6.91	8.95	11.0	13.2	0.131	0.665	1.38	2.18	3.94	5.82	6.79	8.78	10.8	12.9	17.1	20.3
75.0	0.359	1.202	2.16	3.18	5.30	7.50	8.62	10.9	13.2	15.5	0.360	1.20	2.16	3.17	5.27	7.45	8.55	10.8	13.0	15.3	19.9	23.4
50.0	0.863	2.09	3.33	4.57	7.06	9.55	10.8	13.3	15.8	18.3	0.866	2.10	3.34	4.59	7.09	9.59	10.8	13.3	15.8	18.3	23.3	27.1
25.0	1.72	3.33	4.84	6.31	9.14	11.9	13.3	16.0	18.6	21.3	1.73	3.37	4.90	6.39	9.28	12.1	13.5	16.3	19.0	21.8	27.2	31.2
10.0	2.84	4.78	6.52	8.16	11.3	14.2	15.7	18.6	21.4	24.2	2.88	4.86	6.65	8.35	11.6	14.7	16.2	19.3	22.2	25.2	30.9	35.2
5.0	3.68	5.80	7.66	9.39	12.7	15.8	17.3	20.3	23.2	26.0	3.75	5.93	7.87	9.69	13.1	16.4	18.0	21.2	24.3	27.4	33.4	37.8
1.0	5.59	8.00	10.1	12.0	15.6	18.9	20.5	23.6	26.5	29.5	5.76	8.30	10.5	12.6	16.4	20.0	21.8	25.2	28.5	31.8	38.2	42.9
	0.25	1.0	1.5	2.5	4.0	X	6.5	X	10	X	0.25	1.0	1.5	2.5	4.0	X	6.5	X	10	X	15	X
	Acceptable Quality Levels (tightened inspection)																					

Note: All values given in above table based on Poisson distribution as an approximation to the Binomial.

TABLE X-J-2 - SAMPLING PLANS FOR SAMPLE SIZE CODE LETTER: J

Type of sampling plan	Cumulative sample size	Less than 0.15	0.15	0.25	X	0.40	0.65	1.0	1.5	2.5	4.0	X	6.5	X	10	X	15	Higher than 15	Cumulative sample size
Acceptable Quality Levels (tightened inspection)		Less than 0.25	0.25	X	0.40	0.65	1.0	1.5	2.5	4.0	X	6.5	X	10	X	15	X	Higher than 15	
		Ac Re	Ac Re	Ac Re	Ac Re	Ac Re	Ac Re	Ac Re	Ac Re	Ac Re	Ac Re	Ac Re	Ac Re	Ac Re	Ac Re	Ac Re	Ac Re	Ac Re	
Single	80	▽	0 1	Use Letter H	Use Letter L	Use Letter K	1 2	2 3	3 4	5 6	7 8	8 9	10 11	12 13	14 15	18 19	21 22	△	80
Double	50	▽	*				0 2	0 3	1 4	2 5	3 7	3 7	5 9	6 10	7 11	9 14	11 16	△	50
	100						1 2	3 4	4 5	6 7	8 9	11 12	12 13	15 16	18 19	23 24	26 27		100
Multiple	20	▽	*				# 2	# 2	# 3	# 4	0 4	0 4	0 5	0 6	1 7	1 8	2 9	△	20
	40						# 2	0 3	0 3	1 5	1 6	2 7	3 8	3 9	4 10	6 12	7 14		40
	60						0 2	0 3	1 4	2 6	3 8	4 9	6 10	7 12	8 13	11 17	13 19		60
	80						0 3	1 4	2 5	3 7	5 10	6 11	8 13	10 15	12 17	16 22	19 25		80
	100						1 3	2 4	3 6	5 8	7 11	9 12	11 15	14 17	17 20	22 25	25 29		100
	120						1 3	3 5	4 6	7 9	10 12	12 14	14 17	18 20	21 23	27 29	31 33		120
	140						2 3	4 5	6 7	9 10	13 14	14 15	18 19	21 22	25 26	32 33	37 38		140

Acceptable Quality Levels (normal inspection) — top header; Acceptable Quality Levels (tightened inspection) — bottom header.

△ = Use next preceding sample size code letter for which acceptance and rejection numbers are available.
▽ = Use next subsequent sample size code letter for which acceptance and rejection numbers are available.
Ac = Acceptance number
Re = Rejection number
* = Use single sampling plan above (or alternatively use letter M)
\# = Acceptance not permitted at this sample size.

There are other special levels S-1 to S-4 for special conditions. See Section 9.2.

10.3.3. Entering the Standard for a Plan. The following steps are taken to obtain a sampling plan:

1. Decide on the size of lot N, which will be sample inspected. This need not be a production lot size.
2. Decide upon an inspection level, in general II.
3. Using 1 and 2, enter Table I and find the corresponding sample-size code letter A, B, to R, the last calling for largest sample sizes.
4. Decide upon single, double or multiple sampling.
5. Decide whether to start with normal (almost invariably), tightened or reduced sampling.
6. Table IIA, IIIA or IVA will thus be determined by 4 and 5 (if normal inspection is used.)
7. Decide upon the inspection basis of defectives or defects.
8. Decide upon the desired AQL: for defectives (10.0 or less), or defects per 100 units; using only what is available in tables. Note that ".015" is .015%, or decimally .00015.
9. Enter the table determined in 6, using the AQL for the column from 8 and the row for code letter in 3. This commonly gives the acceptance-rejection numbers, the sample size(s) being given to the left of this block.
10. Following the above, we may reach an asterisk or an arrow. An asterisk means to use the single sampling plan for the desired AQL and code letter, instead of double or multiple sampling. If an arrow is encountered, follow it to the first block with acceptance-rejection numbers, using sample sizes to the left of this block, not to the left of the <u>original</u> block.

10.3.4. Further Explanatory Comments. Section 2 in the Standard gives a classification of defects. Three categories of defects are given: critical, major and minor, as defined. For critical defects 100% inspection or testing is commonly used, although sampling may be used after adequate excellent history is built up and then only for very low AQL's. Major and minor defects differ according to seriousness and thus carry lower and higher AQL's respectively. Commonly in any product, there will be a list of major defects and another list of minor defects. It must be born in mind that when there are ten or fifteen different possible major defects, say, it is quite difficult to have the AQL for defectives to be extremely small, because d will be the total number of all major defectives, that is, pieces with one or more major defects.

Note the emphasis in Section 7 on proper choosing of samples. This is essential to sound decisions.

Rules for "switching" are given in Section 8. As regards normal to tightened inspection and vice versa, these have been greatly simplified from earlier Standards, by adopting suggestions of Dodge [5]. It has proven rather difficult in practical use to insure that those in charge of the inspection go to tightened inspection when they should. If tightened inspection is not invoked when called for, then the intended consumer protection provided by the Standard is greatly weakened. This is because the plans are set up with OC curves intentionally lenient toward producers who are cooperating by having $p' \leq$ AQL. But to do this the shape of the OC curves shows that Pa for p''s a bit above the AQL still give quite high Pa's. Hence unless tightened inspection is invoked a producer may fairly comfortably run p' above the AQL for quite a long time. The newer rule calls for tightened inspection whenever 2 out of any 5 consecutive lots are rejected on initial submission.

In order to get back onto normal inspection, 5 consecutive lots must be passed while under tightened inspection. Moreover this transition must be accomplished within 10 consecutive lots, or else sampling inspection is abandoned altogether.

Switching rules for going to reduced inspection are four in number, all of which must be fulfilled to justify the switch. Between them they quite well insure that there will be a low probability of any lot with p' above the AQL being submitted. Thus a lenient plan with small n is feasible. See Section 8.3.3. On the other hand there are four conditions for remaining on reduced inspection. If any one fails, inspection is to revert to the normal plan. See Section 8.3.4. One of these is a special procedure. For example, see Table IIC, AQL 1.5, code letter J. For this single plan n = 32, Ac = 1, Re = 4, which leaves a gap. If $d \geq 4$, of course we reject, or if $d \leq 1$ we accept. But if d is 2 or 3 we still accept that lot under some suspicion, because quality has been excellent, but normal sampling is reinstated.

As pointed out in Section 11.1, the OC curves are calculated using the binomial distribution when both AQL $\leq$ 10.0 and $n_{single} \leq 80$. Thus these are exact Type B curves for percent defective. In the other cases the Poisson distribution was used, and they are thus exact for defects per hundred units, and good approximations for percent defective.

In Section 11.2 it is emphasized that the process average is to be figured on results from lots on original submission only (not after any rectifying 100% inspection). Why?

Section 11.4 indicates the inclusion of AOQL factors in Tables VA normal and VB tightened plans. We do not include the former table, because in order for the AOQL to become meaningful, p' would have to be much worse than the AQL, and under these conditions tightened inspection would have been quickly invoked.

An interesting inclusion is that of relative average sample number ASN curves, Section 11.5. Read all of this Section carefully. The reason that so few ASN curves as in Table IX may be used to cover all cases is that both the sequences of sample sizes and AQL's go up in arithmetic progression (very nearly). Thus we have n's 200, 315, 500, 800, 1250, 2000. Recalling that $\sqrt[5]{10} = 1.5849 = r$, say, then $315 \doteq 200r$, $500 \doteq 200r^2, \ldots, 2000 \doteq 200r^5$. The same is true for the progression of AQL's. This permits the same acceptance-rejection numbers down a diagonal in tables such as IIA. Then all of the ASN curves for double and multiple plans corresponding to the acceptance numbers c = 1, 2,..., 44, for single sampling will be nearly exactly alike, if we use a vertical scale in terms of n_{single}, and as horizontal scale $np'/(100\%)$, that is, the expected number of defectives per <u>single</u> sample. The corresponding double sample size is $(n_{single})/1.5849 \doteq (n_{single}).631$ ar multiple sample size is $(n_{single})/1.5849^3 \doteq (n_{single}).25$. For all <u>normal</u> single sampling plans, indexed by c numbers, the AQL is shown on the horizontal scale by vertical arrows.

Take very special note that these relative ASN curves are only comparable because they have OC curves very nearly alike. Thus the same power of discrimination between lots of differing quality is available by single, double or multiple plans for given AQL and code letter. Hence the relative ASN's can be soundly compared. Table IX shows that the double sampling plans have uniformly more favorable ASN curves than the corresponding single sampling plans, especially so for very good or poor quality levels. Moreover, except for the c = 1 case, the ASN curve for the corresponding multiple sampling plan lies uniformly below that of the double sampling plan. Of course such more favorable ASN curves, for comparable protection, are bought at the price of greater difficulty in teaching and supervision and slower accumulation of history on a producer.

Another point is that Table IX is for non-truncated inspection, that is, any sample started is assumed completely inspected, even if an earlier decision could be reached. See Section 11.5. But in double and multiple inspection, any inspection after completion of the first sample is commonly ended as soon as a rejection decision is reached, if one is. This has little effect if $p' \leq$ AQL or even near it, but has increasing effect as p' increases.

Finally we may take note of sample pages 307 and 308 for plans under code letter J. On page 307 are drawn all of the OC curves, the horizontal scale being applicable to either defects or defectives. Then Table X-J-1 gives precise values of lot qualities for which Pa = .99, .95, ..., .01 for each given AQL. This table is entered from the top for normal inspection or the bottom for tightened inspection. The same method of entry is used in Table X-J-2, which gives the acceptance-rejection criteria for all three kinds of plans, and the AQL's.

10.3.5. Multiple Sampling. Let us now consider multiple sampling which has been already involved in our description of the Standard. For AQL = 1.0, and code letter J, we have n = 20, and the following decision criteria:

Σn	20	40	60	80	100	120	140
Ac	*	0	0	1	2	3	4
Re	2	3	3	4	4	5	5

*Cannot accept on the first sample of 20

For this plan, if we find two or more defectives among the first 20 we reject at once. But if there are zero or one, we draw a new sample of 20 (from all over the lot) and inspect these. Suppose that d_1 was 0, then if d_2 is also 0 we can accept. Similarly if $d_1=0$ and $d_2 \geq 3$ we reject. Otherwise we continue. Or if $d_1 = 1$ and $d_2 \geq 2$ we reject, while if $d_2 = 0$ or 1 we continue. Note that if $d_1 = 1$ the earliest possible acceptance is after 60 more pieces, all of which are good for then we would meet the Ac number for 80. We observe that after 140 a decision is always reached since there is no integer gap between 4 and 5.

10.3.6.* Calculations for Multiple Sampling. If occasion should arise for calculating the OC or ASN curve for multiple sampling, the following approach has been found useful. Consider the plan given in Section 10.3.5 for code letter J and AQL 1.0%. Suppose we figure out from Poisson or binomial distribution the probability of 0, 1, 2, and so on, defectives in n = 20 for given p', $p_0, p_1, p_2 \ldots$ Then let $P_i(j)$ be the probability of having j defectives after i samples, without a decision having been made. Thus $P_1(0) = p_0$, $P_1(1) = p_1$. These are the only cases wherein the decision is still doubtful, since 2 would give rejection.

Now we have after the second sample the incomplete cases where $d_1 + d_2 = 1$ or 2. We find the probabilities of such being the case by using $P_1(j)$ and p_j:

$$P_2(1) = P_1(0)p_1 + P_1(1)p_0 \qquad P_2(2) = P_1(0)p_2 + P_1(1)p_1.$$

These are the only possibilities for incomplete cases at the end of the second sample. After 60 pieces we have

$$P_3(1) = P_2(1)p_o \qquad P_3(2) = P_2(1)p_1 + P_2(2)p_o$$

proceeding recursely to find $P_i(j)$'s from $P_{i-1}(j)$'s. The $P_i(j)$'s are then used for contributions to Pa. The probabilities of acceptance can only occur after 40, 80, 100, 120, 140 pieces inspected. (Why?) Thus

$$Pa = P_1(0)p_o + P_3(1)p_o + P_4(2)p_o + P_5(3)p_o + P_6(4)p_o.$$

Also the ASN is obtained by multiplying these five probabilities by 40, 80, 100, 120, 140, for the contribution under acceptance. One must also calculate the probabilities of rejection at 20, 40, and so forth and obtain the weighted sum to add to the above. This approach can also be used to find the ASN under truncation, via a digital computer.

10.3.7. Isolated Lot Protection. Section 11.6 discusses the case of sampling inspection for an isolated lot. Then we are not in any position to maintain the average outgoing quality and our problem becomes that of making a decision on this one lot. We want to accept it if p' is sufficiently low and to reject it if p' is sufficiently high. There will be some gap in these p''s where it is somewhat immaterial whether we accept or reject and sort; an indifference band. For such a decision we need to use an appropriate OC curve. Convenient information is given in Tables VIA and VIB, which provide the values of p' or c'100% for which Pa = .10. Thus, choosing the column for a desired AQL value of p' in Table VIA we can look down it until we find a sufficiently low p'_{10} for our purposes. Then this determines the code letter and we can use a single, double or multiple plan, as desired.

10.4. What Next? Since its issuance in 1963 the MIL-STD 105D has been studied and many possible items of revision considered. In 1967 Stephens and Larson [6] published an evaluation of the standard as regards its operating characteristics, treated as a system. The idea is to take account of the proportion of the lots for which sampling is on the normal plan, the tightened plan and on the reduced plan, for a given process quality p'. Then one works out the "composite" operating characteristic curve and the composite average sample size curve. This gives insight into the working of the system chosen (code letter and AQL) when p' is fixed. It is also possible to let p' vary according to some dynamic model. Two cases were considered: Case I, using normal, tightened and reduced inspection and Case II, using only the first two (since reduced is not mandatory). The composite OC curve, as one would expect, is more discriminating than the OC curves for either the normal or tightened plans. Moreover the inclusion of reduced inspection with its very lenient OC curve has virtually no eroding effect on the composite OC curve.

A distinguished Japanese Group made substantial studies of 105D and fostered adoption of the standard by the Japanese Standards Association, with some modifications of switching rules [7,8]. The Group's chief concern was with a calculated "considerable" switching back and forth between normal and tightened inspection (and even dropping of inspection altogether) when a producer is running exactly at p' = AQL. In [8] a letter from the Standards Group of the Standards Committee of the American Society for Quality Control, made several points. The first point of difference is in the definition of AQL, where reference [12] of Chapter 9, defines it as the maximum satisfactory process average, not as a perfectly satisfactory quality level at which the producer can run and expect virtually no trouble with going onto tightened inspection. If running right at the AQL, he should expect to have some such trouble, since that is natural in a plan designed to provide the consumer with protection against p''s somewhat worse than the AQL.

The second point is that much of the calculated frequency of switching was for single sampling plans with Ac = 1. But for these plans Pa(AQL) = .910, whereas for those with Ac > 1, Pa(AQL) = .953 and above. For the latter plans switching while p' = AQL would be much less frequent than if Ac = 1.

The third point is that the proposed use of both the 2 out of 5 criterion in 105D Section 8.3.1 and the earlier (105B, 105C) use of an upper limit to $\bar{p}$ ($\bar{p}$ out of control relative to the specified AQL) requiring frequent calculation of $\bar{p}$, makes the tightened inspection criterion quite complicated. This has the calculable effect of giving the consumer considerably less protection against p' > AQL, and gives the producer of such quality too much leniency. Moreover, it would have the practical effect of discouraging the use of tightened inspection at all, which was often avoided under the earlier rule, but avoided much less often under the 2 out 5 rule. To require both criteria to occur before going onto tightened inspection is a substantial stumbling block to the use of tightened inspection. But unless tightened inspection is invoked when called for, the consumer loses much of his protection.

At the time this is being written the Standards Group of the ASQC Standards Committee is working toward a revision of the American National Standards Institute Z1.4 which is currently the same as the 105D. One objective is to make the standard more oriented to industrial production of consumer goods, MIL-STD 105D being worded more to military procurement. Changes under study are

1. To make the probability of acceptance at the AQL as nearly uniform at .95 as feasible for all Ac numbers 1, 2,... This would have the effect, to quite an extent, of removing the objection of the Japanese Group toward frequency of switching. Currently these Pa's go from about .91 at Ac = 1, .95 at Ac = 2, uniformly up to about .995 at Ac = 22.
2. It is difficult to fit Ac = 0 plans into the picture in 1, because the OC curve for such plans is always concave upwards. Hence in 105D, Pa(AQL) is only about .88. To bolster this Pa, consideration is being given to using Chain Sampling as an alternative to Ac = 0 plans. See Chapter 13.
3. By making the acceptance number for tightened inspection one less than that for normal, except when Ac(normal) = 0, one obtains tightened inspection plans with AOQL's fairly close to the AQL's specified. This gives desired consumer protection, and at least in some places in the system, less of a penalty in tightened inspection.
4. Adjustment of sample size in Ac = 0 toward lower n's will raise the Pa(AQL) from .88.
5. Corresponding adjustments can be made for double and multiple sampling to match the OC curves for single plans.

6. Provision of relative ASN curves for double and multiple sampling under truncation after the initial samples, when rejection occurs.
7. Elimination of Table VA, AOQL Factors for normal inspection, since as pointed out earlier, when the level of p' is poor enough to approach that needed in normal sampling to make the AOQ rise to the AOQL, the sampling would have quickly gone onto tightened inspection. Thus AOQL has no meaning for normal inspection (unless one violates 105D completely and never invokes tightened inspection).

10.5. Summary. In this chapter we have discussed two sampling systems with differing aims. These are collections of sampling plans, with instructions as to when to go from one plan to another. The Dodge-Romig has single and double plans to provide specified consumer protection, either AOQL or LTPD, and to give this protection on a minimum ATI. Thus it is specific protection on minimum inspection cost.

The ABC-105D system aims to maintain material at a quality level of the specified AQL or better, being lenient toward quality at this AQL or better, but tightening up when there is evidence of quality deterioration. Single, double and multiple plans are provided.

Both plans are consumer oriented. Although they can be used at the end of the producer's production line, a more natural and desirable approach for a producer would seem to be process control, for example, control charts. In this way, assignable causes of bad product may best be sought out and eliminated and the product made to conform to requirements.

Other approaches are possible. For example, H. C. Hamaker and associates, published in the Philips Technical Review (Eindhoven, Netherlands) in 1949 and 1950 three papers leading to a simple sampling system, indexed on the point of indifference, p'_{50}. The plans were developed to equalize the producer and consumer risks in a sense by keeping nearly constant

$$\frac{\text{slope Pa(at } p'_{50})}{p'_{50}}.$$

Then the producer's risk PR decreases from 50% as p' decreases from p'_{50}, at about same rate as the consumer's risk CR decreases from 50% as p' increases from p'_{50}.

PROBLEMS

Find Dodge-Romig sampling plans for the following conditions and list the amount of the opposite kind of consumer protection:

10.1. Double sampling, AOQL=1.0%, N=3500, $\bar{p}$=.5%, defectives

10.2. Double sampling, AOQL=1.0%, N=600, $\bar{p}$=.9%, defectives

10.3. Single sampling, AOQL=1.0%, N=3500, $\bar{p}$=.5%, defectives

10.4. Single sampling, AOQL=1.0%, N=600, $\bar{p}$=.9%, defectives

10.5. Double sampling, LTPD=4.0%, N=3500, $\bar{p}$=.5% defectives

10.6. Double sampling, LTPD=4.0%, N=600, $\bar{p}$=.9% defectives

10.7. Find a few points to check the AOQL in 10.3, Type B OC curve.

10.8. Find a few points to check the AOQL in 10.4, Type B OC curve.

10.9. Check Pa at p'=4.0% in 10.5, Type B OC curve.

10.10. Check Pa at p'=4.0% in 10.6, Type B OC curve.

10.11. In Tables 10.1 and 10.3, for a fixed lot size, what happens to the p_t protection as $\bar{p}$ increases? Why does this occur? Similarly what happens to p_t as $\bar{p}$ remains fixed and N increases? Why does this occur?

10.12. In Table 10.2, for a fixed $\bar{p}$ and increasing lot size, what happens to the AOQL? Why does this occur?

10.13. Comparing Tables 10.1 and 10.3, one sees that in the AOQL table the $\bar{p}$ classes go up to the AOQL, whereas in the LTPD table the $\bar{p}$ classes only go up to $p_t/2$. Why does this make sense?

10.14. In the Dodge-Romig double sampling plans $Re_1 = c_2+1$ and $Re_2 = c_2+1$ also. This sometimes leaves quite a gap between $Ac_1=c_1$ and Re_1. What are the implications of this, without and with truncation of inspection under rejection on the second sample?

Find the MIL-STD105D sampling plans for the following sets of conditions, giving cumulative sample sizes and corresponding acceptance and rejection numbers. Also give approximate LTPD at Pa=10% for normal plans, and, for tightened plans, the AOQL:

	Lot size N	Inspection Level	Number of samples	Severity	AQL
10.15.	1000	II	Single	Normal	.65
10.16.	1000	II	Single	Tightened	.65
10.17.	1000	II	Single	Reduced	.65
10.18.	1000	II	Double	Normal	.65
10.19.	1000	II	Multiple	Normal	.65
10.20.	1000	II	Multiple	Tightened	.65
10.21.	1000	I	Single	Normal	.65
10.22.	400	II	Double	Normal	.10
10.23.	400	II	Multiple	Normal	.10
10.24.	400	II	Double	Tightened	.65
10.25.	4000	II	Double	Normal	40.
10.26.	4000	II	Multiple	Normal	40.
10.27.	25	II	Multiple	Normal	15.

10.28. Which of the plans in 10.15-10.27 are available for defectives? Which for defects?

10.29. On the same axes show the OC curves for 10.15, 10.16, 10.17. Comment

10.30. On the same axes show the OC curves for 10.15, plus those under the same conditions, except N=200, N=10,000.

10.31. For 10.16, find the AOQL and compare it with the plan's AQL. What is the point of this comparison?

10.32. Why is reduced inspection an optional feature of 105D, whereas tightened inspection is an obligatory feature?

10.33. Why can such relatively lenient OC curves be permitted under reduced inspection?

10.34. Why does Section 11.2 stipulate the omission of results from resubmitted lots when figuring process averages?

10.35. How would a control chart be used for acceptance sampling results under 105D, and how would it be useful?

10.36. What are some of the important considerations in choosing between single, double and multiple sampling?

10.37. The AQL's in 105D are usable for defectives, only up to 10.0, while for defects, all the way up to 1000. Why is it desirable to have some such cut-off?

10.38. Why do you think that the single sample size n is <u>not</u> directly proportional to the lot size N in either Dodge-Romig or 105D?

10.39. Find Pa for 10.15 and 10.18 at p'=.02=2%, and at .04 or 4%.

10.40. Find Pa for 10.19 (multiple) at p'=.02. Use Section 10.3.6.

10.41. Suppose that c'=.10 defect per unit, that is, 10 defects per 100 units. Then in a sample of 100 units we expect 10 <u>defects</u> on the average. How many <u>defectives</u> can we expect in the 100? For this, use Table VII to find $q'=P(0 \text{ defects} | c'=.10)$ then p' and 100 p'. Do similarly for c' = .40 per unit and c' = 1.00 per unit? Does this help on 10.37?

10.42. In the double sampling Dodge-Romig table for AOQL=1.0% find a plan having p_t=4.0%. Compare this plan with the corresponding one in the double sampling table having p_t=4.0%. ($\bar{p}$ classes do not correspond exactly.)

References

1. Statistical Research Group, Columbia University, "Sampling Inspection." McGraw-Hill, New York, 1948.

2. H. F. Dodge, Notes on the evolution of acceptance sampling plans, Parts I to IV. *J. Quality Techn.*, 1, 77-88, 155-162, 225-232 and 2, 1-8, 19 (1969,1970).

3. H. F. Dodge and H. G. Romig, "Sampling Inspection Tables." Wiley, New York, 1959.

4. W. R. Pabst, Jr., MIL-STD 105D. *Indust. Quality Control*, 20 (No. 5), 4-9 (1963).

5. H. F. Dodge, A general procedure for sampling inspection by attributes — based on the AQL concept. Technical Report No. 10, The Statistics Center, Rutgers the State University, New Brunswick, New Jersey (1959).

6. K. S. Stephens and K. E. Larson, An evaluation of the MIL-STD-105D system of sampling plans. *Indust. Quality Control*, 23 (No. 7) 311-319 (1967).

7. T. Koyama, Y. Ohmae, R. Suga, T. Yamamoto and T. Yokoh, assisted by W. R. Pabst, Jr., MIL-STD-105D and the Japanese Modified Standard. *J. Quality Techn.*, 2, 99-108 (1970).

8. Review and letters on reference 7 and letters relative to JIS Z 9015, Sampling Inspection Plans by Attributes with Severity Adjustment. *J. Quality Techn.*, 3, 87-94 (1971).

CHAPTER 11

ACCEPTANCE SAMPLING BY MEASUREMENTS

11.1. Potential Greater Efficiency by Variables than by Attributes.

In general the measurement for any characteristic for a part or an assembly can be converted to an attribute. This is accomplished by choosing specification limits for the characteristic. Then the part or assembly is a "good" or a "defective" according to whether the measurement lies between the limits or outside. For example, one often sees a green sector on a dial gage, such that if the pointer lies in this sector the part is a good one, otherwise it is a defective.*

The reverse is not true, since there are many attributes or defects which a part or assembly can have, to which we can give no measurement. In this case we _must_ use "inspection by attributes" to describe lot or process quality performance.

When either type of inspection may be used, we have a choice to make. In general a measurement provides more information about the part or assembly, than does a mere determination that it lies within or outside limits. It is thus often possible to obtain the same discriminating power on a smaller sample size n, when using measurements than when using attributes. But, in order to make full use of measurements and thus obtain this greater efficiency we commonly need to know the _type_ of distribution of the individual x's, and the standard deviation σ_x. Somewhat less power is still obtainable if σ_x is unknown. But there are other factors which may sway the balance in favor of the method of attributes, namely

*We may also have a band just outside of L to U, such that x's within are minor defectives, and those further out major defectives.

(a) the cost of a single measurement vs. that of finding whether it is in limits or not, (b) the cost of recording, and (c) the cost of analysis. So use of a larger n with attributes might still possibly be cheaper.

11.2. Dependence Upon Knowledge of Distribution of Individual x's. If our objective in sampling a lot is to determine whether or not the percentage outside of specification limits is excessive, then we must know what type of distribution the x's have. Is it normal or is it somewhat unsymmetrical or skewed? Is it one homogeneous distribution, even if not normal, or is it a composite of several distributions? The latter often occurs when a process has been let run at one setting for a while then reset at a new level, then perhaps at a third level, etc. And then too, there is the question as to whether the distribution of x's has been multilated by 100% sorting. This often chops off one or both natural tails of the distribution if it originally lay outside of the limits. Sometimes this is quite a drastic truncation. And there is even the case where we receive the outer tails of a distribution, the central part having been sorted out and sent to another customer with tighter limits than ours!

Unless we can build up evidence on the distribution of measurements, we had better stay with the method of attributes.

On the other hand if we can be reasonably sure of an approximately normal distribution of x's then we are in a strong position, even if we do not know σ_x. Our position is further enhanced if σ_x is known.

11.3. Use of Control Charts and Frequency Distributions for Past Data. When a series of lots is being inspected, $\bar{x}$ and R or s charts form an invaluable record. They provide a picture of the producer's quality level and stability of performance. And, as we have seen, comparison can be made to specifications. Frequency tabulation of the x's obtained should also be made. If performance is consistently satisfactory we may well consider reducing the sample size per lot. Use may also be made of process control chart data obtained by the producer, if he is willing to send them along.

However, when lots are only sent from time to time our past records are helpful, but the process may well have changed in the interval between lots, and larger samples with a frequency distribution may be called for.

We shall now study what can be done in acceptance sampling by measurements, when we can assume a normal distribution of x's.

11.4. Decision-Making on Lot Means, Known σ_x. We assume that the x's are normally distributed. Also in this chapter we shall only consider single sampling. "Sequential sampling", in which the maximum sample size is not fixed in advance, is taken up in the next chapter.

The present problem is basically significance testing on the mean, with σ_x known. We may have one-way protection (e.g. high μ is good, low μ is bad), or two-way protection. There are also two approaches: (a) set n and an α risk of erroneous rejection if μ is actually "good" and find the criterion of decision, and sketch an OC curve, and (b) set an α risk as above and a β risk of erroneous acceptance if μ is actually at some "bad" level and find n and the acceptance-rejection criterion. We shall adopt the latter approach as being perhaps a more useful and direct approach in industry. If it leads to too large an n, modifications can be made.

11.4.1. One-way Protection on the Mean, Known σ_x. Let us consider an example to illustrate this method. For tensile strength of a certain casting, σ_x was known to be about 3500 pounds per square inch. The lower specification was 58,000 psi. Let us set a safe lot mean and an unsafe one. These depend upon the importance of the specification limit. Suppose that 2% of the castings lying below 58,000 psi would be undesirable. Then we might take $\mu_1 = 58{,}000 + 2.05(3500) \doteq 65{,}000$ as an undesirable lot mean. On the other hand if only .2% are below, this might be called fully satisfactory. So $\mu_2 = 58{,}000 + 2.88(3500) \doteq 68{,}000$. In a sense we still have said nothing until we decide upon respective risks. Again this will depend upon engineering considerations. How likely is it that a lot comes along with μ around 58,000, etc? Suppose that we take the following:

$$P(\text{acc}|\mu_1) = \beta = .05, \qquad P(\text{rej}|\mu_2) = \alpha = .10 \tag{11.1}$$

We are thus specifying two points on the operating characteristic curve. See Figure 11.1(a). Now the problem is to find a sample size n and a critical value, say, K such that if $\bar{x} \geq K$ we accept, or if $\bar{x} < K$ we reject. Thus

$$P(\bar{x} \geq K|\mu_1) = \beta \qquad P(\bar{x} < K|\mu_2) = \alpha. \tag{11.2}$$

We need n large enough so as to cut down $\sigma_{\bar{x}} = \sigma_x/\sqrt{n}$ sufficiently that we can have the situation shown in Figure 11.1(b).

Suppose we define z_α for the normal curve (2.25) by

$$\int_{z_\alpha}^{\infty} \phi(v)dv = \alpha \tag{11.3}$$

that is, z_α cuts off a tail area of α. Then we have from (2.24) and (2.27)

$$\frac{K-\mu_1}{\sigma_x/\sqrt{n}} = z_\beta \quad \text{and} \quad \frac{K-\mu_2}{\sigma_x/\sqrt{n}} = -z_\alpha. \tag{11.4}$$

Substituting in what we know, gives equations

$$\frac{K-65{,}000}{3500/\sqrt{n}} = +1.645 \qquad \frac{K-68{,}000}{3500/\sqrt{n}} = -1.282.$$

Subtracting the second from the first gives

$$\frac{3000\sqrt{n}}{3500} = 2.927 \qquad \sqrt{n} = 3.415 \qquad n = 11.7 \text{ or } 12.$$

If a sample size of 12 tensile tests seems reasonable in view of the size of a lot and the risks involved we can proceed to find K. But if n = 12 appears excessive, then we must increase one or both risks, and/or widen the gap between μ_1 and μ_2. Either will give a less discriminating or "lazier" OC curve. Suppose that n = 12 is deemed economically feasible.

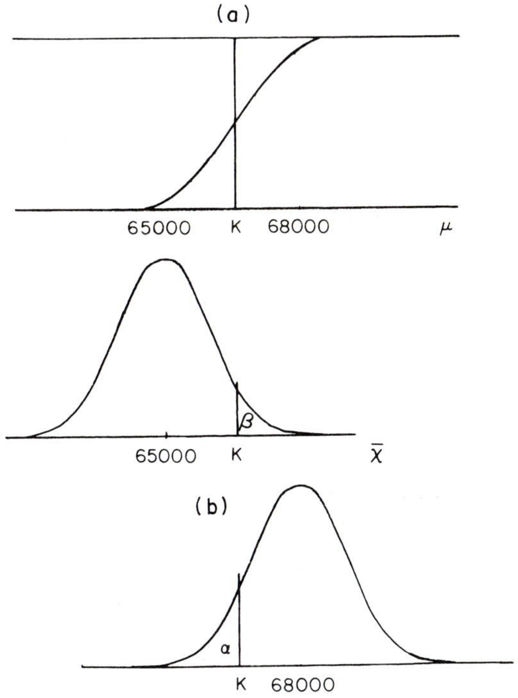

FIG. 11.1. One-way protection on mean. OC curve shown in (a), two $\bar{x}$ distributions in (b).

Then we may find K from either equation. Substituting into the first will preserve β and slightly decrease α, and vice versa, if we substitute into the second. Let us preserve β.

$$K = 65{,}000 + 1.645(3500)/\sqrt{12} = 66{,}662.$$

Then $z = (66{,}662 - 68{,}000)/(3500/\sqrt{12}) = -1.324$

which gives α = .0928.

Thus the plan is:

Take 12 tensile tests and find $\bar{x}$

Then $\bar{x} < 66{,}662$ reject lot, $\bar{x} > 66{,}662$ accept lot*.

---

*The only advantage to retaining so much precision on K is to make $\bar{x} = K$ unlikely.

The OC curve may be easily sketched by using the above two Pa's and the third point

$$P(acc|\mu=66,662) = .50.$$

If desired we may easily find a few more points on the OC curve by using different μ's and proceeding as for the actual α risk.

Summarizing the procedure we have

One-way Protection on Mean, σ_x Known

1. Given AQL = μ_2 > RQL = μ_1, and risks $P(acc|\mu_1) = \beta$, $P(rej|\mu_2) = \alpha$.
2. Find sample size n, rounding up to an integer

$$n = \left[\frac{\sigma_x(z_\alpha + z_\beta)}{\mu_2 - \mu_1}\right]^2 \tag{11.5}$$

3. Using the integer n, find K by either

$$K = \mu_2 - (z_\alpha\sigma_x/\sqrt{n}) \quad \text{or} \quad K = \mu_1 + (z_\beta\sigma_x/\sqrt{n}) \tag{11.6}$$

4. Plan: $\bar{x}_n < K$ reject, $\bar{x}_n \geq K$ accept
5. If AQL = μ_1 < RQL = μ_2, that is, μ_1 is still the lower one, but now low μ's are preferred, then use α at μ_1, β at μ_2. (11.5) still gives n. But for K use

$$K = \mu_2 - (z_\beta\sigma_x/\sqrt{n}) \quad \text{or} \quad K = \mu_1 + (z_\alpha\sigma_x/\sqrt{n}). \tag{11.7}$$

The above procedure is quite satisfactory even if we do <u>not</u> have a normal distribution but it <u>is</u> homogeneous, perhaps skewed, provided we can settle on acceptable and rejectable μ values. This is because $\bar{x}$'s are quite normally distributed even if the x's are not, in line with the central limit theorem. This is also justified by (2.26) and (2.27) still holding under non-normality.

11.4.2. Two-way Protection on the Mean, Known σ_x. Let us illustrate this case with an actual example, which saved considerable money. The application was to the doubled thickness of rubber gaskets, cut from tubes which were extruded continuously and cut off at 30" lengths and cured. The gaskets, somewhat resembling rubber bands, were for caps for food jars. Standard practice had been to measure the doubled thickness of each tube in about five places, form a mental average and then put the tube into one of three piles. The three piles of tubes were supposed to yield gaskets when cut running (a) .096" to .100", (b) .100" to .106" and (c) .106" to .110". But because of the considerable variability within a single tube due to the curing, there was large overlapping. It was finally decided to ask the producers to run an entire lot at one process setting. If they thought that it was running too far off from the desired level, then to stop that lot, and after re-setting to begin a new lot number. Then a decision as to which specification to use the rubber for would be made on the basis of a sample of tubes. This program brought three benefits: (a) the distribution of gaskets became normal, (b) the standard deviation in a lot was considerably decreased, and it was known to be .002", and (c) far less inspection time was needed.

Now let us see the statistics involved. This was a case of σ_x known, namely .002". To test for specification (b) the AQL = μ_0 was set at .103". Then the alternative lot averages or RQL's were set at .098" and .108", that is, symmetrically placed around .103". Taking an α risk of .001 and a β risk of .002, what would be the plan? It is of the form

$\bar{x}$ between .103" - k and .103" + k, accept for (b)

$\bar{x}$ below .103" - k, accept for (a)

$\bar{x}$ above .103" + k, accept for (c).

(There was no trouble with rubber too thin for (a) nor too thick for (c).)

Let us therefore seek n and k. Four equations could be set up, but owing to symmetry, only two are independent:

$$\frac{.103'' + k - .103''}{.002''/\sqrt{n}} = z_{.0005} \qquad \frac{.103'' + k - .108''}{.002''/\sqrt{n}} = -z_{.002}$$

Note particularly that $\alpha = .001$ is split two ways, because we can reject for (b) usage by either too high or low an $\bar{x}$. Also notice the sign of $z_{.002}$. We then have

$$\frac{k\sqrt{n}}{.002''} = 3.30 \qquad \frac{(k - .005'')\sqrt{n}}{.002''} = -2.88$$

giving by subtraction

$$\frac{.005''\sqrt{n}}{.002''} = 6.18 \qquad \sqrt{n} = 2.47 \qquad n = 7.$$

Then using the second equation we find $k = .005'' - 2.88(.002'')/\sqrt{7} = .00282''$ Thus for samples of 7 tubes

$\bar{x}$ between .10018" and .10582" accept for (b)

$\bar{x}$ below .10018" accept for (a)

$\bar{x}$ above .10582" accept for (c).

Actually the company decided to make the decision for lots of several hundred tubes by a random sample of 20 or 25 tubes which is being very conservative, but still brought a large saving.

Let us summarize as follows:

Two-way Protection on the Mean, Known σ_x

1. Set an AQL = μ_0 and two symmetrically placed RQL's on opposite sides of μ_0, say, μ_1, μ_2 so that $\mu_0 - \mu_1 = \mu_2 - \mu_0$.
2. Set risks $P(\text{rej}|\mu_0) = \alpha$ and $P(\text{acc}|\mu_1) = P(\text{acc}|\mu_2) = \beta$
3. Solving $\dfrac{\mu_0 + k - \mu_0}{\sigma_x/\sqrt{n}} = z_{\alpha/2}$ $\dfrac{\mu_0 + k - \mu_2}{\sigma_x/\sqrt{n}} = -z_\beta$

   gives

$$n = \left[\frac{\sigma_x(z_{\alpha/2} + z_\beta)}{\mu_2 - \mu_0}\right]^2 . \tag{11.8}$$

4. Using the integer value of n, find k by

$$k = z_{\alpha/2}\sigma_x/\sqrt{n} \quad \text{or} \quad k = \mu_2 - \mu_0 - z_\beta\sigma_x/\sqrt{n}. \tag{11.9}$$

5. Then for $\bar{x}$ for n x's:

   $\bar{x}$ between $\mu_0 - k$, $\mu_0 + k$ accept

   Otherwise reject.

Again, the normality assumption is not too important as long as σ_x is known.

11.5. Decision Making on Percent Beyond Specification, Known σ_x. This problem is very straight forward for the one-way protection case. Suppose first that we are interested in an upper specification U. Then in terms of fraction defective p', let

$$\text{AQL} = p_1' < p_2' = \text{RQL}$$

and the respective risks be α and β. The assumption of normality now becomes much more crucial, because we are using x distributions, not $\bar{x}$ distributions, for the set-up. Now we set μ_1 so as to have fraction defective p_1' beyond U and μ_2 so as to have p_2' beyond U. Then

$$\mu_1 = U - \sigma_x z_{p_1'} \qquad \mu_2 = U - \sigma_x z_{p_2'}. \tag{11.10}$$

We may now substitute these into (11.5) obtaining

$$n = \left[\frac{z_\alpha + z_\beta}{z_{p_1'} - z_{p_2'}}\right]^2. \tag{11.11}$$

Then for K some care must be exercised. In (11.10) μ_1 is the AQL and μ_2 the RQL. To reach K we must go <u>up</u> from μ_1 by an amount $z_\alpha\sigma_x/\sqrt{n}$. This gives

$$K = U - \sigma_x(z_{p_1'} - z_\alpha/\sqrt{n}). \tag{11.12}$$

Or we must go <u>down</u> from μ_2 by an amount $z_\beta\sigma_x/\sqrt{n}$:

$$K = U - \sigma_x(z_{p_2'} + z_\beta/\sqrt{n}). \tag{11.13}$$

At a lower limit L, formula (11.11) still follows, but now

$$K = L + \sigma_x(z_{p_1'} - z_\alpha/\sqrt{n}) \quad \text{or} \quad K = L + \sigma_x(z_{p_2'} + z_\beta) \tag{11.14}$$

Probably the best approach is not to use these formulas (except as a check) and instead to think through the distributions of x and $\bar{x}$ involved and perform the necessary arithmetic.

The student should compare the formulas (11.12) and (11.13) with those for modified limits (7.9), (7.10).

For two-way protection on the mean, the first step is to see whether σ_x is sufficiently small to permit as little as p_1' outside when $\mu = (U+L)/2$, that is, perfectly centered. Thus

$$U - (U+L)/2 \geq \sigma_x z_{p_1'/2}$$

or

$$(U-L)/(2\, z_{p_1'/2}) \geq \sigma_x. \tag{11.15}$$

Now one possibility is to run a one-way test at whichever specification $\bar{x}$ is nearest. This approach is recommended if σ_x is considerably less than the maximum given by the left side of (11.15), for then the lot mean μ has some latitude to move around in, and $\bar{x}$ must be some little distance away from the nominal (U+L)/2 toward, say U, before we might reject. But then the fraction defective below L will surely be negligible. So we can use p_1' (not $p_1'/2$) and p_2' in (11.11) and (11.12).

If, however, σ_x is close to the maximum given in (11.15), then the method given in the Military Standard 414 [1] seems preferable. There, the percent beyond U and that beyond L are estimated using $\bar{x}$, and the sum of these estimated percents compared with a predetermined maximum allowable estimate. If below, the lot is accepted, if above, it is rejected.

11.6. Decision Making on Percent Beyond a Specification, Unknown σ_x. This case is a bit more complicated than that when σ_x is known, and also places more demands upon the assumption of normality. We shall follow the method given in [2].

The test is quite similar to that given in Section 11.5 for one specification limit. There, for example at an upper specification U, we reject if

$$\bar{x} > U - \sigma_x(z_{p_1'} - z_\alpha/\sqrt{n}) \text{ or } \bar{x} + \sigma_x(z_{p_1'} - z_\alpha/\sqrt{n}) > U.$$

Now we do not know σ_x in the present case, so, taking an analogy of the latter inequality we use the criterion

$$\bar{x} + ks > U \text{ reject}, \bar{x} + ks \le U \text{ accept}.$$

The latter means that we are apparently safely below U. The problem then is to find n and k from p_1', α; p_2', β. We shall shortly derive the formulas (11.16) and (11.17) below. But first we give the general plan and illustrate with an example.

Percent Beyond One Specification, Test on $\bar{x}$, Unknown σ_x

1. Case of protection against an upper specification U

   a. Set fraction defectives relative to U: $AQL = p_1' < p_2' = RQL$.

      and respective risks $\alpha = P(rej|p_1')$ and $\beta = P(acc|p_2')$

   b. Find the two constants k and n determining the plan:

$$k = \frac{z_\alpha z_{p_2'} + z_\beta z_{p_1'}}{z_\alpha + z_\beta} \tag{11.16}$$

$$n = \frac{k^2 + 2}{2} \cdot \left(\frac{z_\alpha + z_\beta}{z_{p_1'} - z_{p_2'}}\right)^2 \tag{11.17}$$

      If n is non-integral, round up to the next integer [3].

   c. Then the plan is to take n measurements, finding $\bar{x}$ and s:

$$\bar{x} + ks \le U \quad \text{accept} \tag{11.18}$$

$$\bar{x} + ks > U \quad \text{reject} \tag{11.19}$$

2. Case of protection against a lower specification L

   a. Identical to Section 1a

   b. Use (11.16) and (11.17) just as in 1b

   c. Then the plan is to take n measurements, finding $\bar{x}$ and s:

$$\bar{x} - ks \geq L \quad \text{accept} \tag{11.20}$$

$$\bar{x} - ks < L \quad \text{reject} \tag{11.21}$$

Comparing (11.17) with (11.11), it is easily seen that the ratio of the required n's, with σ_x unknown and known is $(k^2 + 2)/2$, which is always above 1, and increases as we try to discriminate between smaller p_1' and p_2' fractions.

Example 11.6. Suppose that the maximum time of blow of fuses is 150 seconds under a certain marginal current, and that 1% is considered satisfactory, but 6% not. Call these p_1' and p_2'. Let us also take corresponding risks $\alpha = .05$ and $\beta = .10$. Then we have from Table I,

$$z_{p_1'} = z_{.01} = 2.326, \quad z_{p_2'} = z_{.06} = 1.555, \quad z_\alpha = z_{.05} = 1.645, \quad z_\beta = z_{.10} = 1.282$$

Thus

$$k = \frac{1.645(1.555) + 1.282(2.326)}{1.645 + 1.282} = 1.893$$

$$n = \frac{1.893^2 + 2}{2} \cdot \left(\frac{1.645 + 1.282}{2.326 - 1.555} \right)^2 = 40.23 \text{ or } 41$$

Since the test is destructive, this is likely to be too many to be practical, so some compromise is needed. If σ_x were known, we could use (11.11), giving $n = 15$. Also Table 9.5 would give $Ac = 3$, $np_2' = 6.68$ yielding $n = 112$ for a pure attribute plan, that is, blowing time below or above 150 seconds.

11.6.1*. Derivation of (11.16), (11.17). The following facts are available in most mathematical statistics texts:

1. If the population of x's is normal, then $\bar{x}$ and s are independent and

   a. $\bar{x}$ has mean μ and standard deviation $\sigma_{\bar{x}} = \sigma_x/\sqrt{n}$, and the $\bar{x}$'s are normally distributed.

b. s has mean approximately σ_x (actually $c_4\sigma_x$) and standard deviation σ_s approximately $\sigma_x/\sqrt{2(n-1)}$ (actually $c_5\,\sigma_x$), and the s's are fairly normally distributed, increasingly so as n increases.

2. If two variables u and v are independent, then $\mu_{au+bv} = a\mu_u + b\mu_v$ and $\sigma^2_{au+bv} = a^2 \cdot \sigma^2_u + b^2 \cdot \sigma^2_v$ and au + bv is more normal than the less normal of au and bv.

Now consider the variable

$$w = \bar{x} + ks.$$

From 1 and 2 above we have that

1. w has mean approximately $\mu + k\sigma_x$
2. w has standard deviation

$$\sigma_{\bar{x}+ks} \doteq \sigma_x\sqrt{\frac{1}{n} + \frac{k^2}{2(n-1)}}.$$

3. w is approximately normally distributed.

Let us consider the problem of distinguishing between the two critical means of x's (given fraction defectives p_1', p_2')

$$\mu_1 = U - z_{p_1'}\sigma_x = \text{AQL and } \mu_2 = U - z_{p_2'}\sigma_x = \text{RQL}.$$

These give respective means for the variable $w = \bar{x} + ks$ of

$$\mu_{w(1)} = U - z_{p_1'}\sigma_x + k\,\sigma_x = U - \sigma_x(z_{p_1'} - k),\ \mu_{w(2)} = U - \sigma_x(z_{p_2'} - k).$$

Recalling that U is the cut-off point for the criterion of acceptance or rejection, we have in direct analogy to (11.4):

$$\frac{U - [U - \sigma_x(z_{p_1'} - k)]}{\sigma_x\sqrt{\frac{1}{n} + \frac{k^2}{2(n-1)}}} = z_\alpha \tag{11.22}$$

$$\frac{U - [U - \sigma_x(z_{p_2'} - k)]}{\sigma_x\sqrt{\frac{1}{n} + \frac{k^2}{2(n-1)}}} = -z_\beta. \tag{11.23}$$

Simplifying and further approximating by changing the n-1 to n gives

$$\frac{z_{p_1'} - k}{\sqrt{(2+k^2)/2n}} = z_\alpha \tag{11.24}$$

$$\frac{z_{p_2'} - k}{\sqrt{(2+k^2)/2n}} = -z_\beta . \tag{11.25}$$

Dividing (11.24) by (11.25) gives

$$\frac{z_{p_1'} - k}{z_{p_2'} - k} = -\frac{z_\alpha}{z_\beta} . \tag{11.26}$$

Solving (11.26) for k gives (11.16). On the other hand, if we subtract (11.25) from (11.24) we have

$$z_{p_1'} - z_{p_2'} = (z_\alpha + z_\beta) \sqrt{(2+k^2)/2n}$$

which when solved for n gives (11.17).

11.7. Single Sampling for Variability. A moderately common problem is to test the variability of a process or measurement technique. In the former case we are concerned with process capability, while in the latter case the interest is in measurement error. Typically low population standard deviations σ_x are good and higher σ_x poor. Accordingly a one-way test is needed.

We may set up the test by deciding upon

$$\text{AQL} = \sigma_1 < \sigma_2 = \text{RQL}$$

We might proceed in either of two ways:

1. Decide upon risks of wrong decisions

$$\alpha = P(\text{rej} | \sigma_1) \text{ and } \beta = P(\text{acc} | \sigma_2)$$

and then find n and the acceptance - rejection criterion.

2. Decide upon the sample size n and one of the two risks, then find the acceptance - rejection criterion and evaluate the other risk.

The former approach seems particularly natural and by use of Table VIII, proves to be very easy. In either approach assuming the x's are normally distributed, s^2 is the best test statistic and we would use the criteria

$$s^2 \leq K \quad \text{accept,} \quad s^2 > K \quad \text{reject.} \tag{11.27}$$

Also in either case it is wise to at least sketch the OC curve.

In order to use Table VIII, we need to use $\alpha = \beta$ set at one of .10, .05, .02 or .01. If one should desire unequal α, β or other risk values, this can be done, but Table VIII , is not applicable.

Example 11.7. Suppose that we are concerned with the consistency in muzzle velocity of ammunition. σ_1 = 30 ft/sec is regarded as satisfactory and σ_2 = 60 ft/sec as unsatisfactory. (The test is unconcerned with average muzzle velocity.) Let us settle upon risks $\alpha = \beta = .05$. Then we first find $\sigma_2/\sigma_1 = 2.00$. Now entering Table VIII, in the left side at the .05 column, we seek 2.00. We find for n = 13, 2.01 and for n = 14, 1.95. So we take n = 14 (unless we permit the risks to slightly exceed .05). Then looking at the right side of the table under .05, opposite n = 14 we find the multiplier for σ_1^2 to give K. It is 1.72, giving

$$K = 1.72 \cdot (30 \text{ ft/sec})^2 = 1548 \text{ (ft/sec)}^2$$

Then this K value is used to make the decision as in (11.27).

Thus if lots have σ_x = 30 ft/sec, they will be accepted 95% of the time, but if 60 ft/sec they will be rejected 95% of the time. Probabilities for other σ's appear on the OC curve.

Protection Against Excessive Variablity

1. Assume normal population
2. Choose an acceptable lot or process standard deviation σ_1, and a rejectable standard deviation σ_2.

3. Decide upon the risk α that a lot or process having $\sigma = \sigma_1$ will be erroneously rejected, and a risk β that one with $\sigma = \sigma_2$ will be erroneously accepted. If these are made equal and at one of .10, .05, .02 or .01, Table VIII can be used.

4. To use the table, form σ_2/σ_1 and in whichever column (2) to (5) corresponds to $\alpha = \beta$, seek the desired ratio. If it does not occur, take the next lower entry (interpolating if $n > 31$). This gives n.

5. Now find K by finding the entry in the appropriate column (6) to (9) opposite n. Multiply this entry by σ_1^2, for K.

6. Then plan is $s^2 \leq K$ accept lot or process, otherwise reject.

7. This plan will match α, but slightly cut β.

8. An economic balance should be struck between sample size and the discriminating power of the test. This is easily done by using the table.

9. An OC curve can be sketched as below.

OC Curve. The probabilities for other σ's than σ_1 and σ_2 are easily found as follows. A standard theorem from mathematical statistics is that if a random sample of n x's is chosen from a normal population having standard deviation σ, then

$$\frac{(n-1)s^2}{\sigma^2} = \chi^2_{n-1} \tag{11.28}$$

follows the chi-square distribution with n-1 degrees of freedom. Now from (11.27)

$$Pa = P(s^2 \leq K|\sigma). \tag{11.29}$$

In order to build up (11.28) within the parenthesis multiply both sides of the inequality by $(n-1)/\sigma^2$, obtaining

$$Pa = P[(n-1)s^2/\sigma^2 \leq (n-1)K/\sigma^2].$$

But by (11.28) we have

$$Pa = P[\chi^2_{n-1} \leq (n-1)K/\sigma^2]. \qquad (11.30)$$

Hence

$$\chi^2_{(Pa\ below)} = (n-1)K/\sigma^2. \qquad (11.31)$$

To use (11.31), we might substitute desired σ's and then try to find Pa by interpolation in Table II, at the row corresponding to the degrees of freedom. But this is a nuisance. Instead let us find $\chi^2_{(Pa\ below)}$ for values of Pa available in the table. Then

$$\sigma^2 = \frac{(n-1)\cdot K}{\chi^2_{(Pa\ below)}}. \qquad (11.32)$$

Thus in Example 11.7, n = 14, df = 13 and we have

$$\sigma^2 = \frac{13(1548)}{\chi^2_{(Pa\ below)}}.$$

From Table II, with df = 13 we have

Pa = .25	$\chi^2_{.25} = 9.299$	$\sigma^2 = 2164\ (ft/sec)^2$	$\sigma = 46.5$ ft/sec
Pa = .75	$\chi^2_{.75} = 15.984$	$\sigma^2 = 1259\ (ft/sec)^2$	$\sigma = 35.5$ ft/sec

These supplement the σ_1, σ_2 points. The full curve is shown in Figure 11.2.

11.7.1*. Derivation of Method. Making use of (11.28) we have the following two equations for the plan which is to have (σ_1, Pa = 1 - α), (σ_2, Pa = β).

$$P(s^2 \leq K|\sigma_1) = 1 - \alpha \qquad P(s^2 \leq K|\sigma_2) = \beta. \qquad (11.33)$$

Now as before, form χ^2_{n-1} on the left sides:

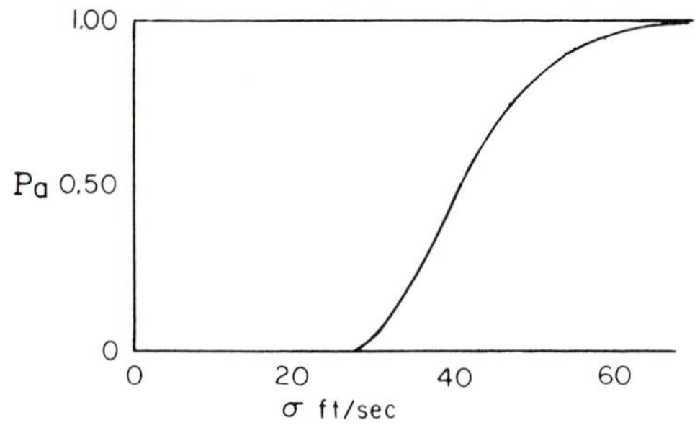

FIG. 11.2. OC curve for plan n=14, $s^2 \leq 1548(\text{ft/sec})^2$ accept, otherwise reject.

$$P[(n-1)s^2/\sigma_1^2 \leq (n-1)K/\sigma_1^2] = 1 - \alpha \tag{11.34}$$

$$P[(n-1)s^2/\sigma_2^2 \leq (n-1)K/\sigma_2^2] = \beta. \tag{11.35}$$

Thus we have

$$\chi^2_{n-1\text{ df, }1-\alpha\text{ below}} = (n-1)K/\sigma_1^2 \tag{11.36}$$

$$\chi^2_{n-1\text{ df, }\beta\text{ below}} = (n-1)K/\sigma_2^2 \tag{11.37}$$

Now we would like to solve for n and K, but this cannot be done directly, and of course again we shall go to an integer n for sample size. Dividing (11.37) into (11.36) and taking the square root yields

$$\frac{\sigma_2}{\sigma_1} = \sqrt{\frac{\chi^2_{n-1\text{ df, }1-\alpha\text{ below}}}{\chi^2_{n-1\text{ df, }\beta\text{ below}}}}, \tag{11.38}$$

This gives the entries in columns (2) to (5) in Table VIII, tabulated against n. Risks α and β were taken equal in order to cut down the multiplicity of cases. Then we solve (11.36) for K giving

$$K = \frac{\chi^2_{n-1\text{ df, }1-\alpha\text{ below}}}{n-1} \cdot \sigma_1^2. \tag{11.39}$$

The multiplier of σ_1^2 is what is given in columns (6) to (9) in Table VIII, tabulated against n.

11.8. Military Standard 414 Plan. We will now describe MIL-STD 414, [1], which is a collection of sampling plans to use measurement statistics, such as, $\bar{x}$, to control the percentage of pieces beyond specifications. It is thus concerned with the types of problems we have considered in Sections 11.5 and 11.6. As this is being written there is considerable discussion of Standard 414 in many quarters around the world. Because of this we shall content ourselves with a description of the plan. Many cases can be handled adequately by the methods already discussed here.

Early work on such techniques was done by Romig [4]. Military Standard 414 is an outgrowth of [2], [3], [5]-[7]. All of these plans, whenever they are concerned with the percentage beyond specification limits, make strong use of the assumption of normality, because lack of normality tends to affect the tails of distributions most strongly. The normality assumption is often insufficiently emphasized in practice.

Now let us proceed with a description of the 414 plans.

A. All are single sampling plans; making decisions on the basis of n measurements, rather than upon counts of the number outside of specifications. All assume the distribution of measurements x in the lot to be normal.

B. The plans are nevertheless concerned with controlling the percentage outside of limits L to U, or L alone, or U alone. These are two-way or one-way plans.

C. Plans are available for two general cases

   1. σ known. Section D, using σ.

   2. σ unknown. Section B, using sample s, or Section C using sample ranges, R's.

D. Plans are to control fraction defective. They include normal, tightened and reduced plans.

E. There are two forms, 1 and 2 (unfortunately, for it makes the standard more bulky and complicated). Decisions on a lot would be the same under either form. Since Form 2 is used whenever a two-way decision (both L and U given) is made, we here recommend using Form 2 throughout, for one-way decisions too.

F. OC curves for Pa vs. lot fraction defective are provided for normal and tightened inspection plans. They are in Table A-3 of the Standard, a page for each sample size code letter, listed for normal inspection; but also correct for tightened inspection by finding what normal inspection plan corresponds to the desired tightened plan.

G. To start to find a plan, specify an AQL (acceptable quality level) in percent defective. Table A-1 converts this to one of the index AQL's, for example, .65%. Then for inspection levels I to V (smallest to largest n's), and lot size N, Table A-2 gives a sample size code letter B, C,..., Q. Use level IV if none is specified. Thus we have a code letter and an AQL <u>in per cent</u>.

H. Now choose Section B, C or D according to desire and knowledge as in our Section C above.

I. Form 2. Using sample standard deviation s. Section B.

   1. One-way, say, U

      a. For the AQL and code letter from G, find in Table B-3 the required sample size n. Then for a random sample of n measurements x, find $\bar{x}$ and s. Also note the maximum allowed estimated percent defective M in Table B-3. (This M will be larger than the AQL, so that lots with p' = AQL will have a high Pa.)

b. Form the "quality index"

$$Q_U = (U - \bar{x})/s. \qquad (11.40)$$

(This is much like (2.24), $z = (x-\mu)/\sigma$ for obtaining the percentage beyond x, but uses instead the sample $\bar{x}$ and s.) From Q_U and n, enter Table B-5 to find estimated percentage above U, say p_U.

c. Decision:

$$p_U \leq M \text{ accept} \qquad p_U > M \text{ reject}. \qquad (11.41)$$

d. If protection is relative to a lower specification L, use Section a as it is, but in Section b use

$$Q_L = (\bar{x} - L)/s. \qquad (11.42)$$

Then use Table B-5 as above to find p_L and make decision by

$$p_L \leq M \text{ accept} \qquad p_L > M \text{ reject}. \qquad (11.43)$$

e. Tightened plans are found in Table B-3 by entering from the AQL's at the bottom rather than those at the top as in normal plans. Reduced plans are found in Table B-4.

2. Two-way, L and U both given

a. Same as Section 1a.

b. Form Q_L by (11.42) and Q_U by (11.40) and use Table B-5 to find estimates p_L and p_U.

c. Decision

$$p_U + p_L \leq M \text{ accept lot} \qquad p_U + p_L > M \text{ reject lot} \qquad (11.44)$$

d. Same as 1e.

3. Switching rules are based in considerable part upon the estimated fraction defective from a series of lots, that is, upon $\bar{p}_U$, or $\bar{p}_L$, or $\bar{p}_U + \bar{p}_L$, usually for ten lots. The rules are rather similar to those in MIL-STD 105D.

J. Form 2. Using sample ranges, R. Section C.

1. If n = 3,4,5 or 7, use R. Or if n = 10,15, etc. break up the sample into subsamples of 5 each, finding the range for each and use $\bar{R}$.

2. Proceed as in I1 and I2, except using R or $\bar{R}$ and Tables C-3 to C-5, and

$$Q_U = c(U - \bar{x})/\bar{R}, \quad Q_L = c(\bar{x} - L)/\bar{R}. \tag{11.45}$$

which provide estimates p_U and p_L, from Table C-5. Tables C-3 and C-4 provide the c values for each n; they are like d_2 values for control charts.

K. Form 2. Using known population σ. Section D.

1. Steps very similar to Section I, but use Tables D-3 to D-5, the first two providing n, M and a v-quantity. Then use

$$Q_U = v(U - \bar{x})/\sigma \qquad Q_L = v(\bar{x} - L)/\sigma. \tag{11.46}$$

and find estimates p_L and p_U from Table D-5.

L. If there is real doubt as to the normality of the distribution, then one possibility is to accept only by MIL-STD 414. But if the variables approach 414 would call for rejection, then do not reject yet, but continue onward, counting the number of pieces outside of specifications, until the n for an appropriate attribute plan is completed. Such a plan might be from MIL-STD-105D. This approach is called "variables-attributes sampling." It particularly protects a producer who may have a process running outside of L to U, but has sorted his product carefully to these limits.

M. If the shape of the distribution of x's is not normal, but the curve shape is known, then one might use [8]. See also [9] and [10]. If the long tail in a skewed curve is toward U, one can use a smaller n than for the normal, but if the short tail is toward U a larger n is needed.

11.9. Checking a Process Setting and a Process Capability. In general it is recommended that process control be maintained by measurement control charts, $\bar{x}$ and R or s. This is because they supply the dual role of helping find assignable causes thereby improving and stabilizing the process, and secondly of telling when to adjust the process, for example resetting the level. If the object is only that of periodically checking the process level μ, in relation to specifications, and σ_x is known from past records, then we may wish to make a one-way or a two-way check of the process level μ.

Two-way Check of Level. Suppose that the specifications are rather tight so that the tolerance T = U - L is only six or seven σ_x's and we must therefore maintain μ close to the nominal $(L+U)/2 = \mu_0$, say. Then a good, practical plan is the following one, with risks below 10%:

"Safe process level" μ_0: P(approval) = .942 $\doteq$ .95

"Unsafe process levels" $\mu_0 \pm 1\sigma_x$: P(rej) = .897 $\doteq$ .90

Take n = 10, and find $\bar{x}$. Then

$\bar{x}$ between $\mu_0 - .6\sigma_x$ and $\mu_0 + .6\sigma_x$, approve setting

$\bar{x}$ outside, reject and reset process.

This plan follows from (11.8), using $\alpha = .10 = \beta$, but then using n = 10 instead of n = 9 for simplicity in finding $\bar{x}$'s, and also making the critical value simpler.

One-way Check of Level. If, contrary to the above, the tolerance T = U - L is eight or more times σ_x, then there is some latitute for the process mean μ to move around in. This is a case like Section 7.2. A practical one-way check is the following one with approximate 10% risks:

1. Level nearing upper specification U

   Take n = 7 measurements and find $\bar{x}$. Then

   $\bar{x} \leq U-2.5\sigma_x$ approve setting $\bar{x} > U-2.5\sigma_x$ reject setting

   $P(\text{approval} \mid \mu=U-3.0\sigma_x) = .907$ p' = .0013 above U

$P(\text{approval}|\mu=U-2.5\sigma_x) = .500 \quad p' = .0062$ above U

$P(\text{approval}|\mu=U-2.0\sigma_x) = .093 \quad p' = .0228$ above U

2. Level nearing lower specification L.

Take n = 7 measurements and find $\bar{x}$. Then

$\bar{x} \geq L + 2.5\sigma_x$ approve setting $\quad \bar{x} < L + 2.5\sigma_x$ reject setting.

These two tests may be combined by using n = 7 and approving setting if $\bar{x}$ lies between $L + 2.5\sigma_x$ and $U - 2.5\sigma_x$.

Checking Process Capability. When buying machine tools for a particular production requirement, or testing the capabilities of present equipment for a job, we want to test the variability. In general, as in Section 7.2 we think of σ_x as the short-term process standard deviation. Now if the needed tolerance T = U - L is only $6\sigma_x$, then μ must be held very closely to the nominal (U+L)/2. Thus in order to permit any degree of drift of μ (for tool wear for example) or inaccuracy of set-up, we need T to be well above $6\sigma_x$. One possibility for checking the process capability is to call

$\sigma_1 = T/10$ and $\sigma_2 = T/6$.

Then we may proceed as in Section 11.7. We may well use $\alpha = \beta = .05$ for the test. Then using Table VIII, we find

$\sigma_2/\sigma_1 = 1.67, \ n = 23$

and then since $\sigma_1 = .1T$,

$K = 1.54(.1T)^2 = .0154T^2$

Thus approve process if $s^2 \leq .0154T^2$, etc. This plan, however, will reject the process about half of the time when $T = 8\sigma_x$. Hence one might alternatively let

$\sigma_1 = T/8.33, \quad \sigma_2 = T/5$

with the same risks. Then still use n = 23, but instead use

$K = 1.54(T/8.33)^2 = .0222T^2$

Then $T = 8.33\sigma_x$ will only be rejected about 5% of the time.

11.10. Summary. This chapter has been concerned with sampling decision-making on the basis of measurement data. The decision can be on a lot or a process. In the latter case, the use of control charts on the process is recommended in order to seek out assignable causes as well as to take action on the process. The use of sampling acceptance tends to concentrate only on the second, taking action, such as adjusting the level up or down.

For decision-making on a particular lot from a producer, or a series of lots, control charts can still be of much use, but the primary job is to take action on this one lot. Then acceptance sampling is a natural decision-making tool.

If we cannot tell what kind of distribution of measurements we have in the lot, then we are much safer to use a random sample of parts or pieces and measure each, comparing each x with specification limits L and U, and taking action on the basis of some attribute plan. This will call for a substantial sample size in general. The measurements can well be recorded too, and thus build up knowledge on the producer's frequency distribution. If this proves to be normal or nearly so we may begin to try the normal curve methods described in this chapter. These would be unknown-σ_x plans, Section 11.6. In a series of lots, plotting of R or s charts can quickly lead to a good estimate of σ_x, if there is control. Then we can use this in known-σ_x plans, Sections 11.4, 11.5, and thereby further decrease the required sample size.

Military Standard 414 is an effective system of integrated plans, providing normal, tightened and reduced sampling, for protection against one or two specification limits. It is concerned with fraction or per cent defective. The plans assume normality, but one hedge, if this is not an available assumption, is to use variables - attributes inspection, permitting quick acceptance via measurements, but only rejecting via attributes.

Checks on process level and process capability were also given, for use when normality of x's can be assumed.

PROBLEMS

11.1. Verify the risks given in the two-way check of level in Section 11.9, for Pa vs. μ.

11.2. Verify the risks given in the one-way check of level in Section 11.9, for Pa vs. μ. Also verify the three corresponding fractions defective for Pa vs. p'.

11.3. A minimum specification for individual tensile strength tests is 90,000 pounds per square inch (psi) and σ_x is known to be about 4000 psi. The distribution of tests is approximately normal. For lot average μ, AQL = 102,000 psi, and RQL = 98,000 psi, respective risks are set $\alpha = \beta = .05$. Determine an appropriate sampling plan for a decision on a lot and sketch the OC curve Pa vs. μ, labeling axes.

11.4. For "acid number" for lead battery material the minimum specification is 8, while σ_x is known to be .39. There is approximate normality. Set risks as follows $P(\text{acc}|\mu = 8.8) = .05$ and $P(\text{rej}|\mu = 9.2) = .05$. Set up an appropriate sampling plan, cutting β if necessary and sketch the OC curve Pa vs. μ.

11.5. For an inner axle carrier, specifications of 12.742" to 12.745" are uncomfortably tight, so control of the mean μ essential. The standard deviation is .000,834". Taking RQL's at 12.743" and 12.744" and risks $\alpha = .10 = \beta$ determine a two-way protection plan for testing μ.

11.6. For a hydraulic stop light switch, specifications for operation are 45 to 100 pounds per square inch (psi). σ_x is known to be just about

6 psi. Set up two one-way checks of the process level as in Section 11.9. Why not use one two-way test?

11.7. For a distributor valve bushing, carrying specifications of 1.1760" to 1.1765", σ_x is known to be about .000,040". Set up two one-way process setting checks as in Section 11.9. Why not use one two-way test?

11.8. Suppose that the maximum time of blow of a fuse under a certain current is 150 secs. (a) If σ_x is unknown, set up a looser plan than that in Section 11.6, with $p_1' = .005$, $p_2' = .080$ and respective risks $\alpha = .05$, $\beta = .10$. (Assume normality.) (b) If σ_x is known to be 25.0 secs. what does the plan become? (c) What attribute plan would provide the protection desired?

11.9. Weight of fill of a container of insecticide is subject to a minimum of 427 g. Suppose that the proportions below 427 g. are taken as $p_1' = .005$ and as $p_2' = .050$. Respective risks are $\alpha = .05$, $\beta = .10$. (Assume normality.) (a) Set up an appropriate test for $\bar{x}$ with σ_x unknown. (b) Set up an appropriate test on $\bar{x}$, if σ_x is known to be 8.7 g. (c) What is the attribute plan for the desired protection?

11.10. For the "Scott value" for a material for a battery the maximum specification is 26. $\sigma_x = .56$, and assume normality. Set up a sampling plan if $\mu = 25$ is to be rejected 98% of the time and $\mu = 24.5$ accepted 90% of the time.

11.11. An analytical technique is supposed to have a standard error of measurement of 2 parts per million. Set $\sigma_1 = 1.5$ ppm and $\sigma_2 = 2.5$ ppm and risks both .05, and determine an appropriate test for the technique.

11.12. A standard gage has had a measurement error of $\sigma_1 = .000,025"$. A less expensive one is being considered. Set $\sigma_2 = .000,035"$ and risks of .05. Find an appropriate test for the gage.

11.13. For a lot of thermometers, a standard deviation of .2°C is considered satisfactory, while .3°C is unsatisfactory. Using risks of .10 respectively on each, set up an appropriate sampling plan. (Thermometers are to be placed at constant temperature and read.) What is assumed?

11.14. Given $\sigma_1 = 1.715$, $\sigma_2 = 3.43$ as AQL and RQL respectively, with $\alpha = \beta = .10$, find a single sample test on variability. Draw a sample of the required size from distribution A of Table 4.1 and make the test. Do the same for a sample from distribution C or D.

11.15. Specifications for overall length of an inner roller bearing race are 1.2475" ± .0025". Set up a process capability test, T/10 vs. T/6 for the process. How would you choose the races to be measured? $\alpha=\beta=.05$.

11.16. For the inside diameter of a transmission main shaft bearing retainer the specifications are 2.8341" - 2.8351". Set up a process capability test, T/10 vs. T/6 case. How would you choose the required retainers for the sample?

References

1. Military Standard MIL-STD 414, "Sampling Procedures and Tables for Inspection by Variables for Percent Defective." U. S. Govt. Printing Office, Washington, D.C., 1957. Also Amer. Nat. Standards Inst. Z1.8.
2. Statistical Research Group, Columbia Univ., "Techniques of Statistical Analysis." McGraw-Hill, New York, 1947.
3. W. A. Wallis, Lot quality measured by proportion defective. In "Acceptance Sampling, A Symposium," Amer. Statist. Assoc., 1950, 117-122.
4. H. G. Romig, "Allowable Average in Sampling Inspection." Ph.D. Thesis Columbia Univ., New York, 1939.
5. Naval Ordnance Standard OSTD 80(1952). "Sampling Procedures and Tables for Inspection by Variables."

6. A. H. Bowker and H. P. Goode, "Sampling Inspection by Variables." McGraw-Hill, New York, 1952.

7. Ordnance Inspection Handbook, ORD-M608-10, "Sampling Inspection by Variables." 1954.

8. W. J. Zimmer and I. W. Burr, Variables sampling plans based on non-normal populations. Indust. Quality Control 20 (No. 1), 18-26 (1963).

9. D. B. Owen, Summary of recent work on variables acceptance sampling with emphasis on non-normality. Technometrics 11, 631-637 (1969).

10. J. N. K. Rao, J. Subrahmaniam and D. B. Owen, Effect of non-normality on tolerance limits which control percentages in both tails of a normal distribution. Technometrics 14, 571-575 (1972).

CHAPTER 12

SEQUENTIAL ANALYSIS

12.1. Introduction. Work by Harold Dodge and Harry Romig, and Walter Bartky on double and multiple sampling respectively supplied background to the development of sequential analysis. Experimentation in successive stages was also suggested and used by Harold Hotelling and P. C. Mahalanobis. The problem of sequential analysis first arose in the Statistical Research Group of Columbia University in some comments of Captain G. L. Schuyler of the Bureau of Ordnance. Following this idea up, Milton Friedman and W. Allen Wallis conjectured that sequential procedures might be developed having as good control of errors of misclassifying lots as single or double plans, but which would require less average inspection. This was brought to the attention of the late Abraham Wald, who then developed the probability ratio sequential test, and worked out the general theory of the test and its application to many important problems. Hence, sequential analysis is in large part a contribution of Abraham Wald. The foregoing account is largely from reference [1].

Fundamentally the aim of sequential analysis is to obtain decisions of requisite security on the basis of a smaller average number of tests or determinations than is possible even with the very best single sampling plans possible. The saving is often as much as 50% at critical levels of quality and considerably more at very good or poor quality. In this chapter we will discuss briefly the fundamental principles of sequential analysis and how these work out in the various cases in practice. Further, we want to show how unified and comparatively simple the sequential plans are.

In general, a sequential test is one in which, after each measurement or determination, we may either accept a hypothesis (or lot or process) reject the hypothesis, or request additional evidence. As such, the sample size is, in general, a variable; sometimes it is very small, while at other times it is quite large. Single and double sampling are special cases of sequential analysis, in which neither acceptance nor rejection is possible until n (or n_1) pieces are measured or tested. Thus the only decision on the earlier pieces is a request for more evidence. The sequential test terminates according to definite rules, that is, when sufficient evidence for a decision has accumulated.

12.2. The General Approach. For either sequential or single sampling (as we have been seeing in the latter case) we may proceed as follows. Let q stand for any population quality characteristic (for example μ, σ_X or p'), and let the AQL be q_1, RQL (rejectable quality level) be q_2. Then we consider the "null" hypothesis H_1: $q = q_1$, and alternate hypothesis H_2: $q = q_2$. Corresponding to these quality levels we set risks α and β, that is, $\alpha = P(\text{rej } H_1 | q = q_1)$ and $\beta = P(\text{acc } H_1 | q = q_2)$, just as usual. Commonly rejection of H_1 means rejection of a lot or adjustment of the production process, and acceptance of H_1 means acceptance of a lot or approval of a process.

Given the four quantities q_1, α; q_2, β it is possible to find single and sequential plans, and in some cases double or multiple plans, to meet these conditions. One must always keep in mind that to compare two sampling plans as to cost and usefulness, we must first make sure that they supply comparable protection, that is, have similar operating characteristic, OC curves. If the four quantities just given, which determine two points on the OC curve, are the same, then the curves will be sufficiently alike for sound comparison. Knowing that the OC curves are alike, we can compare the average sample number ASN curves for single and sequential plans, to see whether the gain in using the latter is worth the increasing complexity of the sequential plan. The gain tends to be greater for very good or very poor q values and least for q values somewhere between q_1 and q_2. (ASN's for double plans commonly lie between those for single and sequential, but not always.)

Sequential sampling plans are not really very complicated at all, once one gets acquainted with the approach.

12.3. The Sequential Probability Ratio Test. This approach seems quite straightforward and makes excellent practical sense. Let us form the following relative probabilities for each sample size n.

P_{2n} = relative probability of observed sample, if H_2 is true (12.1)

P_{1n} = relative probability of observed sample, if H_1 is true. (12.2)

Now form the probability ratio criterion P_{2n}/P_{1n}. Then

If $P_{2n}/P_{1n} \leq B$ accept H_1 (12.3)

If $P_{2n}/P_{1n} \geq A$ reject H_1 (12.4)

If $B < P_{2n}/P_{1n} < A$ continue sampling. (12.5)

Criterion (12.3) makes sense, because in this case we have continued our sampling until P_{2n} is a great deal smaller than P_{1n}. Therefore the explanation H_2 is so much poorer as an explanation of the observed sample than is H_1 that we are justified in accepting H_1. (B is of course considerably smaller than 1 in any practical case.) On the other hand in (12.4) we have continued sampling until H_2 has become a sufficiently better explanation than H_1, and we are justified in taking the action indicated. (Likewise A is in general much above 1.)

We now need to be able to set A and B to develop a specific plan. These are largely determined by the risks α and β, but also to a slight extent by the type of case in question. In practice, however, α and β alone would determine A and B very closely. In fact we set, in general

$$A = (1-\beta)/\alpha \qquad B = \beta/(1-\alpha) \tag{12.6}$$

Sequential plans with A and B so chosen actually will not have precisely the requested risks α and β, but instead, say, risks α' and β' according to the case. But it can be shown [1] that:

$$\alpha' \leq \alpha/(1-\beta), \qquad \beta' \leq \beta/(1-\alpha) \tag{12.7}$$

$$\alpha' + \beta' \leq \alpha + \beta. \tag{12.8}$$

It could still happen that one of α', β' obtained by using A and B from (12.6) could be a very small amount above the respective α or β, but their sum is always less than or equal to the sum of the desired risks $\alpha + \beta$. So we use (12.6) to set A and B in practice.

In the special cases we shall see that the use of criteria (12.3) to (12.5) can be codified into quite simple rules, providing acceptance and rejection numbers for each n. At any given n, the test statistic for the case in question will lie in one of three intervals (a) accept, (b) reject, and (c) continue.

In many cases we can find points on the OC curve besides $q=q_1$, $Pa=1-\alpha$; $q=q_2$, $Pa = \beta$. And thus we would have a spectrum of conditional probabilities $P(acc|q)$. In most cases the average sample number ASN curve is available too, providing ASN_q as a function of q. This is the average number of observations needed to reach a decision, whatever it may be.

One other point which bothers some persons is the possibility of a sequential test going on indefinitely and never terminating. The probability of this can be shown to be zero. In practice we do set a maximum sample size and curtail the sampling there.

12.4. Six Special Cases of Sequential Analysis Summarized. We can easily handle the following cases by sequential analysis:

(a). Fraction defective, hypotheses: $p' = p_1' = \text{AQL}$ and $p' = p_2' = \text{RQL}$. Binomial population for defectives d, assumed.

(b). Number of defects per unit, hypotheses: $c' = c_1' = \text{AQL}$ and $c' = c_2' = \text{RQL}$. Poisson population for defects c, assumed.

(c). Mean of measurements, one-way, σ known, hypotheses: $\mu = \mu_1 = \text{AQL}$, $\mu = \mu_2 = \text{RQL}$. Normal population of x's assumed. Here we assume $\mu_1 < \mu_2$ and that low values are good, with α risk at μ_1, β risk at μ_2. See later on, the modification if high values are good, subsection (7).

(d). Mean of measurements, two-way, σ known, hypotheses: $\mu = \mu_0 = \text{AQL}$ and $\mu = \mu_0 \pm d = \text{RQL}$. Normal population of x's, assumed. Here there is one AQL and two symmetrically placed RQL's with respective risks α and β.

(e). Standard deviation, one-way, μ known, hypotheses: $\sigma = \sigma_1 = \text{AQL}$ and $\sigma = \sigma_2 = \text{RQL}$, $\sigma_1 < \sigma_2$. Normal population of x's, assumed.

(f). Standard deviation, one-way, μ unknown, hypotheses: $\sigma = \sigma_1 = \text{AQL}$, $\sigma = \sigma_2 = \text{RQL}$, $\sigma_1 < \sigma_2$. Normal population of x's, assumed.

12.4.1. Particulars of the Six Cases. The following criteria and formulae contrast and compare the various cases, and show their essential unity.

(1). Constants a and b from A and B of (12.6). For case (a), defectives, we use base 10 logarithms, "log":

$$a = \log[(1-\beta)/\alpha] = \log A \qquad b = \log[(1-\alpha)/\beta] = \log (1/B). \tag{12.9}$$

But for cases (b) through (f) we use base e logarithms, "ln":

$$a = \ln[(1-\beta)/\alpha] = \ln A \qquad b = \ln[(1-\alpha)/\beta] = \ln (1/B). \tag{12.10}$$

Thus a and b reflect the specified risks at AQL and RQL. Note that a and b are both positive, in general.

(2). The sequential test for each n is a cumulated function of the n observations in the sample, as follows:

(a). Total defectives, d, in n pieces or parts

(b). Total defects, c, in n units

(c). Total measurements, Σx, or often coded from k and using $\Sigma(x-k)$ for n measurements

(d). Total deviations from μ_0, summed algebraically, then absolute value taken, $|\Sigma(x-\mu_0)|$ for n measurements

(e). Total variation from known mean μ, $\Sigma(x-\mu)^2$ for n measurements

(f). Total variation from sample mean $\bar{x}$, $\Sigma(x-\bar{x})^2 = \Sigma x^2 - [(\Sigma x)^2/n]$ for n measurements

(3). Criteria of acceptance, rejection and continuation at sample size n are all based upon acceptance number $-h_1 + sn$ and rejection number $h_2 + sn$, where $-h_1$ and h_2 are intercepts and s is the slope of two parallel lines for the test criterion plotted against n. If the test criterion lies above the upper line we reject, if below the lower line we accept, if between we continue. See Figure 12.1. (However, in case (d) there is a slight curving in the lower line for small n's.)

(4). Formulas for h_1, h_2 and s for the cases. Be sure to use the correct formula for a and b, that is, (12.9) for case (a), (12.10) for the others.

(a). $$h_1 = \frac{b}{\log \dfrac{p_2'(1-p_1')}{p_1'(1-p_2')}} \qquad h_2 = \frac{a}{\log \dfrac{p_2'(1-p_1')}{p_1'(1-p_2')}} \qquad (12.11)$$

$$s = \frac{\log [(1-p_1')/(1-p_2')]}{\log \dfrac{p_2'(1-p_1')}{p_1'(1-p_2')}}. \qquad (12.12)$$

(b). $$h_1 = \frac{b}{\ln c_2' - \ln c_1'} \qquad h_2 = \frac{a}{\ln c_2' - \ln c_1'} \qquad (12.13)$$

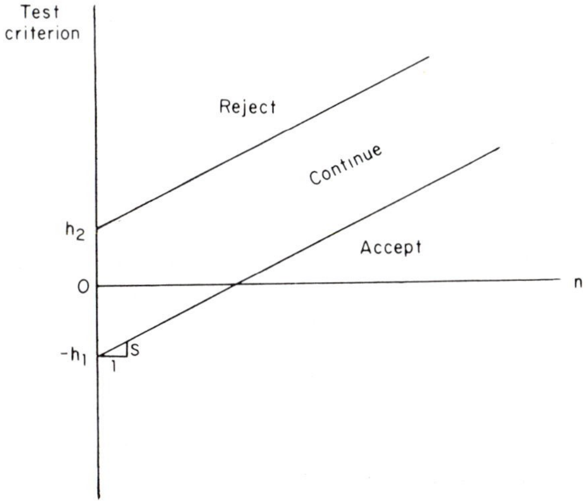

FIG. 12.1. General sequential test criteria regions.

$$s = \frac{c_2' - c_1'}{\ln c_2' - \ln c_1'} \tag{12.14}$$

(c). $h_1 = b\sigma^2/(\mu_2-\mu_1) \qquad h_2 = a\sigma^2/(\mu_2-\mu_1)$ (12.15)

$$s = (\mu_1 + \mu_2)/2 \tag{12.16}$$

(d). $h_1 = (\sigma^2/d)(b-.693) \qquad h_2 = (\sigma^2/d)(a + .693)$ (12.17)

$$s = d/2 \tag{12.18}$$

Technically, for small n the acceptance criterion is

$$\text{Accept if } \ln \cosh\,[d|\Sigma(x-\mu_0)|/\sigma^2] \le -b + (nd^2/2\sigma^2) \tag{12.19}$$

But it usually suffices to plot the intercept

$$n = 2b\sigma^2/d^2 \tag{12.20}$$

and draw a curve to the acceptance straight line, $-h_1 + sn$.

(e). $h_1 = \frac{2b}{(1/\sigma_1^2)-(1/\sigma_2^2)}$ $\qquad h_2 = \frac{2a}{(1/\sigma_1^2)-(1/\sigma_2^2)}$ (12.21)

$$s = \frac{\ln\ (\sigma_2^2/\sigma_1^2)}{(1/\sigma_1^2)-(1/\sigma_2^2)} \tag{12.22}$$

(f). Use same h_1, h_2 and s as in (e) but instead of plotting $-h_1 + sn$ and $h_2 + sn$ vs. n, plot $-h_1 + s(n-1)$, $h_2 + s(n-1)$ vs. n.

(5). Plotting of three points and using Pa = 0 and Pa = 1 as asymptotic lines is usually enough for a sketch. These are called "five point OC curves." By appropriate formulas they may be supplemented [2].

(a). $p' = 0$ $\qquad$ $Pa = 1$
$p' = p_1'$ $\qquad$ $Pa = 1 - \alpha$
$p' = s$ $\qquad$ $Pa = h_2/(h_1 + h_2)$
$p' = p_2'$ $\qquad$ $Pa = \beta$
$p' = 1$ $\qquad$ $Pa = 0$

(b). $c' = 0$ $\qquad$ $Pa = 1$
$c' = c_1'$ $\qquad$ $Pa = 1 - \alpha$
$c' = s$ $\qquad$ $Pa = h_2/(h_1 + h_2)$
$c' = c_2'$ $\qquad$ $Pa = \beta$
c' large $\qquad$ $Pa \doteq 0$

(c). μ low $\qquad$ $Pa \doteq 1$
$\mu = \mu_1$ $\qquad$ $Pa = 1 - \alpha$
$\mu = s$ $\qquad$ $Pa = h_2/(h_1 + h_2)$
$\mu = \mu_2$ $\qquad$ $Pa = \beta$
μ high $\qquad$ $Pa \doteq 0$

(d). μ low $\qquad$ $Pa \doteq 0$
$\mu = \mu_0 - d$ $\qquad$ $Pa = \beta$
$\mu = \mu_0$ $\qquad$ $Pa = 1 - \alpha$
$\mu = \mu_0 + d$ $\qquad$ $Pa = \beta$
μ high $\qquad$ $Pa \doteq 0$

(e). $\sigma = 0$ $\quad$ $Pa = 1$

$\sigma = \sigma_1$ $\quad$ $Pa = 1 - \alpha$

$\sigma = \sqrt{s} = \sqrt{\text{slope}}$ $\quad$ $Pa = h_2/(h_1 + h_2)$

$\sigma = \sigma_2$ $\quad$ $Pa = \beta$

σ high $\quad$ $Pa \doteq 0$

(f). Same as (e).

(6). Likewise we may draw the average sample number ASN curves. These too are often sketched by five points, but intermediate points may be found [2].

(a). $p' = 0$ $\quad$ $ASN = h_1/s$ round up to whole number

$p' = p'_1$ $\quad$ $ASN = \dfrac{(1-\alpha)\cdot h_1 - \alpha\cdot h_2}{s-p'_1}$

$p' = s$ $\quad$ $ASN = h_1h_2/[s(1-s)]$

$p' = p'_2$ $\quad$ $ASN = \dfrac{(1-\beta)h_2 - \beta\, h_1}{p'_2 - s}$

$p' = 1$ $\quad$ $ASN = h_2/(1-s)$ round up to whole number

(b). $c' = 0$ $\quad$ $ASN = h_1/s$ round up to whole number

$c' = c'_1$ $\quad$ $ASN = \dfrac{(1-\alpha)h_1 - \alpha\, h_2}{s - c'_1}$

$c' = s$ $\quad$ $ASN = h_1h_2/s$

$c' = c'_2$ $\quad$ $ASN = \dfrac{(1-\beta)h_2 - \beta\, h_1}{c'_2 - s}$

$c' = \infty$ $\quad$ $ASN = 0$ (we reach rejection number before completing inspection of even one unit)

(c). μ low $\quad$ $ASN \doteq 1$

$\mu = \mu_1$ $\quad$ $ASN = \dfrac{(1-\alpha)h_1 - \alpha\, h_2}{s - \mu_1}$

$\mu = s$ $\quad$ $ASN = h_1h_2/\sigma^2$

$\mu = \mu_2$ ASN $= \dfrac{(1-\beta)h_2 - \beta\, h_1}{\mu_2 - s}$

μ high ASN $\doteq 1$

(d). ASN curve not available

(e). $\sigma = 0$ ASN $= h_2/s$ round up to whole number

$\sigma = \sigma_1$ ASN $= \dfrac{(1-\alpha)h_1 - \alpha\, h_2}{s - \sigma_1^2}$

$\sigma = \sqrt{s} = \sqrt{\text{slope}}$ ASN $= h_1 h_2/2s^2$

$\sigma = \sigma_2$ ASN $= \dfrac{(1-\beta)h_2 - \beta\, h_1}{\sigma_2^2 - s}$

σ high ASN $\doteq 1$

(f). ASN precisely 1 greater than for case (e).

(7). Comments on particular cases.

(a). A great many particular cases are worked out in [2] providing h_1, h_2, s and the five ASN points, for p_1', p_2' and with $\alpha = .05$ and β either .10 or .50. The high β's give "reduced" sampling plans.

(b). Note that test criterion c is the total defects on n units. It might be thought of as $c_1 + c_2 + \ldots + c_n$ if the c''s are above 1 (or even below).

(c). If high values of x are good and low values bad, then it is convenient to retain μ_1 as the low mean but now call it an RQL; μ_2 as the high mean but now call it an AQL. Also we retain α as the risk at μ_1, namely for acceptance, and β as the risk at μ_2 for rejection (just contrary to common practice). This is done so that all formulas in subsections (4), (5) and (6) follow through, except that Pa in (5) is now interpreted as P(rej).

Another point is that coding around some number near μ_1 and μ_2 is desirable, so as to avoid huge s's which make the acceptance-rejection graph unmanageable. In particular if we use coding from the nominal, $\mu_0 = (\mu_1 + \mu_2)/2$, then s becomes zero and the test criterion $\Sigma(x-\mu_0)$ may be compared to two horizontal straight lines. That is, the acceptance and rejection numbers are constant.

(d). Great care must be used with signs in accumulating $|\Sigma(x-\mu_0)|$.

(e), (f). It is fairly easy to accumulate the test criteria $\Sigma(x-\mu)^2$, but it is quite a chore to make direct use of $\Sigma(x-\bar{x})^2$, so the form $\Sigma x^2 - [(\Sigma x)^2/n]$ is recommended. See also the sequential use of ranges in Section 12.6.

12.5 Some Examples. Let us now illustrate the preceding material with a few examples.

12.5.1. Defectives. One company in the author's experience changed over from large amounts of 100% sorting of purchased piece parts to the use of sequential analysis. This dropped the cost per 1000 parts received, from \$7.00 to \$.28, that is, a 96% saving. Moreover there was no more trouble with bad parts than usual. The few rejected lots could be more carefully sorted than the former routine sorting, and still obtain the 96% saving. The plan used was $p_1' = .01$, $\alpha = .05$; $p_2' = .02$, $\beta = .15$, and truncation was used.

Another company making carburetors developed and made wide use of three sequential sampling plans in receiving inspection. They used the rather tight risks of $\alpha = .05$, $\beta = .01$ on three plans:

(a) Critical $p_1' = .01$, $p_2' = .03$

(b) Major $p_1' = .01$, $p_2' = .05$

(c) Minor $p_1' = .02$, $p_2' = .10$

To use the plans they made three tables of acceptance and rejection numbers, by classes of n's going up till n was about equal to (a) 1.5 (ASN at p_1'), (b) 2.3(ASN at p_1'), and (c) 3(ASN at p_1'). If no decision was reached by then, they sorted the lot 100%.

Let us illustrate with a sequential sampling plan for

$$p_1' = .01,\ \alpha = .05;\ p_2' = .05,\ \beta = .10.$$

A corresponding single sampling plan for these requirements may be found from Table 9.5 to be Ac = 3, Re = 4, n = 134.

From (12.9), (or we could use Table XI), we find for this case of $\alpha = .05$, $\beta = .10$: a = 1.25527 b = .97772. Then we use the formulas for case (a), (12.11) and (12.12)

$$h_1 = \frac{.97772}{\log \frac{.05(.99)}{.01(.95)}} = 1.3638,\quad h_2 = \frac{1.25527}{\text{same}} = 1.7510$$

$$s = \frac{\log \frac{.99}{.95}}{\text{same}} = .02500.$$

Next we use subsections (5a) and (6a):

$$Pa(0)=1,\ Pa(.01)=.95,\ Pa(.025)=\frac{1.7510}{3.1148} = .562,\ P(.05)=.10,\ Pa(1)=0.$$

$$ASN(0)=1.3638/.02500=54.6 \text{ or } 55$$

$$ASN(.010)= \frac{.95(1.3638) - .05(1.7510)}{.02500 - .010} = 80.5$$

$$ASN(.025)= \frac{1.3638(1.7510)}{.02500(.975)} = 98.0$$

$$ASN(.050)= \frac{.9(1.7510) - .1(1.3638)}{.050 - .02500} = 57.6$$

$$ASN(1)= 1.25527/(1-.02500)=1.29 \text{ or } 2.$$

Thus even when the ASN is at its approximate maximum there is a considerable gain over the single n = 134. Much more gain is obtained as p' moves away from the value $p' = .025 = s$.

The acceptance-rejection regions are shown in Figure 12.2. These are drawn up to three times the larger of $ASN(p_1')$ and $ASN(p_2')$, as it is

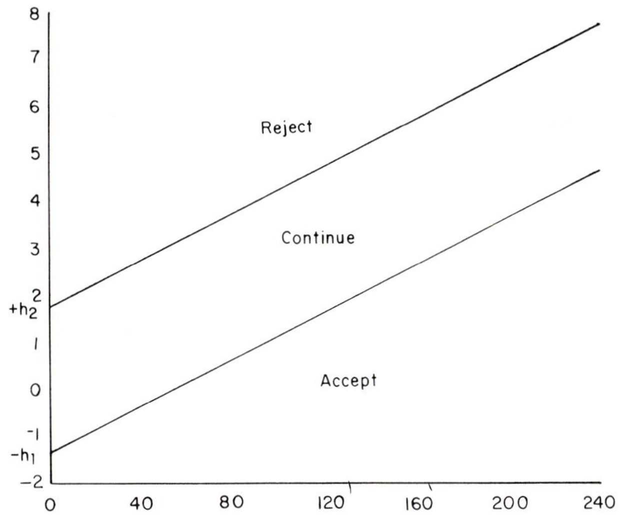

FIG. 12.2. Acceptance-rejection regions for defectives sequential plan: $p_1' = .01$, $\alpha = .05$; $p_2' = .05$, $\beta = .10$.

common and quite sound practice to truncate at such an n and make the decision according to which region the test criterion is closest to. Such a procedure does not do much violence to the α and β risks. Thus we go up to n = 240. Then if at n = 240 we have 7 defectives in 240, we would reject, or if 5 or 6 we accept. In fact just as soon as a seventh defective is found we could stop and reject. This could occur as early as n = 210.

Although a figure such as 12.2 could be used to plot observed results and to make decisions, it is commonly easier to make up a table of acceptance and rejection numbers from

$$\text{Acc. No.} = -h_1 + sn \qquad \text{Rej. No.} = h_2 + sn. \tag{12.23}$$

Thus here we have

Acc. No. = -1.3638 + .02500n   Rej. No. = 1.7510 + .02500n.

A desk calculator can easily be used to show that for the integral numbers of defectives we have:

Acc. No.	Class of n's	Rej. No.	Class of n's
(Negative)	1 - 54	2	2 - 9
0	55 - 94	3	10 - 49
1	95 - 134	4	50 - 89
2	135 - 174	5	90 - 129
3	175 - 214	6	130 - 169
4	215 - 254	7	170 - 209
		8	210 - 249

For this we find the Acc. No. by (12.23) to be negative up through n = 54 where it is still -.0138. But for n = 55, it becomes +.0112, whence d = 0 is acceptable. But then the Acc. No. remains less than 1 till n = 94 when it is .9862, but at n = 95, it is 1.0112, whence d = 1 is now acceptable. And so on. On the other hand continuous calculation of the Rej. No. by (12.23) gives 1.8010 at n = 2, so d = 2 is rejectable (if the first two pieces are both defectives). The Rej. No. remains below 2 till n = 9. But at n = 10 it becomes 2.0010 so that it now takes d = 3 to exceed the Rej. No. The last remains below 3 through n = 49 (2.9760). But at n = 50 the Rej. No. becomes 3.0010 and it then takes d = 4 to reject, etc. Such classes are obtained continuously without clearing the machine. They can now be put together as follows:

Class of n	Acc. No.	Rej. No.
2 - 9	*	2
10 - 49	*	3
50 - 54	*	4
55 - 89	0	4
90 - 94	0	5
95 - 129	1	5
130 - 134	1	6

135 - 169	2	6
170 - 174	2	7
175 - 209	3	7
210 - 214	3	8
215 - 249	4	8

* Cannot accept on such small n's.

12.5.2. Test on Means, Lower Specification. Let us take as an illustrativ example, that of Section 11.4.1 for tensile strength. Assuming it is feasible to prepare and test samples one at a time sequentially, quite a bit of sample-size saving is available using a sequential test. We here use the comments under Section 12.4.1, subsection (7c). We take μ_1 = 65,000 psi, μ_2 = 68,000 psi (the higher average). Also σ = 3500 psi. Now in Section 11.4.1, since μ_1 is rejectable, the probability of erroneously accepting such a lot was called β = .05, and of erroneously rejecting a lot with $\mu = \mu_2$ was called α = .10. Now in order to make direct use of the formulas in Section 12.4.1, subsections (1), (2c), (4c), (5c) and (6c), we rename the risk at μ_1 = 65,000 psi to be α = .05, and that at μ_2 = 68,000 psi to be β = .10. We are now ready to use the formulas for a, b, h_1, h_2 and s. We may find a and b from Table XI, or use (12.10) obtaining

$$a = 2.890 \quad b = 2.251.$$

Next for s, if we use the formula in (4c) we find

$$s = \frac{(65{,}000 + 68{,}000)\text{ psi}}{2} = 66{,}500 \text{ psi}$$

We would find it awkward to use such a large slope, so suppose we code in 100 psi units since the original measurements were to the nearest 100 psi. Now in line with (2c) we could choose u = (x - k)/100 psi where k is a convenient number of psi. k could well be μ_1 = 65,000 psi. But the most convenient one is $\mu_o = (\mu_1 + \mu_2)/2$ = 66,500 psi. Then we would measure the tensile strengths in terms of u = the number of 100 psi from 66,500 psi, and $\mu_1 = -15$, $\mu_2 = +15$ and s becomes zero! Meanwhile $\sigma_x = 35$. Next let us find h_1, h_2:

$$h_1 = b\sigma^2/(\mu_2 - \mu_1) = 2.251(35^2)/30 = 91.92$$

$$h_2 = a\sigma^2/(\mu_2 - \mu_1) = 2.890(35^2)/30 = 118.01$$

Hence for the test criterion

$$\Sigma u_i = \Sigma(x_i - 66{,}500)/100$$

we have the very simple rule

$$\text{"Acc. No."} = -h_1 = -91.92$$

$$\text{"Rej. No."} = +h_2 = +118.01.$$

The quotation marks were used because in reality if

$$\Sigma u_i \leq -91.92 \quad \text{Reject lot}$$

$$\Sigma u_i \geq +118.01 \quad \text{Accept lot.}$$

Now how much sampling do we do on the average (ASN)? We have from (6c):

$$\text{ASN}(\mu_1) = \frac{(1-\alpha)h_1 - \alpha h_2}{s-\mu_1} = \frac{.95(91.92)-.05(118.01)}{0-(-15)} = 5.43$$

$$\text{ASN}(\mu_2) = \frac{(1-\beta)h_2 - \beta h_1}{\mu_2 - s} = \frac{.9(118.01)-.1(91.92)}{+15-0} = 6.47.$$

Meanwhile the approximate maximum ASN occurs at $\mu = 66{,}500$, giving

$$\text{ASN}[(\mu_1+\mu_2)/2] = h_1h_2/\sigma^2 = 91.92(118.01)/35^2 = 8.86$$

These ASN's compare favorably with the single n = 12.

12.5.3. Test on Means, Two Specification Limits. Let us consider a test on hardness Rockwell C. Suppose that $\sigma = 1.7$ points and $\mu_o = 65$ is desired (specifications perhaps being 60 and 70). Let d = 1.7 also matching σ. Then suppose that we set risks $\alpha = \beta = .10$. Then by the appropriate Section 12.4.1, subsections (1), (2d), (4d) and (5d) we can find the desired criteria and sketch the OC curve:

$$a = 2.197 = b$$

$$h_1 = (\sigma^2/d)(b-.693) = (1.7^2/1.7)(2.197-.693) = 2.56$$

$$h_2 = (\sigma^2/d)(b+.693) = (1.7^2/1.7)(2.197+.693) = 4.91$$

$$s = d/2 = .85.$$

These suffice to set the acceptance-rejection regions, except for the short curved section for acceptance. The intercept of this curve on the n axis is at $2b\sigma^2/d^2 = 2(2.197) = 4.4$. See Figure 12.3, which shows a curve with vertical slope at n = 4.4, rounding into the acceptance line as an asymptote. The OC curve can only be sketched from the α and β points in this case, and the ASN curve is not available at all.

The quality manager or engineer in charge would have to decide whether risks of .10 are too large. A single sampling plan giving comparable protection has an n = 9.

12.6. Use in Checking a Process Setting. A frequently occurring problem in industry, is that of determining when to adjust or reset the process level and when to leave it alone. Of course the short-term process variation

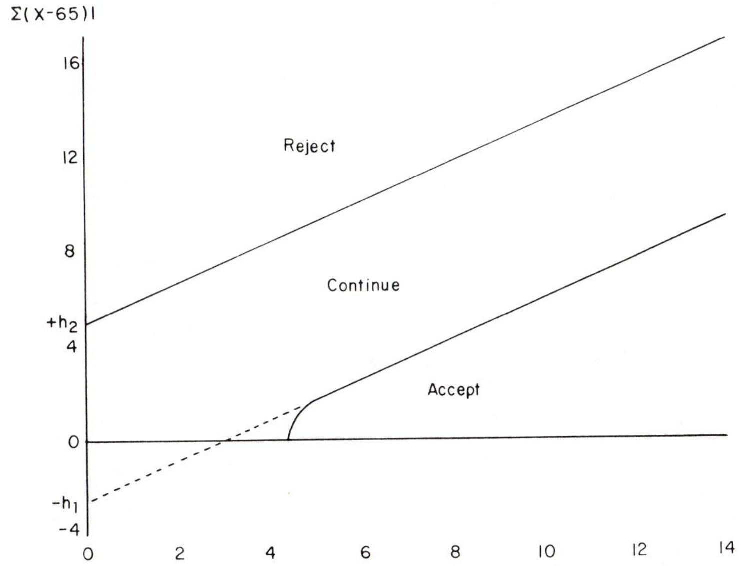

FIG. 12.3. Acceptance and rejection regions for two-way test on means. Test criterion $\sum(x-65)$. $\mu_0 = 65$, d = 1.7, $\sigma = 1.7$, $\alpha = .10$, $\beta = .10$.

σ_x is what complicates this decision. In the early stages of using any process, we recommend the use of $\bar{x}$ and R or s charts to seek out and eliminate undesirable assignable causes. Eventually the variability chart, at least, stabilizes, and we may estimate the short-term process standard deviation, σ_x. It then becomes a question as to whether to continue use of an $\bar{x}$, R chart set-up, or to go to periodically checking the process level. In any case, some way of checking the process average level at the beginning of a production run is standard practice. We have already presented a pair of single sampling checks of process average in Section 11.9, for one-way and two-way cases.

Two sequential checks of process average were developed in [3] for this problem, and tables provided so that no calculation is needed prior to making the process check, as long as σ_x is known well enough to place it in one of the given ranges of σ_x. See Tables XIII and XIV, given in the back of this book. Knowing σ_x from past data (R or s chart in control), we find which of the 12 class-intervals contains our σ_x, coding the x's if necessary.

Consider first a one-way check. Table XIII, gives the acceptance and rejection numbers for $\Sigma_{i=1}^{n}(x_i-\mu_1)$, where μ_1 is safely below an upper specification, U. This is for protection against a level $\mu_2 = \mu_1 + \sigma_x$ which is unsafely close to U. Respective risks are $\alpha = \beta = .10$. If at a lower specification, L, we still call μ_1 the lower mean which is now "unsafe", and $\mu_2 = \mu_1 + \sigma_x$ now "safe", and make the test on $\Sigma_{i=1}^{n}(x_i-\mu_1)$. But now the numbers called "Acc." in the table are rejection numbers. That is, $\Sigma(x_i-\mu_1) \leq$ "Acc." then reject the setting as being too low, etc.

For a two-way check of μ_o vs. $\mu_o \pm \sigma_x$ with risks both at .10, use Table XIV, and use the test criterion $|\Sigma_{i=1}^{n}(x_i-\mu_o)|$. Signs must be carefully watched, then the numerical value of the sum taken, to compare against the appropriate pairs of acceptance and rejection numbers in the table.

These sequential checks of the process setting reach a decision much more quickly on the average, than do single sampling plans with the same discriminating power, but they are harder to teach and to administer. This requires a decision!

12.7. A Sequential Test for Process Capability, Using Ranges. Corresponding to the process capability check as given in Section 11.9, there is a very simple check involving ranges [4]. As pointed out in Section 11.9, if the tolerance T = U - L must be met, then $\sigma_1 = T/10$ provides comfortable meeting of this tolerance, but $\sigma_2 = T/6$ makes it difficult to maintain the tolerance T. (Both σ's are for the short-term variability.) Two sequential plans are given with risks (a) $\alpha = \beta = .05$ and (b) $\alpha = \beta = .01$. To use the plan one has but to multiply the entries in columns (3) and (4), or (5) and (6) in Table XII by the required T. Then take a random sample of n = 8 pieces from production (over a short period of time) and find the range R_1. If, for example, with risks .05, $P_1 \leq .19T$, accept the process as being capable. But if $R_1 \geq .54T$, reject it. But if R_1 lies between, take another sample of n = 8 and find R_2. Then if $R_1 + R_2 \leq .55T$ accept, if $\geq .90T$ reject, otherwise take an additional sample, etc. A maximum of 8 samples of n = 8 might be required for a decision, if $\alpha = \beta = .05$. This plan is simple, and yet it has a much more favorable ASN curve than does the comparable single sampling plan. In fact its ASN curve is very nearly as good as the sequential plan for σ's using $\Sigma(x_i - \bar{x})^2$, μ unknown, which is quite a chore to use, with the necessity of refiguring $\Sigma x_i^2 - [(\Sigma x_i)^2/n]$ after each observation.

12.8. Use in the Laboratory. Sequential analysis for measurements can be quite an aid to laboratory experiments. This is especially true where observations or outcomes, such as runs, naturally occur in succession rather than in parallel. The latter would suggest use of single sampling.

One can run an effective significance of difference test on two materials, two formulations, two methods, etc., if one can estimate the respective σ_1 and σ_2 (or a reasonable upper limit). All we need do is to start taking pairs of observations (x_{11}, x_{21}), (x_{12}, x_{22}), (x_{13}, x_{23}), ... Then form

$$y_1 = x_{11} - x_{21},\ y_2 = x_{12} - x_{22},\ y_3 = x_{13} - x_{23},\ \ldots$$

Now one reasonable test is as to whether the two materials differ or are reasonably alike. We can set

$$\mu_o \text{ for an AQL for } y\text{'s}$$

and

$$\mu_o \pm d \text{ for two RQL's for } y\text{'s.}$$

Using appropriate risks α and β and $\sigma_y = \sqrt{\sigma_1^2 + \sigma_2^2}$, we may then use case (d) of Section 12.4.1 for a two-way test on μ_y. Decisions would be to take action appropriate if (a) $\mu_1 \geq \mu_2 + d$, or (b) $\mu_1 \leq \mu_2 - d$, or if (c) $\mu_1 = \mu_2$ (no appreciable difference).

It is easy to modify the foregoing for a one-way check on μ_y by deciding upon two critical levels, so that there is no alternative such as (c) above.

12.9*. Two Derivations of Acceptance-Rejection Constants. In order to give some of the flavor of sequential analysis proofs we derive formulas for h_1, h_2 and s for cases (a) and (c) of Section 12.4.1, using (12.1) through (12.6).

For the case of defectives, we have observed some ordered sample of good (G) and defective (D) pieces, which has not yet lead to a decision. For example, for n pieces we might have GG ... GDG ... GDG ... G where there are two D's in n pieces. The probability for this particular <u>ordered</u> sequence of outcomes is $p'^2(1 - p')^{n-2}$. Now form P_{2n}/P_{1n}. We have for d D's, n - d G's:

$$\frac{P_{2n}}{P_{1n}} = \frac{p_2'^d(1 - p_2')^{n-d}}{p_1'^d(1 - p_1')^{n-d}} \geq A \text{ gives rejection.} \tag{12.24}$$

The criterion is to be solved for the rejection numbers for d, that is, for each n, what d would reject. Taking logs to the base 10 of (12.24) gives

$$d \log p_2' + (n-d)\cdot\log(1-p_2') - d \log p_1' - (n-d)\cdot\log(1-p_1') \geq \log A.$$

Collecting terms and using the laws of logs gives

$$d \log \frac{p_2'(1-p_1')}{p_1'(1-p_2')} - n \log \frac{1-p_1'}{1-p_2'} \geq a \qquad \text{rej.} \tag{12.25}$$

where log A = a, as defined in (12.9). Solving (12.25) for an inequality for d gives

$$d \geq \frac{a}{\log \dfrac{p_2'(1-p_1')}{p_1'(1-p_2')}} + n\cdot\frac{\log \dfrac{1-p_1'}{1-p_2'}}{\log \dfrac{p_2'(1-p_1')}{p_1'(1-p_2')}} \qquad \text{rej.} \tag{12.26}$$

since the coefficient of d is always positive with $p_2' > p_1'$. Defining (12.26) to be

$$d \geq h_2 + ns \qquad \text{rej.} \tag{12.27}$$

we find the formulas for h_2 and s given in Section 12.4.1, subsection (4a).

For h_1 we first recall that from (12.9) b = log(1/B) or log B = -b, and then everything goes through as for the rejection case. The quantities $h_2 + sn$ and $-h_1 + sn$ are called "rejection" and "acceptance numbers," respectively.

For the case of a one-way test on the mean, say with AQL = $\mu_1 < \mu_2$ = RQL and with σ known, we assume a normal distribution. We call P_{1n} and P_{2n} relative probabilities because we actually use density functions.* Now for the first observation we have

---

*They can be made approximate ordinary probabilities by multiplying by products of Δx's, which then proceed to cancel out.

$$\frac{P_{21}}{P_{11}} = \frac{e^{-(x_1-\mu_2)^2/2\sigma^2}/\sigma\sqrt{2\pi}}{e^{-(x_1-\mu_1)^2/2\sigma^2}/\sigma\sqrt{2\pi}} = e^{-(x_1-\mu_2)^2/2\sigma^2 + (x_1-\mu_1)^2/2\sigma^2}. \qquad (12.28)$$

Now consider the exponent of e. We have

$$\frac{-(x_1^2 - 2x_1\mu_2 + \mu_2^2) + (x_1^2 - 2x_1\mu_1 + \mu_1^2)}{2\sigma^2} = \frac{2x_1(\mu_2 - \mu_1) - (\mu_2^2 - \mu_1^2)}{2\sigma^2}$$

So $$\frac{P_{21}}{P_{11}} = \exp\left[\frac{2x_1(\mu_2 - \mu_1) - (\mu_2^2 - \mu_1^2)}{2\sigma^2}\right]$$

where "exp" means the exponential function. Now if P_{21}/P_{11} lies between A and B we continue with a new x_2. Then

$$\frac{P_{22}}{P_{12}} = \exp\left[\frac{2(x_1 + x_2)(\mu_2 - \mu_1) - 2(\mu_2^2 - \mu_1^2)}{2\sigma^2}\right]$$

because we merely add in the new exponent. In general then reject as soon as

$$\frac{P_{2n}}{P_{1n}} = \exp\left[\frac{2(\Sigma x_i)(\mu_2 - \mu_1) - n(\mu_2^2 - \mu_1^2)}{2\sigma^2}\right] \geq A.$$

Now take base e logarithms "ln," giving

$$\frac{2(\Sigma x_i)(\mu_2 - \mu_1) - n(\mu_2^2 - \mu_1^2)}{2\sigma^2} \geq \ln A = a$$

by (12.10). Solving for Σx_i as test criterion yields

$$\Sigma x_i \geq \frac{a\sigma^2}{\mu_2-\mu_1} + n\left[\frac{\mu_2+\mu_1}{2}\right]$$

if we recall that $\mu_2 > \mu_1$. Thus, since this is h_2 + ns we have the formulas for h_2 and s as given in Section 12.4.1, subsection (4c). The formula for h_1 follows similarly.

Derivations for other cases and for OC and ASN curve points are given in [1].

12.10. Summary. Sequential analysis, as we have seen, can handle the various cases of significance testing or decision-making normally done by single sampling (or double sampling). Moreover, sequential sampling is the usual starting point for developing a multiple sampling plan for defectives, with a desired OC curve.

The various cases are compared in Section 12.4. Two other cases can be handled: (a) differences between two proportions, for example, two processes, [1], and (b) a one-way test on the mean with σ_x unknown. The latter requires use of a new t-test calculation after each new observation [5].

Use of sequential analysis is most natural when measurements are relatively expensive and can best be done in sequence, rather than several all at once. This may frequently be the case in laboratories and in research. Another opportunity, which is quite feasible, occurs when we can work out the test for routine usage, as in Sections 12.6 and 12.7, so that we do not need to start calculating the criteria all over again for each new decision situation.

Again it is emphasized that average sample number curves are only to be compared if the protection (OC) curves are quite similar.

PROBLEMS

12.1. Set up a sequential sampling plan for defectives: $p_1' = .005$, $\alpha = .05$; $p_2' = .030$, $\beta = .10$. Find h_1, h_2, s and sketch OC and ASN curves, labeling axes. Make up a partial table of acceptance and rejection numbers for classes of n.

12.2. Set up a sequential sampling plan for defectives: $p_1' = .010$, $\alpha = .05$; $p_2' = .080$, $\beta = .10$. Find h_1, h_2, s and sketch the OC and ASN curves, labeling axes. Make up a partial table of acceptance and rejection numbers for classes of n.

12.3. Set up a sequential sampling plan for defectives as in problem 11.9 (c). Find h_1, h_2, s for the plan, sketch acceptance-rejection regions, and the OC and ASN curves, labeling axes.

12.4. Set up a sequential sampling plan for defectives for the case in Problem 11.8(c). Find h_1, h_2, s, sketch acceptance-rejection regions and the OC and ASN curves, labeling axes.

12.5. Corresponding to Problem 11.3, use σ_x = 4000 psi, AQL = 102,000 psi, α = .05; RQL = 98,000 psi, β = .05, and set up a sequential test assuming normality. Code the x's in convenient fashion, find h_1, h_2, s and sketch acceptance-rejection regions, OC and ASN curves, labeling axes.

12.6. Corresponding to Problem 11.4, set up a one-way sequential test of means. Code the x's in convenient fashion, if desired, find h_1, h_2, s and sketch acceptance-rejection regions, OC and ASN curves, labeling axes.

12.7. Set up a sequential test, as in Problem 11.9(b), first finding the two critical mean weights of fill having $p_1' = .005$ and $p_2' = .050$. Then find h_1, h_2, s using suitable coding, sketch acceptance-rejection regions, and the OC and ASN curves, labeling axes.

12.8. Set up a sequential plan of process level for blowing time of fuses as in Problem 11.8(b), finding h_1, h_2, s sketching acceptance-rejection regions and the OC and ASN curves, labeling axes.

12.9. Corresponding to Problem 11.7, set up a sequential check of process level for the lower specification 1.1760". Use μ_1 = 1.17608" = RQL, μ_2 =

1.17612" = AQL, $\alpha = \beta = .10$, σ_x = .000,040". Find h_1,h_2,s using a convenient coding, sketch acceptance-rejection regions, and the OC and ASN curves, labeling axes.

12.10. Set up a one-way test on means, corresponding to Problem 11.10. Find h_1,h_2,s and sketch acceptance-rejection regions, and the OC and ASN curves, labeling axes.

12.11. Corresponding to Problem 11.5, set up a two-way sequential test of means. Use coded variable z = (x-12.7435")/.0001", and find h_1,h_2,s, sketch acceptance and rejection regions, and OC curves, labeling axes.

12.12. Corresponding to Problem 11.6 for pressure on hydraulic stop light switch, set up a one-way sequential test at the upper specification 100 pounds per square inch, using $\alpha = \beta = .10$ at $\mu_1 = 82$ vs. $\mu_2 = 88$ psi. Find h_1,h_2,s, sketch acceptance-rejection regions, and the OC and ASN curves, labeling axes.

12.13. Find a sequential sampling plan to correspond with the single plan on variability, μ unknown, in Problem 11.11. Find h_1,h_2,s and sketch acceptance-rejection regions, and the OC and ASN curves, labeling axes.

12.14. Find a sequential sampling plan to correspond with the single plan on variability, μ unknown, in Problem 11.12. Find h_1,h_2,s and sketch acceptance-rejection regions and the OC and ASN curves, labeling axes.

12.15. Corresponding to Problem 11.13, find a sequential plan to make the test $\sigma_1 = .2°C$, $\sigma_2 = .3°C$, $\alpha = \beta = .10$. Find h_1,h_2,s, sketch acceptance-rejection regions, and the OC and ASN curves, labeling axes. Use μ known case. Note that (a) μ known as the exact temperature, then we are testing "accuracy", which includes bias and repeatability, and (b) μ unknown, then we are testing repeatability or "precision".

12.16. Corresponding to Problem 11.14, set up a sequential test for the case $\sigma_1 = 1.715$, $\sigma_2 = 3.43$, $\alpha = \beta = .10$. Sketch the acceptance-rejection regions, and the OC curve and ASN curve, labeling axes, for the μ-known case. Test once each with distributions A and C of Table 4.1, using given μ's. (Class results, decisions and n's may be pooled.)

12.17. Corresponding to Problem 11.15, set up a sequential test of process capability for specifications 1.2450" to 1.2500" or T = .0050" for $\sigma_1 = T/10$ vs. $\sigma_2 = T/6$, $\alpha = \beta = .05$. Find h_1, h_2, s, sketch acceptance-rejection regions, and the OC and ASN curves, labeling axes for the case μ known. Also give scales for μ unknown.

12.18. Corresponding to Problem 11.16, set up a sequential test of process capability for specifications 2.8341" to 2.8351" or T = .0010" for $\sigma_1 = T/10$, $\sigma_2 = T/6$, $\alpha = \beta = .05$. Find h_1, h_2, s and sketch the acceptance-rejection regions, the OC and ASN curves, labeling axes, for the case μ known. Also give scales for μ unknown.

12.19. Set up a sequential range test of process capability with T=17, as in Section 12.7, with $\alpha = \beta = .05$. Test once with population A or B of Table 4.1. Test also once with population C or D. What would you guess Pa is for the latter distributions?

12.20. Set up a sequential sampling plan for defects: $c_1' = 4$, $\alpha = .05$; $c_2' = 6$, $\beta = .10$. Find h_1, h_2, s and sketch the acceptance-rejection regions, and the OC and ASN curves, labeling axes.

12.21. Set up a sequential sampling plan for defects: $c_1' = 6$, $\alpha = .05$; $c_2' = 10$, $\beta = .10$. Find h_1, h_2, s and sketch the acceptance-rejection regions and the OC and ASN curves, labeling axes.

12.22. For a Poisson distribution, derive the formulas for h_1, h_2 and s from criteria on P_{2n}/P_{1n}.

12.23. In what respect is a sequential sampling plan more efficient than a corresponding single sampling plan?

12.24. Suppose that a sequential sampling plan is worked out for a one-way test on process means, for a given σ_x. If the resulting plan is used when σ_x is actually larger than assumed, what does this do to the OC curve for μ_1 vs. μ_2?

12.25. Why must we always have $\alpha + \beta < 1$? Explain in relation to the OC curve for q_1 vs. q_2.

References

1. A. Wald, "Sequential Analysis." Wiley, New York, 1947.

2. Statistical Research Group, Columbia Univ., "Sequential Analysis of Statistical Data: Applications." Columbia Univ. Press, New York, 1945.

3. I. W. Burr, A new method for approving a machine or process setting, Parts I to III. Indust. Quality Control 5 (No. 4), 12-18, 6 (No. 2), 15-19, 6 (No. 3), 13-16 (1949).

4. H. M. Wies, Jr. and I. W. Burr, A simple capability acceptance test, a sequential test on ranges. *Indust. Quality Control* 21 (No. 5),266-268, (1964).

5. National Bureau of Standards, Tables to Facilitate Sequential t-tests. Dept. of Commerce, Applied Mathcs. Series 7 (1951).

CHAPTER 13

SOME OTHER SAMPLING PLANS

13.1. Sampling Inspection of Continuous Production. When it is desired to use sampling-inspection control of a continuous flow of discrete units, parts, items or assemblies, the situation is different from that in which product appears in separate lots, such as was the case in Chapters 9 and 10.

Of course one approach is to use a control chart set-up on such product, with a view to finding causes of sub-standard material and improving the production processes. Control charts could be for p, c, or for $\bar{x}$ and R if a measurable characteristic is the criterion. Moreover the chart could be for 100% of the units, segregated into samples of n in succession or could be samples of n chosen unbiasedly from larger segments of production. The objective would be to improve the processes and thereby obtain adequate and dependable quality. The chart could be thus used for action on the process, and if sampling is being employed, for action on the present product, such as, sorting 100% if danger is signaled.

An alternative to this approach is to use a sampling inspection plan for continuous production, some of which plans we will describe. These plans can well be used during production or at a final inspection station, especially under the following conditions.

Conditions for Sampling Inspection of Continuous Production

1. There are no natural lots submitted for inspection and decision. Instead there is a continuous stream of product as on a conveyor belt.

2. There are discrete items of product each of which is either a good one or a defective, that is, has none or has one or more defects.
3. Production is not so rapid but what 100% inspection is possible.
4. The quality of production has an acceptably low fraction defective for at least much of the time.

Continuous sampling plans for a particular defect, or for a class of defects, call for alternating periods of 100% sorting and of sampling inspection. The relative amount time spent on each depends upon the quality level of the defect in question: if quality is excellent, most of the time is spent under sampling inspection, and so forth.

13.1.1. The Original CSP-1 Plan. This plan was developed by Dodge [1]. The procedure is as follows:

1. At the outset, inspect, for the defect(s) in question, 100% of the units consecutively as produced, and continue such inspection until i units in succession are found clear of the defect(s).
2. When i units in succession are found clear of defects, discontinue 100% inspection, and inspect only a fraction f of the units, selecting the sample units one at a time from the flow of units, in such a way as to insure an unbiased sample.
3. Whenever, under sampling, a defective is found, revert immediately to 100% inspection of succeeding units, and continue until again i units in succession are found clear of defects as in section 1, when sampling inspection is resumed.
4. Correct or replace with good units, all defective units found.

The protection provided by the plan is determined by the two chosen constants i and f. The larger i is, the more difficult it is to qualify, or requalify, for sampling inspection, and thus there is more protection against relatively poor quality. Likewise the larger the proportion f we sample, the more quickly we will find a defective and thus be returned to 100% inspection, especially when quality is relatively poor. Suppose that the units being inspected come from a controlled process at some fraction defective p'. Then the constants i and f will determine the long-run proportion of the time (a) the inspection is on sampling with the outgoing fraction defective still at about p', and (b) the inspection is 100% with the outgoing fraction defective zero. Thus the outgoing fraction defective or $AOQ_{p'}$, is a weighted average of p' and zero, the weights for conditions (a) and (b) depending upon p' and of course i and f.

Now over the whole spectrum of possible p' fractions defective there will be some maximum AOQ, called the AOQL (average outgoing quality limit), just as in lot by lot acceptance sampling. This will not be a function of p' like the AOQ is, but will depend upon i and f only. Of course it will occur at some critical p'. This p' will lie between relatively good p''s for which there will be mostly sampling inspection with the AOQ close to p', and relatively poor p''s with most of the inspection 100% thus also giving low AOQ's. The AOQ curves for such plans would be much like those for typical lot by lot plans. See Figure 9.3. Thus specified constants i and f provide an AOQL protection, whenever the process is controlled.

But what if the process is not controlled, which is the more likely case? Here the AOQL still applies, if the shifts of p' are not directly correlated with whether the plan is on sampling or 100% inspection. Since such would seem to be very unlikely the AOQL is still meaningful. Moreover, the actual average AOQ is likely to be considerably below the AOQL, because the latter can only occur when p' is right at the critical p'.

Now how are the AOQL, i and f related? This relationship is shown graphically in Figure 13.1, which uses logarithmic scales. One might

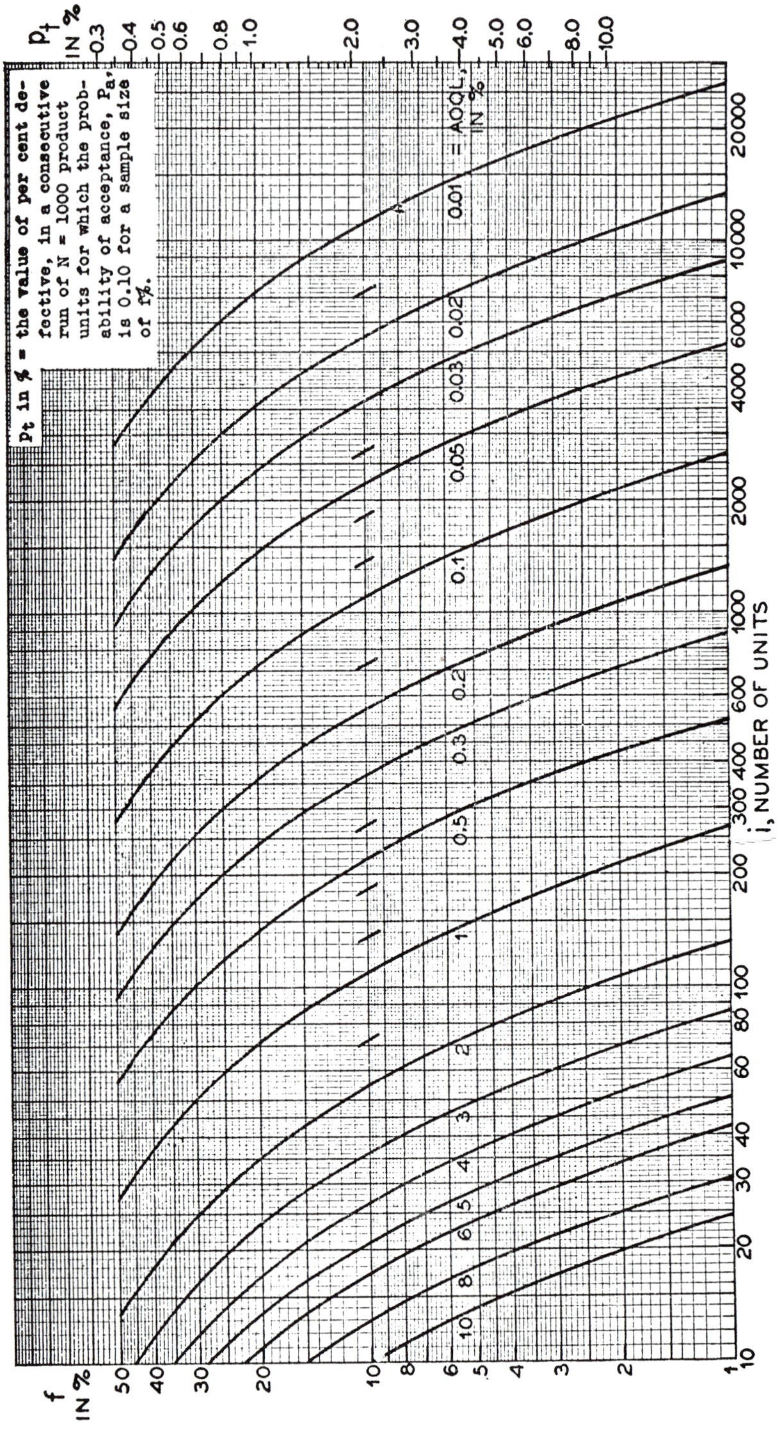

FIG. 13.1. Curves for determining values of i and f for a given value of AOQL [1]. Reproduced from reference [1] with the kind permission of the Editor of the Annals of Mathematical Statistics.

specify any two of the three and read the third from the graph. The usual thing, however, is to decide upon the desired AOQL and a sampling proportion f, for example 10%. Then read out the required number i to qualify for sampling inspection. For example, if f = 12.5%, or one out of eight, and the AOQL = 2% we find i = 48. Or if f = 5%, or one out of 20, with the same AOQL = 2%, then we find i=76. That is, if we are going to sample inspect fewer pieces once onto sampling, we must make it harder to qualify, if we are to have the <u>same</u> AOQL protection. Thus for <u>fixed</u> AOQL, i and f are somewhat inversely related. Take note that great care must be used in reading the logarithmic scales in Fig. 13.1.

With the CSP-1 plan there is really no operating characteristic curve in the sense of those in Chapters 9 and 10. Thus we cannot speak of an "acceptable quality level," AQL for the plan. But we can concern ourselves with the proportion of the time we are on sampling inspection, that is, with the percent of the total production accepted on a sampling basis.

Also we can describe the probability of remaining on sampling throughout a production run of 1000 units. See the note in the upper right corner of Figure 13.1 and the right hand scale. This p_t is not analogous to p_t in lot by lot acceptance sampling, Section 10.2. Here p_t is that fraction defective for which the probability is only .10 that the next 1000 units produced will all be accepted while under sampling. Of these 1000, we shall only inspect 1000f/(100%) = K of them. Then the binomial probability of finding no defectives among the K and thus of not going back to 100% inspection is

$$P(0 \text{ def. in } K|p_t) = (1-p_t)^K. \tag{13.1}$$

Equating this to .10,

we may find p_t by

$$.10 = (1-p_t)^{1000f/(100\%)} \tag{13.2}$$

Note that p_t is a function of f only, not i nor AOQL. If f = 5% we solve $.10 = (1-p_t)^{50}$, using logarithms, for p_t = .0450 or 4.5%.

Reference [1] gives appropriate formulas for finding the proportion of the time on sampling and 100% inspection from which the AOQ's and AOQL's may be obtained.

A practical problem arises as to how to sample at a rate of f%. Using the reciprocal of f, we may have to sample one out of five, eight, ten or 20. Then it is a good practice to choose which of each five, eight, ten or 20 consecutive pieces is to be inspected, by using a table of random numbers such as Table IV. This avoids bias.

Another practical problem lies in the shifting inspection load. If for example f = 10%, changing from sampling to 100% inspection involves quite a jump in load. One possibility which is used is to have the Production department take over the 100% inspection whenever it is called for.

Yet another practical matter is that of using CSP-1 on more than one class of defects. For example, we may wish to have AOQL's of .5% for major defects and 2.0% for minor defects. For this it seems good practice to use the same value of f for both classes of defects, and let the difference in treatment come via i, the qualification number. It is then quite easy to use the same sample of units to inspect for both classes of defects. For example, if f is 10% in the above we would find the i values of 220 and 55. Once qualified on one or both, we take one unit randomly out of each ten, for inspection for the defects qualified for sampling.

Properly used, CSP-1 can be utilized for <u>outgoing quality control</u>, and <u>process control</u> as well as by bringing attention to defect-producing conditions when they occur.

As proven in [1], if p' is the process fraction defective, then for given i and f, one has

u = average number of units inspected following a defective, till qualifying

$$= \frac{1-(1-p')^i}{p'(1-p')^i} \qquad (13.3)$$

v = average number of units passed while under sampling

$$= 1/fp'. \qquad (13.4)$$

It can then be shown that the average fraction of the units which are passed without inspection (that is, while on sampling inspection at a rate f) is

$$1-F = 1 - \frac{u+fv}{u+v}. \qquad (13.5)$$

This proportion passed, for fixed i and f, is thus a function of the process p'. It of course decreases as p' increases. It would seem that the use of CSP-1 would tend to force process correction until p' is sufficiently low to give 1-F above some critical value, say, 80% to 90%. That is, p' should be improved until the great majority of the time the units are on sampling inspection. (The AOQ will then be well below the AOQL of the plan.)

Further discussion of CSP-1 is given in Dodge [2].

13.1.2. Modifications of CSP-1. Two variations of CSP-1 were developed to reduce the frequency of alternation between 100% sorting and sampling, of continuous production. These are given in Dodge and Torrey [3]. In the first of these, CSP-2, the qualification procedure is the same as in CSP-1, that is, i consecutive good units must be observed under 100% inspection to go onto sampling. However, when a defective unit is found under sampling, 100% inspection is <u>not</u> reinstituted at once. Instead, careful note is made of the ensuing sample-inspected units (still at the rate f). As soon as any of the next i sampled units proves defective, 100% sorting _is_ reinstituted. But if no defective occurs within the next i sampled units, then the product is again "invulnerable" to a single defective, having requalified, so to speak, through the i consecutive good sampled units. If at any time after this run of i units, a new defective occurs, a careful watch of the next i sampled units is made. A second defective within these i causes the plan to revert immediately to 100% inspection, and so on.

Now is not this a more lenient plan _for given_ i and f, than is CSP-1? Yes it is, because it will stay on sampling inspection for longer periods, thus reducing the protection and giving a higher AOQL. Hence in order to obtain the _same_ AOQL for the _same_ frequency of sampling f, we must _increase_ i.

For example from [3]:

AOQL	.5%			1.0%			2.0%		
f	5%	10%	20%	5%	10%	20%	5%	10%	20%
i CSP-1	305	220	142	151	108	71	76	54	36
i CSP-2	390	293	200	194	147	100	96	72	50

This brief table shows the price we pay for greater freedom from alternation between sampling and 100% inspection. In these cases, it is roughly a third more pieces required to qualify for sampling.

Another variation from CSP-1 is called CSP-3 [3]. Its objective is to provide additional protection against a sudden run, perhaps brief, of bad quality, that is, "spottiness" of production. It is probabilistically identical to CSP-2, when inspecting units from a controlled process. It starts off with the same qualifying number i as in CSP-2. Then it goes onto sampling at the specified rate f. Whenever the first defective occurs, the very next four units produced are all inspected. If any of these four are defective, 100% inspection is reinstituted. If none of the four is defective, then these count as the first four of the i units all of which must be good ones to remain on sampling inspection. (The choice of units to inspect after the four proceeds as usual at a rate f.) Then if the i units are all good ones, a new defective found does not cause reversion to 100% inspection. This occurs only if one is found in the next four units produced or in the i-4 succeeding sampled units. And so on.

Thus we use the same i and f for a given AOQL in CSP-3 as in CSP-2, but in usage have the additional protection against spottiness of production.

13.1.3. Other Plans. One variation on continuous sampling plans is to use more than one sampling frequency. Such a plan is aptly called multi-level continuous sampling. It proceeds as follows: After qualifying by i consecutive good units, then sampling is instituted at a rate f. Then if the next i units inspected under sampling are all good, the rate of sampling goes to f^2. If the next i inspected at this rate are all good the rate goes to f^3, and so on. But if there occur conditions while sampling at a rate f^k, which under CSP-1, CSP-2 or CSP-3 would call for a return to 100% inspection, then under the multi-level plan the rate becomes f^{k-1}, that is, one level of a greater proportion sampled. Or if at f already, then 100% inspection is reinstituted, requiring requalification. Such plans may be designated CSP-M.

Another variation making use of "production intervals," operates like CSP-1 (or CSP-2), but has a maximum number a of defectives to be found in any one production interval, including both 100% and sampling inspections.

There is a military standard [4] which makes use of CSP-1, CSP-2, CSP-A and CSP-M plans. In the first two there is a maximum number of units L, before the completion of which the plan must have permitted sampling inspection, otherwise inspection is terminated pending improvement by the producer. The standard has specified AOQL's. But in an attempt to work closely with MIL-STD-105D it lists corresponding AQL's as index values. This seems to be a misnomer, since there really is no p' which can properly be called an AQL. If one finds the p' at which, say, 95% of the pieces will be accepted on a sampling basis, this might be a useful thing to list for the AOQL plan. But such a p' can hardly be called an AQL in the sense of lot sampling (Chapters 9 and 10).

The standard is currently being revised as this is being written. It is hoped that the above defect will be remedied, and also that the standard will be much simplified and streamlined. Moreover, as it currently stands, there is inadequate guidance for the user in the many needed decisions. Supervision of usage of the standard would seem to be very difficult.

The author is not greatly impressed by the practicability of multilevel plans. They require low levels of f (1/2 or 1/3) in order to avoid extremely low fractions on higher powers of f. This makes the saving small at f or even f^2, and the multi-levels make scheduling of work loads difficult. But undeniably multi-level plans are interesting and mathematically pretty. Multi-level plans are the outgrowth of work by Lieberman and Solomon [5].

13.2. Chain Sampling Plan, ChSP-1. In the case of lot by lot sampling, where small samples are called for and the acceptance number is c = 0, we obtain rather undiscriminating OC curves. Moreover, with an acceptance number of zero, the curve shape is always concave up like those in Figure

9.7, instead of having an inflection point such as the curves in Figures 9.2 and 9.6. The c=0 curves thus make it difficult to pass consistently, even quite high quality (low p') lots. The chain sampling plan was devised to obtain a more discriminating OC curve by cumulating the results from several lots. It was developed by Dodge [6]. Mostly the plan gives greater probability of acceptance of lots at relatively low p' than does the c=0 plan, yet giving about the same protection against accepting lots at relatively poor p''s.

The conditions and procedure are given in [6] for using ChSP-1:

"1. Conditions for Application

(a) Interest centers on an individual quality characteristic that involves destructive or costly tests, such that normally only a small number of tests per lot can be justified.

(b) The product to be inspected comprises a series of successive lots (of material or of individual units) produced by an essentially continuing process.

(c) Under normal conditions the lots are expected to be of essentially the same quality (expressed in percent defective).

(d) The product comes from a source in which the consumer has confidence.

"2. Procedure

(a) For each lot, select a sample of n units (or specimens) and test each unit for conformance to the specified requirement.

(b) Acceptance number of defects, c=0; except c=1 if no defects are found in the immediately preceding i samples of n. (i=1,2,3...)"

Hence a lot is accepted if the sample of n units shows no defects, or rejected if two or more defects. If there is only one defect, then the lot can still be accepted if no other defect has occurred within the samples for the immediately i preceding lots. Thus under the good history of no defects in the preceding i small samples, these count somewhat as though they were from the current lot. But if one of these samples contained a defect, then the enlarged sample (of i n's) contains two defects and calls for rejection of the current lot.

Our Figure 13.2 is reproduced from [6]. All of the ChSP-1 plans have the elongated S shape characteristic of sampling plans with $c > 0$, showing a point of inflection. Figure 13.2 is useful in picking out an appropriate OC curve. In the revision of MIL-STD 105D, the provision of ChSP-1 as an alternative to ordinary single sampling plans with $c=0$, deserves consideration and is a current possibility.

The OC curve has slightly different meaning than in lot by lot sampling, wherein we can talk of the probability of acceptance of just the one present lot. This is without regard to what has gone before. Under ChSP-1, however, the vertical scale, as seen in Figure 13.2 is the expected proportion of lots to be accepted, given p'. Under this plan if there is no defect in the previous i lot-samples then <u>for the current lot</u> $Pa = P(d \leq 1 | n, p')$, while if there was one defect among the i samples then $Pa = P(d=0 | n, p')$. (There could not be two defects in the i samples, why?)

Dodge suggests that the i=1 plan is not a preferred choice. It gives too little previous history to draw upon, and may well be too lenient.

The author regrets the confusion in the foregoing between "defects" and "defectives." In the preceding, reading "defectives" instead of "defects" is likely to be more what is meant, but the original source used "defects."

13.3. Skip-lot Sampling Plan, SkSP-1. The basic idea of this plan is to authorize omission of receiving inspection altogether on some fraction of the lots being submitted in a series, when quality in the past has been sufficiently good and consistent. It is an application of continuous sampling plan CSP-1 to <u>lots</u> (rather than to individual units), Dodge [7].

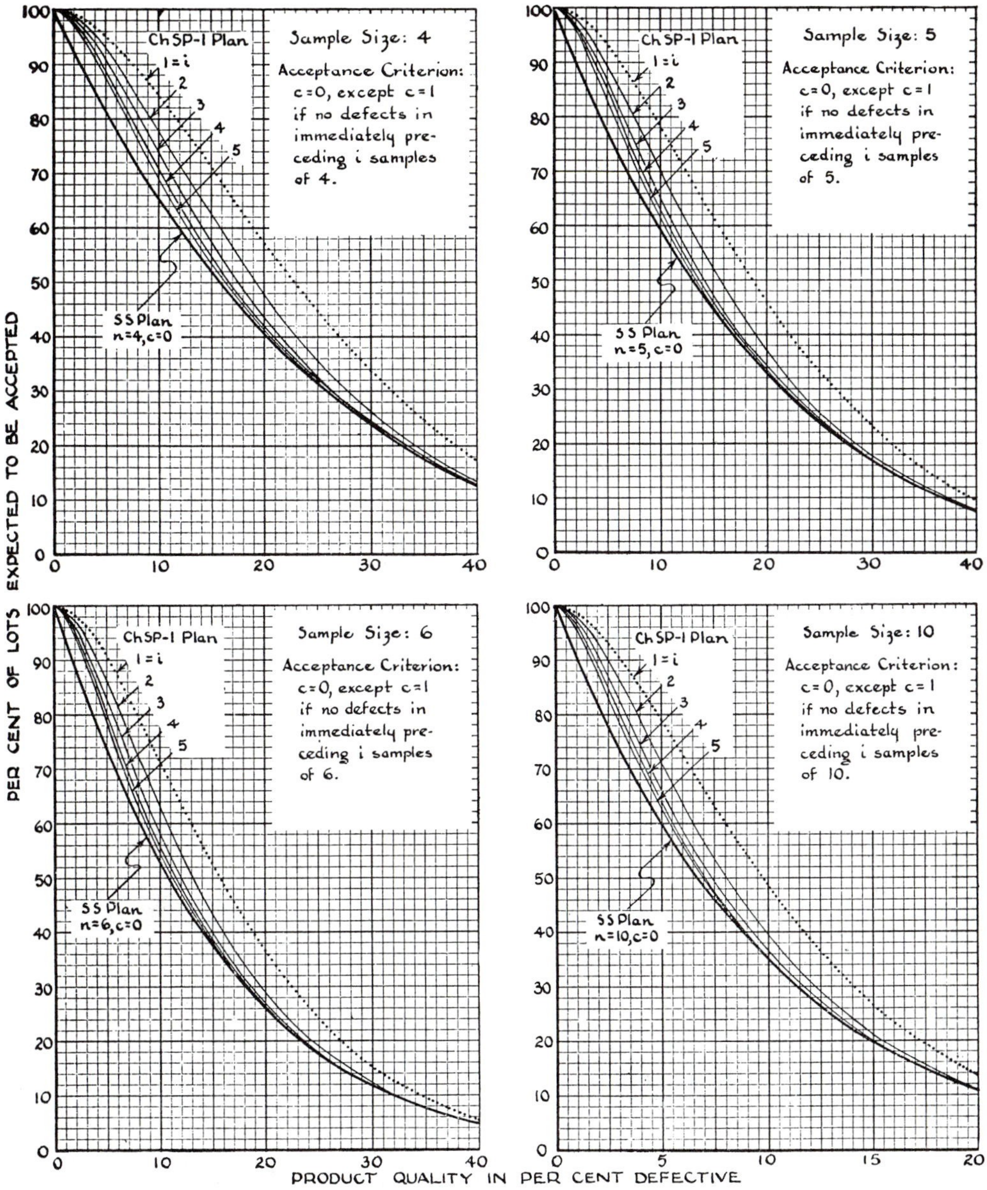

FIG. 13.2. OC curves for ChSP-1 Plans with values of i from 1 to 5. For comparison single sampling plans (n = 4, 5, 6 and 10, with c = 0) are shown. Reproduced from reference [6] by kind permission of the American Society for Quality Control.

It can be applied to either attribute inspection of a lot, or to measurement characteristics as, for example, chemical or physical analyses of bulk product. Under the system each lot or batch is examined or tested against a specific requirement. This could be, for example, in attributes inspection, relative to some single sampling plan n, Ac-Re; or for a measurement characteristic the value lying within, or outside, of a specified range of approved values. The latter could also be on the basis of an $\bar{x}$, or $\bar{x}$ and s, decision.

A series of lots or batches is thus examined until i in succession have been passed. One then goes onto sampling, examining a fraction f of the lots. As soon as a lot is rejected, the plan reverts to examination of every lot until again i in succession are all passed.

Under SkSP-1, the meaning of the AOQL is somewhat changed [7]. In the continuous plans, AOQL referred to the average fraction defective of individual <u>units</u>, including sampled and sorted units. This AOQ was a function of the incoming p' for units. In the skip-lot plan the AOQL refers to the proportion of <u>lots</u> conforming to requirements in the outgoing lots, after rejection (and possibly correction) of some of the lots, that is, those examined and judged to be non-conforming. Now the AOQ refers to the proportion of <u>lots</u>, which when examined, would be passed. In either plan CSP-1 or SkSP-1, the AOQL is the maximum of all AOQ's for whatever the incoming quality level might be.

There are two cases [7]. Procedure 1 calls for reprocessing or correcting each rejected lot until it is conforming, or replacing the rejected lot with a conforming one, then entering it into the stream of outgoing lots. For this we use the value of i in a CSP-1 table (Figure 13.1). In Procedure 2, however, rejected lots are removed and not corrected nor replaced. Then to qualify we must have i+1 accepted lots in succession to yield the AOQL protection specified.

Some plans given in [7] are:

AOQL	f	i: Proc. 1	i: Proc. 2
Standard Plan			
2%	1/2	14	15
Other Plans			
3%	1/2	9	10
5%	1/2	5	6
5%	1/3	9	10
5%	1/4	12	13

Careful consideration should be given to the legal and quality control implications of making no test on a characteristic on some lots. If the characteristic can prove at all hazardous, if a lot is non-conforming, use of the plan seems doubtful or dangerous. But when the characteristic is not so critical, production is steady and the quality level consistently high the plan can well be useful.

A number of plans, SkSP-2 by Dodge and Perry [8], are worked out for the case where inspection is by attributes, for example, n, Ac-Re, and the probability of lot acceptance is given for i and f. Use of f=2/5 provides a possible alternative to reduced inspection in MIL-STD-105D, using the normal plan, but only sample-inspecting two lots out of five when qualified for reduced inspection.

13.4. Cusum Charts. This is the commonly used name for cumulative sum charts. The idea is to follow any quality characteristic, say Q, in relation to some standard value Q_0, by cumulating the deviations from Q_0.

Q could be x, $\bar{x}$, R, s^2, d defectives or c defects. We analyze the current situation through plotting

$$\text{cusum}(Q) = \sum_{i=1}^{k} (Q_i - Q_0). \qquad (13.6)$$

Then if Q_0 is actually the average value of the characteristic Q, the cusum will tend to remain relatively close to zero, or only drift slowly away. But if the average value of Q is quite a bit different from Q_0, the cusum will rather rapidly rise from zero (or go below if $E(Q) < Q_0$). Now, one could put upper and lower limits to the total amount of drifting from zero. But it is also important to know how long (number of points) it took to reach such a limit. Thus the steepness of ascent is important too. This led those who developed the technique to use a V-shaped mask to make a test after each new point arose. See Figure 13.3. This is the usual test presently used. See Johnson and Leone [9] and Duncan [10] for expositions of the approach, and references to the early papers.

The dimensions of the V-shaped mask, specifically θ and d, are determined from the α risk of a decision that $E(Q) \neq Q_0$ when E(Q) is in fact Q_0, and the β risk of not concluding $E(Q) \neq Q_0$ when in fact $E(Q) = Q_0 + D$ (or $Q_0 - D$). Also d and θ depend on the quality statistic in question. The setting up of the mask and chart is something of a chore and must be done for each new application.

The cusum chart is in reality a type of sequential analysis, since it relies upon past data for each decision, which can be (a) conclude $E(Q) > Q_0$, (b) conclude $E(Q) < Q_0$, or (c) continue with new data. Cusum charts perform the same general test as do Shewhart control charts for the case "standards given." Thus Q_0 is a standard reference value, and for cusum charts for $\bar{x}$ or x, σ is needed, or a close estimate of it. Cusum charts do not correspond to Shewhart control charts for the case "analysis

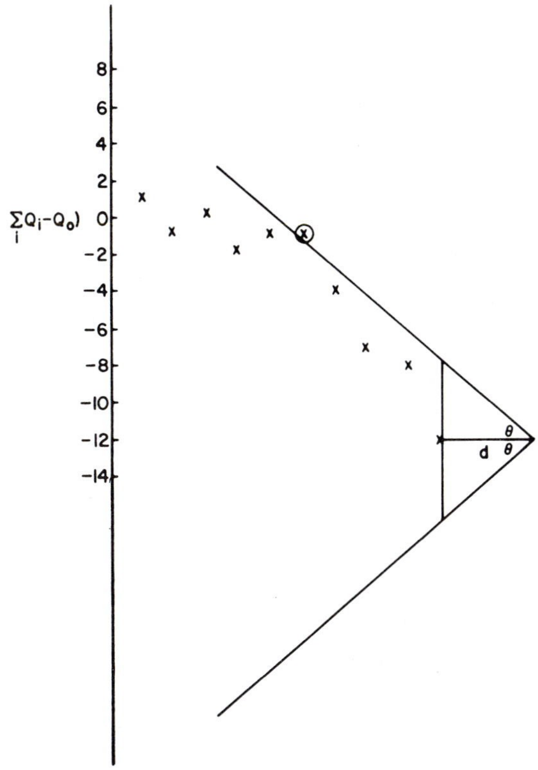

FIG. 13.3. A typical V-mask, showing d and the angles θ. After each point is plotted the mask is placed so that the left end of the d-segment coincides with the point. If any previous points lie outside the V, a decision is reached that $E(Q) \neq Q_0$ at this point. In this figure $E(Q) < Q_0$.

of past data," wherein the aim is to watch for statistical control or homogeneity, and for assignable causes disrupting such control.

The objective of cusum charts is to obtain an indication of departure from the standard Q_0, and to reach such an indication quickly. Specifically we can design a cusum chart which has a similar operating characteristic (OC) curve to the standards-given control chart in the following sense: If $E(Q) = Q_0$ then there should be a long "average run length," ARL, of points before a false warning is given. For example, on an $\bar{x}$ chart with $\pm 3\sigma_{\bar{x}}$ limits, the probability of a point out is .0027. And thus the ARL is 1/.0027 = 370 to a false warning. (See Section 7.6.) But if the sample size for $\bar{x}$'s is n=4, say, and μ should shift up $1\sigma_x$, that is, $2\sigma_{\bar{x}}$, the probability of an $\bar{x}$ point lying above the upper control limit is found from Figure 13.4 to be .1587. Then the ARL under such a process level shift is short, namely 1/.1587 = 6.3. Now if we set up a cusum chart to have an ARL = 370, when $E(\bar{x}) = \mu_0$, the ARL, under a shift in μ of $+1\sigma_x$ will be less than 6.3. Thus the cusum chart is more powerful in this sense. We note

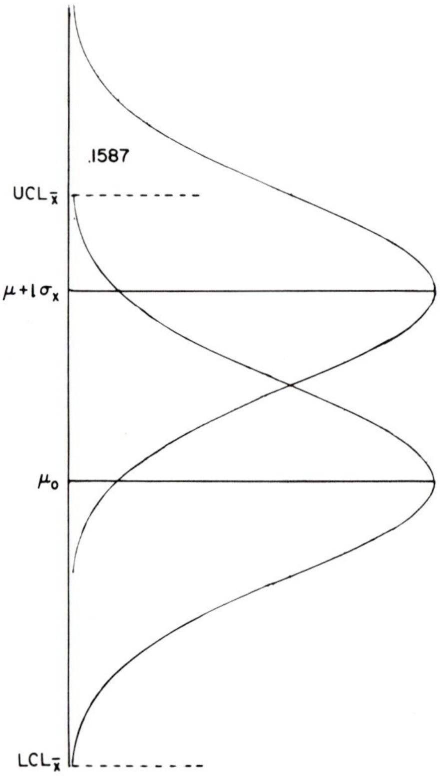

FIG. 13.4. Two distributions of $\bar{x}$ for n=4. One is centered at μ_0 with control limits of $\mu_0 \pm 3\sigma_{\bar{x}}$. The upper one is centered $1\sigma_x = 2\sigma_{\bar{x}}$ higher. It has .1587 of its $\bar{x}$'s above the upper control limit for $\bar{x}$'s from μ_0.

that this comparison is made for the cases (a) where the process stays at the standard and (b) where the average level jumps to some specified new level and remains there till apprehended.

For practical application, the worker must decide whether the greater effort in setting up a cusum chart is worth the gain in power. The present author is not convinced that it is very often worth the additional complications.

13.5. Summary. In this chapter we have introduced additional sampling plans. The first group (CSP) was for sampling control of continuous production. The plans are applicable to a continuous flow of units or pieces which can be inspected 100% when necessary. But if quality is sufficiently good then much of the time sampling inspection can be used, meanwhile providing protection if quality should deteriorate. The savings in inspection time can be considerable if, as should be the case, quality is good enough to stay on sampling 80% to 90% of the time.

Next there were presented two plans for inspection of a continuous flow of lots. The chain sampling plan (ChSP) has as its objective the attainment of a more discriminating OC curve than can be obtained by c=0 plans. The plan is to not reject on just one defective if there have been no other recent defectives (in the last i lots). The application is to conditions under which very small samples are needed, for example destructive or very expensive testing.

The other plan for inspection of a flow of lots is the skip-lot plan (SkSP). It makes use of the principles of CSP-1, but applies them to lots rather than single units. Under the plan, after a sufficient run, i, of lots, all of which were accepted, the plan goes to omission of all inspection in some proportion of the lots. Bolstering conditions would seem to make the plan usable, unless the characteristic(s) are of critical nature.

The fourth sampling plan was the cusum chart. It has resemblance to Shewhart control charts for the case, standards given. Also it has close relations to acceptance sampling for attributes or variables, and to sequential analysis. The typical application is to a flow of units or continuous material, inspecting or testing periodically for evidence of a process change. The objective is to achieve desired discriminating power on minimum average run lengths (ARL).

PROBLEMS

13.1. Find a CSP-1 plan for the following cases, filling blanks:

i	f	AOQL	p_t
	10%	3%	
50		5%	
		2%	7%
60	25%		

13.2. Find a CSP-1 plan for the following cases, filling blanks:

i	f	AOQL	p_t
	10%	1.5%	
100		2%	
200		.5%	
		.3%	4%

13.3. Find a CSP-1 plan for f=10% AOQL=1.5%. What would you expect i to be for the corresponding CSP-2 or CSP-3 plan?

13.4. Find a CSP-1 plan for f=20%, AOQL=.4%. What would you expect i to be for the corresponding CSP-2 or CSP-3 plan?

13.5. Explain why i and f are inversely related for <u>fixed</u> AOQL, in practical terms.

13.6. Why is p_t only related to f and not to i nor the AOQL in CSP-1?

13.7. If two classes of defects for the same piece are to be put under CSP-1 plans, with f=12.5%, find the respective i's for AOQL=1% and 3%. Why not use the same i instead of the same f?

13.8. What administrative difficulties would you foresee in using a multi-level continuous sampling plan? What is the potential gain?

13.9. Consider chain sampling plans ChSP-1 for n=4. What is the percent of lots expected to be accepted, using i=2 and i=5, p'=5%? Compare these with the probability for the single plan n=4, c=0. Do the same at p'=20%.

13.10. Consider chain sampling plans ChSP-1 for n=10. What is the percent of lots expected to be accepted, using i=2 and i=5, at p'=2.5%. Compare these with the probability for the single plan n=10, c=0. Do the same at p'=10%.

13.11. In what sense does ChSP-1 give a more favorable OC curve?

13.12. Consider a skip-lot sampling plan for AOQL=2%, for f=1/2. Explain the meaning of the two i's and of the AOQL.

13.13. Find a SkSP-1 for AOQL=2% and f=1/3. For f=1/4.

13.14. Find a SkSP-1 for AOQL=1% and f=1/2. For f=1/3.

References

1. H. F. Dodge, A sampling inspection plan for continuous production. Ann. Math. Statist. 14, 264-279 (1943).

2. H. F. Dodge, Sampling plans for continuous production. Indust. Quality Control 4 (No. 3), 5-9 (1947).

3. H. F. Dodge and M. N. Torrey, Additional continuous sampling inspection plans. Indust. Quality Control 7 (No. 5), 7-12 (1951).

4. Military Standard MIL-STD-1235 (ORD), "Single and Multi-level Continuous Sampling Procedures and Tables for Inspection by Attributes." Department of the Army, 1962.

5. G. J. Lieberman and H. Solomon, Multi-level continuous sampling plans. Ann. Math. Statist. 26, 686-704 (1955).

6. H. F. Dodge, Chain sampling inspection plan. Indust. Quality Control 11 (No. 4), 10-13 (1955).

7. H. F. Dodge, Skip-lot sampling plan. Indust. Quality Control 11 (No. 5), 3-5 (1955).

8. H. F. Dodge and R. L. Perry, A system of skip-lot plans for lot by lot inspection. Amer. Soc. Quality Control, Ann. Tech. Conf. Trans. 1971, 469-477.

9. N. L. Johnson and F. C. Leone, "Statistics and Experimental Design." Vol I. Wiley, New York, 1964.

10. A. J. Duncan, "Quality Control and Industrial Statistics." Irwin, Homewood, Illinois, 1965.

Additional References

H. F. Dodge and M. N. Torrey, A check inspection and demerit rating plan. Indust. Quality Control 13 (No. 1), 5-12 (1956).

E. G. Schilling and H. F. Dodge, Procedures and tables for evaluating dependent mixed acceptance sampling plans. Technometrics 11, 341-372 (1969).

K. S. Stephens, A Multi-level AOQL sampling plan with weighted defect classification. The Engineer, Western Electric 6 (No. 4), 39-49 (1962).

CHAPTER 14

STATISTICS OF COMBINATIONS, TOLERANCES FOR MATING PARTS

14.1. Introduction. Virtually all individual piece parts are made to be combined with other parts. The same is basically true of processed materials, such as, chemicals, plastics, rubber, glass, pharmaceuticals and many food products. The main thing of interest is the distribution of the important characteristics of the assembly or combination of the components, not so much the distributions of the characteristics of the components. You see we are here talking in terms of distributions rather than individual values, just as we have been stressing throughout this book. Of course the raw materials, operators and production processes must combine in such a way as to provide a satisfactory distribution to each characteristic of the components. We must deal in terms of distributions, since we cannot produce components all exactly alike. There will always be some variation, if only we measure precisely enough.

In a sense the traditional design engineer is determining the distribution of the characteristic of a component when he sets minimum and maximum specifications, L and U, on each such characteristic. This is still usually being done in a quite conservative manner, so that if each component part is anywhere between its specified limits, the assembly characteristic will be satisfactory. For example, the "stack-up" or sum of four dimensions added together might be the assembly characteristic in question. Suppose this sum is to lie between L and U. Then the traditional approach says set $L_1, \ldots, L_4$ so that if all your components are at these respective lower limits then

$L_1 + L_2 + L_3 + L_4 = L$, and similarly, for $U_1 + U_2 + U_3 + U_4 = U$. Now what could be more logical or safe? What is wrong with this approach?

The main thing wrong with such an approach is that it completely ignores probabilities and distributions. What is the probability of the four components being right at the four lower limits L_1, L_2, L_3, L_4 so that their sum will be right at L? It is extremely remote! (In fact theoretically zero.) Let us consider that the first dimension x_1 lies somewhere in the bottom 10% of the interval L_1 to U_1 and similarly for x_2, x_3, x_4. Then $y = x_1 + x_2 + x_3 + x_4$ lies in the bottom 10% of the interval L to U. Also suppose that, very conservatively, we assume that the distributions between L_i and U_i are rectangular, that is, the probability "mass" is uniformly distributed between the limits. Then there is only one chance in 10,000 of all four components being simultaneously in their respective bottom 10%'s, which gives the sum y in the bottom 10% of L to U.* If the components' distributions meet the specifications L_i to U_i in a $\pm 3\sigma$ sense, with normal distributions, then the probability for y in the bottom 10% of L to U is far rarer than .0001 (actually about .000,016). This is due to compensating variation: if one component is low, others may well be high.

The objective of this chapter is to explain some of the theory of combining distributions and to show how to make use of these laws in practice.

14.2. Statistics of Sums and Differences. We may prove the following facts for simple sums and differences. Let

$$y = x_1 \pm x_2 \pm \cdots \pm x_k. \tag{14.1}$$

for k component dimensions or other characteristics. y is the assembly or combination characteristic we wish to control.

---

*There are other ways for y to be in the bottom 10% of L to U. If the intervals L_i to U_i are all equal, the exact calculation yields a probability of about .0011.

Thm. 1. If each x_i component has finite mean μ_i, than for (14.1)

$$\mu_y = \mu_1 \pm \mu_2 \pm \ldots \pm \mu_k. \tag{14.2}$$

This theorem holds whether or not the x_i's are independent, or normally distributed. For variabilities, however, we find it desirable to assume independence, that is, the random variable (component) x_i is independent of all of the other x_j's; knowledge of the latter tells nothing about the former.

Thm. 2. If each x_i component has finite standard deviation σ_i and the $x_1, \ldots, x_k$ are all mutually independent then

$$\sigma_y^2 = \sigma_1^2 + \sigma_2^2 + \ldots + \sigma_k^2, \quad \sigma_y = \sqrt{\sigma_1^2 + \sigma_2^2 + \ldots + \sigma_k^2}. \tag{14.3}$$

This theorem says that σ_y is in general much less than the sum of the separate σ_i's, but this requires independence of the x_i's. If the x_i's are not independent then (14.3) is affected by the so-called linear correlation coefficients. We then have as the very extreme limits:

$$0 \leq \sigma_y \leq \sigma_1 + \sigma_2 + \ldots + \sigma_k. \tag{14.4}$$

Also to be noted in (14.3) is that although (14.1) has $\pm$'s, (14.3) has +'s only.

Next, we have, besides the mean and standard deviation of y, its distribution shape. Theorems 1 and 2 made no assumption of normality of x's. We have the following theorem.

Thm. 3. For y defined as in (14.1), and the x_i's normally distributed, then y is also normally distributed.

Here the x_i's need not be independent.

Now suppose that instead of (14.1) we have

$$y = a_1 x_1 + a_2 x_2 + \ldots + a_k x_k. \tag{14.5}$$

Then under the same corresponding hypotheses as in Theorems 1 to 3, we have

$$\mu_y = a_1 \mu_1 + a_2 \mu_2 + \ldots + a_k \mu_k \tag{14.6}$$

$$\sigma_y^2 = a_1^2\sigma_1^2 + a_2^2\sigma_2^2 + \ldots + a_k^2\sigma_k^2. \tag{14.7}$$

y is normally distributed.

If the x_i's are not normally distributed in (14.1) or (14.5), then the distribution of y is not normal, but tends toward normality as k increases, provided none of the contributions to σ_y^2 in (14.3) or (14.7) predominates. This is from a more general central limit theorem than that for component x's from the same distribution.

14.2.1.* Derivation of (14.6) and (14.7). Consider first the case of k = 2 in (14.5). Then we must have been given the joint density function of x_1 and x_2, that is, $f(x_1,x_2)$. The "marginal" density functions are found by integration:

$$f_1(x_1) = \int_{-\infty}^{\infty} f(x_1,x_2)\,dx_2, \quad f_2(x_2) = \int_{-\infty}^{\infty} f(x_1,x_2)\,dx_1. \tag{14.8}$$

Then the respective means are given by

$$\mu_1 = \int_{-\infty}^{\infty}\int_{-\infty}^{\infty} x_1\, f(x_1,x_2)\,dx_2\,dx_1 = \int_{-\infty}^{\infty} x_1 \int_{-\infty}^{\infty} f(x_1,x_2)\,dx_2\,dx_1$$

$$= \int_{-\infty}^{\infty} x_1\, f_1(x_1)\,dx_1 = E(x_1) \tag{14.9}$$

where E means "expected value of" or "theoretical average of." Also

$$\mu_2 = \int_{-\infty}^{\infty} x_2 f_2(x_2)\,dx_2 = E(x_2).$$

Then too

$$E(x_1 - \mu_1) = \int_{-\infty}^{\infty} (x_1 - \mu_1)\, f_1(x_1)\,dx_1 = \int_{-\infty}^{\infty} x_1\, f_1(x_1)\,dx_1 - \mu_1 \int_{-\infty}^{\infty} f_1(x_1)\,dx_1$$

$$= \mu_1 - \mu_1 1 = 0. \tag{14.10}$$

Furthermore the definition of population variance is

$$\sigma_1^2 = E[(x_1 - \mu_1)^2] = \int_{-\infty}^{\infty} \int_{-\infty}^{\infty} (x_1 - \mu_1)^2 f(x_1,x_2)dx_2dx_1 =$$

$$= \int_{-\infty}^{\infty} (x_1 - \mu_1)^2 f_1(x_1)dx_1 \qquad (14.11)$$

and similarly for x_2. Finally consider

$$E[x_1 - \mu_1)\cdot(x_2 - \mu_2)] = \int_{-\infty}^{\infty} \int_{-\infty}^{\infty} (x_1 - \mu_1)(x_2 - \mu_2)f(x_1,x_2)dx_1dx_2 \qquad (14.12)$$

$$= \text{covar } (x_1,x_2).$$

This is by definition the "covariance" of x_1 and x_2. It is not in general equal to zero. But if x_1 and x_2 are independent then by definition

$$f(x_1,x_2) \equiv f_1(x_1)\cdot f_2(x_2) \qquad x_1,x_2 \text{ independent} \qquad (14.13)$$

and we have

$$E[x_1 - \mu_1)\cdot(x_2 - \mu_2)] = \int_{-\infty}^{\infty} \int_{-\infty}^{\infty} (x_1 - \mu_1)\ (x_2 - \mu_2)\ f_1(x_1)\ f_2(x_2)dx_1dx_2$$

$$= [\int_{-\infty}^{\infty} (x_1 - \mu_1)\ f_1(x_1)dx_1][\int_{-\infty}^{\infty} (x_2 - \mu_2)\ f_2(x_2)dx_2] = 0\cdot 0=0 \qquad (14.14)$$

if x_1,x_2 independent.

Now with this background consider (14.5) - (14.7) with k=2:

$$\mu_y = E[a_1x_1 + a_2x_2] = \int_{-\infty}^{\infty} \int_{-\infty}^{\infty} (a_1x_1 + a_2x_2)\ f(x_1,x_2)dx_1dx_2$$

$$= a_1E(x_1) + a_2E(x_2) = a_1\mu_1 + a_2\mu_2$$

as in (14.6). Then if x_1 and x_2 are independent we have

$$\sigma_y^2 = E[(y-\mu_y)^2] = \int_{-\infty}^{\infty} \int_{-\infty}^{\infty} [(a_1x_1 + a_2x_2)-(a_1\mu_1 + a_2\mu_2)]^2 f(x_1,x_2)dx_1dx_2$$

$$= \int_{-\infty}^{\infty} \int_{-\infty}^{\infty} [a_1^2(x_1-\mu_1)^2 + 2a_1a_2(x_1-\mu_1)(x_2-\mu_2) + a_2^2(x_2-\mu_2)^2]f_1(x_1)f_2(x_2)dx_1dx_2.$$

Breaking up into three separate double integrals we have respectively

$$a_1^2\sigma_1^2, \text{ zero, and } a_2^2\sigma_2^2$$

yielding (14.7) for the independent case. Now if x_1 and x_2 are not independent, the first and third terms are the same but the middle term will not in general be zero, but instead $2a_1a_2$ covar (x_1,x_2). As usual we may define the "correlation coefficient" by

$$\rho_{x_1x_2} = \frac{\text{covar }(x_1,x_2)}{\sigma_1\sigma_2} \tag{14.15}$$

which lies between -1 and +1, and which is zero if x_1 and x_2 are independent. ρ can be zero when x_1,x_2 are not independent, in which case we call them "uncorrelated." Thus we have for the generalization of (14.7) for k=2:

$$\sigma_y^2 = a_1^2\sigma_1^2 + a_2^2\sigma_2^2 + 2a_1a_2\left(\rho_{x_1x_2}\right)\sigma_1\sigma_2 \tag{14.16}$$

For k = 3, we derive (14.6) and (14.7) by regarding y as the sum of the two random variables $(a_1x_1 + a_2x_2)$ and a_3x_3, then use the results for k = 2. Higher k's are similar .

Note that (14.1) through (14.3) are corollaries with $a_i = \pm 1$.

14.3. Use of Distributions in Assemblies. Let us consider some examples to illustrate the formulas of the previous section.

Example 1. Bearing and Shaft. Suppose that we are producing bearings and shafts to be assembled. In order to insure that the shafts will all be capable of assembly at random into a bearing the following non-overlapping specification limits are set:

Inside diameter (ID) of bearing: x_1 .4022" $\pm$ .0012" or .4010", .4034".

Outside diameter (OD) of shaft: x_2 .4000" $\pm$ .0009" or .3991", .4009".

Taking the appropriate extreme limits, the diametral clearance

$$y = x_1 - x_2 \tag{14.17}$$

would then lie between .0001" and .0043". Now suppose that our production processes are capable of meeting the respective limits for x_1 and x_2 in a $\pm 3\sigma$ sense and with approximately normal distributions. What then can we say about the distribution of y? Using (14.2) and (14.3) we find

$$\mu_y = .4022'' - .4000'' = .0022''$$

$$\sigma_1 = .0012''/3 = .0004'' \qquad \sigma_2 = .0009''/3 = .0003'' \text{ so that}$$

$$\sigma_y = \sqrt{(.0004'')^2 + (.0003'')^2} = .0005''$$

y approximately normally distributed.

Thus practically all (about 99.7%) of the diametral clearances will lie within

$$.0022'' \pm .0015'' = .0007'', .0037''.$$

Note carefully how much closer together these limits are than the extreme possible limits .0001" to .0043".

What can then be done? One possibility is to run μ_1 and μ_2 closer together, so as to decrease the maximum clearance if this is desirable. Thus if a minimum of .0001" is deemed satisfactory then we might use

$$.0016'' \pm .0015'' = .0001'', .0031''.$$

and

$$.4016'' \pm .0012'' = .4004'', .4028'' \qquad .4000'' \pm .0009'' \text{ or } .3991'', .4009''.$$

These limits must be met in a $\pm 3\sigma$ sense and with approximately normal distributions. Notice that they overlap somewhat but that taking into account the distributions, 99.7% of the pairs of bearing and shaft being assembled would have clearance between .0001" and .0031".

Another alternative is to use the full limits .0001" to .0043", if these limits are indeed permissible. To meet them in a $\pm 3\sigma$ sense we can say that for clearance y we can let

$$\mu_y = .0022''$$

and since the limits are .0022" ± .0021", we can let

$3\sigma_y = .0021''$ or $\sigma_y = .0007''$.

Now suppose we decide to allocate $\sigma_1 = (4/3)\sigma_2$. Then using (14.3)

$$.0007'' = \sqrt{(4/3)^2\sigma_2^2 + \sigma_2^2} = (5/3)\sigma_2$$

or

$$\sigma_2 = .00042'' \qquad \sigma_1 = .00056''.$$

Then we could use specification limits for the components (if we can make sure of their being met in a $\pm 3\sigma$, normal curve sense) of

$$.40220'' \pm .00168'' = .40052'', .40388''$$

$$.40000'' \pm .00126'' = .39874'', .40126''.$$

Again note that there is overlapping in these two sets of tolerance. Further we probably should not be setting specifications to such high precision (.00001").

Gains such as this of permitting 40% greater tolerance can mean substantial savings in production, inspection and rework costs, not to mention scrapping costs. But how can we <u>safely</u> obtain this gain? The answer is by controlling the <u>distributions</u> of dimensions through use of measurement control charts and/or acceptance sampling by measurements. See Section 14.4 for one recent systematic approach. The watchword then is: Engineering to permit substantially greater process variability in exchange for guarantees by Production of adequate control of process means. Quality Control must work closely with both.

Example 2. Three resistors are to be connected in series so that their resistances add together for the total resistance. One is a 150 ohm resistor and the other two are of 100 ohms, carrying the same part number. Suppose the respective specifications are

150 ± 7.5 ohms and 100 ± 6 ohms.

Let us take, similar to (14.1)

$$y = x_1 + x_2 + x_3.$$

Then we set $\mu_1 = 150$, $\mu_2 = \mu_3 = 100$ ohms. Now suppose that the specifications are being met in a $\pm 3\sigma$ sense by normal distributions. Then $\sigma_1 = 2.5$, $\sigma_2 = \sigma_3 = 2.0$ ohms. Now what can we say about y? Surely if all three resistors are inside their specifications, then y will lie inside $350 \pm (7.5 + 6 + 6) = 350 \pm 19.5$ ohms. But what of y's distribution? We have

$$\mu_y = 350, \quad \sigma_y = \sqrt{2.5^2 + 2.0^2 + 2.0^2} = 3.77, \text{ y is normal.}$$

Thus very nearly all of the total resistances y will lie inside

$$350 \pm 3(3.77) = 338.69, 361.31.$$

Note that these are far narrower than $350 \pm 19.5 = 330.5, 369.5$. In fact these latter limits are $\mu_y \pm 5.17\sigma_y$. It is therefore apparent that if 350 ± 19.5 is all that is required on the total resistances y, then the specification limits for the component resistors can be considerably widened. In fact we could use

$$150 \pm 7.5(5.17/3) = 150 \pm 12.9 \text{ and } 100 \pm 6.0(5.17/3) = 100 \pm 10.3.$$

But this is true only if we will meet such limits with approximately normal distributions and in a $\pm 3\sigma$ sense.

In practice what often happens is that when the assembly characteristic requires specifications like the ± 19.5 limits, then this 19.5 is split up into pieces whose <u>sum</u> is 19.5, for example, here into 7.5, 6.0 and 6.0. Limits so set, which are unnecessarily tight, may well be beyond the process capabilities, requiring expensive 100% sorting, reworking and scrapping, to be met. On the other hand if we use the tolerance of T = 39 for the sum, y, and set

$$T = \sqrt{T_1^2 + T_2^2 + T_3^2} \tag{14.18}$$

then the component tolerances, T_i, can be much larger than the $2(7.5) = 15$, $2(6.0) = 12$ and 12. In fact as we have seen they could be $2(12.9) = 25.8$, $2(10.3) = 20.6$ and 20.6. It could well be that the production process could have the capabilities of producing to these tolerances in a $\pm 3\sigma$ sense, with reasonable normality; thus by suitable process control all 100% inspection, rework and scrap are eliminated. At least it would be much easier for Production to meet these limits which are 72% wider.

But to make use of (14.18) the Engineering Department must be given assurance that the Production Department will exercise control of the process means of the components to be close to the nominal and that the distribution will be reasonably normal. The latter tends to mean that the limits are not to be met only by heavy truncation through 100% inspection. The next section gives a set of process controls and acceptance sampling plans which, if used, will permit the use of formulas like (14.18) to give maximum tolerances to the various components.

Another point we can well make here is that it might be tempting to use

$$y = x_1 + 2x_2$$

here, as in (14.5) rather than

$$y = x_1 + x_2 + x_3.$$

This first formula would give $\mu_y = 150 + 2(100) = 350$ all right, by (14.6). But for σ_y we would have from (14.7)

$$\sigma_y^2 = 2.5^2 + 2^2(2.0)^2 = 22.25 \text{ or } \sigma_y = 4.72.$$

This is larger than the 3.77 we found previously. The reason it is larger is that it basically assumes that if one "100 ohm resistor" is 105 the other one being assembled in the circuit will also be exactly 105 too. That is, it assumes perfect correlation in the two. There might in fact be some correlation leading to σ_y a bit above 3.77, but not as high as 4.72.

Example 3. In the author's consulting experience a problem of additive tolerances arose in which 10 dimensions (inside and outside diameters, pitch diameters and eccentricities, that is, distances between center-points) added and subtracted, determining whether the parts could be assembled. The company had available frequency distributions of 300 observations on components.

These gave a clear picture of the process capabilities, since control was quite good, as was normality. The desire was to double the tolerance on a pipe thread pitch diameter, so as to eliminate a second, finer threading operation. By direct use of (14.2) and (14.3) the author showed that there was not one chance in 10,000 of a collection of parts being unable to be assembled. This would only occur with all ten dimensions being at their very extreme wrong directions for the parts to fail to assemble. When these facts were presented to the government agency, the approval for the relaxed specification came back in four days. This application saved 8¢ on each of an order of 4,000,000 assemblies.

Example 4. An engineer in the automotive industry made a thorough study of a 12-volt motor for starting an automobile. He found that 38 dimensions added and subtracted along the rotor shaft determining the amount of "end-play" the rotor would have after assembly. Just for illustration let us suppose that a total tolerance of .038" could be permitted, and that each of the 38 dimensions would be equally hard to hold. Then the purely additive approach would say "Give each dimension a total tolerance of .001" and set the nominal mean dimensions so that they combine to give the desired average to end-play." The nominal mean is (U+L)/2. Now suppose that the production processes were each capable of meeting $\pm.0005$ in. in a $\pm 3\sigma$ sense, with reasonable normality. Then for each,

$$3\sigma_i = .0005 \text{ in.} \qquad \sigma_i = .000167 \text{ in.}$$

Thus (14.3) gives

$$\sigma_{\text{end-play}} = \sqrt{38(.000167 \text{ in.})^2} = .00103 \text{ in.}$$

or

$$T_{\text{end-play}} = 6\sigma_{\text{end-play}} = .00618 \text{ in.}$$

Compare this with the permissible tolerance of .038 in! It is only a sixth as great. So although we can permit .038 in. tolerance in the assembly characteristic we have forced a far smaller variation by our use of pure additivity of tolerances.

What size of equal tolerances could we use? They could be $\sqrt{38} = 6.16$, that is, about six times as great each. Then each individual tolerance T_i would be .006 in. or $\sigma_i = .001$ in. Then

$$\sigma_{\text{end-play}} = \sqrt{38(.001 \text{ in.})^2} = .00616 \text{ in.}$$

or

$$T_{\text{end-play}} = 6\sigma_{\text{end-play}} = .037 \text{ in.}$$

But, and this is extremely important, the averages of the production distributions would have to be held close to the various nominals $(L_i + U_i)/2$. Normality should be easier to attain, since the excessively tight tolerance might well be beyond the production capabilities, thereby forcing 100% inspection.

Now in practice, undoubtedly some component part characteristics would be more difficult to hold than others, and therefore should be given greater tolerances. But this can be done subject to (14.2) and (14.3). Also some mechanism is needed to insure that the means are properly controlled, otherwise the whole approach is not very safe. Such an approach is given in Section 14.4.

By using such an approach we can often enlarge the tolerance on those dimensions which are the most difficult to hold. In this way it is often possible to (1) eliminate 100% sorting thereby saving inspection and parts, (2) use a cheaper production process, or (3) avoid the necessity of an extra operation. In these ways huge savings are possible in industry.

14.3.1. The Rectangular Distribution. It is sometimes useful to assume that the x's are uniformly distributed between the specification limits, as a rather extreme alternative to the normal curve with $\pm 3\sigma$ just meeting the limits. The rectangular distribution is defined by

$$f(x) = \frac{1}{U-L} \quad L < x < U \quad \text{Rectangular distribution} \tag{14.19}$$
$$= 0 \quad \text{elsewhere.}$$

For this distribution we have*

$$\mu_x = \frac{U+L}{2} \quad \text{Rectangular distribution} \tag{14.20}$$

$$\sigma_x = \frac{U-L}{2\sqrt{3}} \quad \text{Rectangular distribution} \tag{14.21}$$

Thus for the rectangular distribution filling the interval from L to U uniformly, the range U-L is only $2\sqrt{3}\sigma_x = 3.46\sigma_x$, instead of $6\sigma_x$ as in the normal distribution (in a $\pm 3\sigma_x$ sense). And so for component distributions meeting limits L to U and centered at the nominal (U+L)/2, the range U-L might be thought of as running between 3.46 and 6 σ_x's.

Now suppose we have four components, perhaps resistors with their resistances each meeting 100 ± 6 ohms, and connected in series so they add.

---

*For the density function (14.19) we have

$$\mu_x = \text{expectation of } x = E(x) = \int_L^U \frac{x\,dx}{U-L} = \left.\frac{x^2}{2(U-L)}\right]_L^U = \frac{U^2-L^2}{2(U-L)} = \frac{U+L}{2}$$

$$\text{Expectation of } x^2 = E(x^2) = \int_L^U \frac{x^2dx}{U-L} = \left.\frac{x^3}{3(U-L)}\right]_L^U = \frac{U^3-L^3}{3(U-L)} = \frac{U^2+UL+L^2}{3}.$$

Since it is easily shown that

$$\sigma_x^2 = E(x^2) - [E(x)]^2. \tag{14.22}$$

we have

$$\sigma_x^2 = \frac{U^2+UL+L^2}{3} - \frac{U^2+2UL+L^2}{4} = \frac{1}{12}(U^2-2UL+L^2) = (U-L)^2/12$$

or

$$\sigma_x = \frac{U-L}{2\sqrt{3}}.$$

If they meet the limits with a rectangular distribution, each would have

$$\sigma_x = 12/2\sqrt{3} = 2\sqrt{3} \text{ ohms}.$$

Then for $y = x_1 + x_2 + x_3 + x_4$

$$\sigma_y = \sqrt{4\sigma_x^2} = \sqrt{4(2\sqrt{3})^2} = 4\sqrt{3} = 6.93$$

So that y is meeting

$$400 \pm 3(6.93) = 400 \pm 20.8.$$

Note that this is in a $\pm 3\sigma$ sense with good normality by the Central Limit theorem. For just four components it is already closer than the purely additive tolerance

$$4(100) \pm 4(6) = 400 \pm 24,$$

even though we used very conservative distributions for meeting limits. The gain is greater if k is larger, even for these assumed very conservative distributions.

Consider once again Example 4, on the automobile motor. Again suppose each of the 38 dimensions is given a tolerance of .001 in., but that now the range L_i to U_i is met no better than with a rectangular distribution; a very conservative assumption. Now, what limits is the end-play y meeting? By (14.21)

$$\sigma_i = .001 \text{ in.}/2\sqrt{3} = .000289 \text{ in.}$$

so that by (14.3)

$$\sigma_{\text{end-play}} = \sqrt{38(.000289 \text{ in.})^2} = .00178 \text{ in.}$$

or

$$T_{\text{end-play}} = 6(.00178 \text{ in.}) = .0107 \text{ in.}$$

This is less than a third of the .038 in. tolerance the design engineer was assumed to need, despite the conservatism of the assumption.

Reversing the problem as before

$$6\sigma_{\text{end-play}} = 6\sqrt{38\sigma_i^2} = .038 \text{ in.}$$

yielding

$\sigma_i = .038 \text{ in.}/(6\sqrt{38}) = .00103 \text{ in.}$

Then using $T_i = U_i - L_i = 2\sqrt{3}\sigma_i = .00357$ in.

or about 3.6 times as great a tolerance as the purely additive tolerances yield.

Again we must emphasize that the gain comes about in considerable part only by controlling all distributions to average reasonably close to their nominals, $(U_i + L_i)/2$. Also assumed is independence of the x_i for (14.3), which is quite likely to be the case in industrial production, certainly so if the parts are produced under statistical control.

14.4. Tolerancing Assemblies by Specifying Process Controls and Acceptance Sampling. This section is the outgrowth of a series of works [1]-[6]. These assuredly are not to be thought of as the only potential references. G. Bennett [7] lists 187 references, as of 1964, most of which bear on the general subject of statistics of combinations.

Suppose that virtually all of the distribution of the assembly characteristic is to lie within $\mu_0 \pm (T/2) = L, U$, that is, the tolerance is T. This can be accomplished if L to U is $\mu_0 \pm 3\sigma$ and the distribution is reasonably normal.

Now there are several sources of variation in the dimension or other characteristic of any part: (1) the <u>variation</u> of an individual part characteristic around the lot mean or a short-term segment of production, (2) the variation of the <u>means</u> of the lots or segments of production as in (1), and (3) the mean of the means after applying applicable process controls or acceptance sampling tests. This combined variation is what we wish to control so that the <u>total variation</u> for the <u>assembly</u> shall be T or less and centered close to μ_0.

Appropriate process controls and acceptance sampling tests were developed in [3] which led to simple approaches published in [4]. These were further studied by James in [5]. In this his aim was to obtain sufficiently discriminating operating characteristics so as to avoid, quite largely, the rejection and sorting of satisfactory lots or segments of production while still having quite low probability of passing material needing screening. He also showed that the approach did well on fairly non-normal distributions. In [6] there is a compromise reached between the simplicity and somewhat smaller sample sizes of [3] and [4] and the greater complications and somewhat larger sample sizes of [5]. The author recommends the use of [6], which is herewith summarized.

The general approach is as follows:

1. Set up the equation for the manner in which the characteristics of the components combine to determine the characteristic for the assembly. It is usually of the form (14.1) with some combination of signs:

$$y = x_1 \pm x_2 \pm \cdots \pm x_k. \tag{14.1}$$

2. Determine the desired limits for y, the assembly characteristic, L, U, with tolerance T = U - L and with nominal mean $(L+U)/2 = \mu_y$.
3. Allocate the tolerances, T_i, for the component characteristics entering into (14.1) by

$$T = \sqrt{T_1^2 + T_2^2 + \cdots + T_k^2} \tag{14.23}$$

according to their supposed difficulty of manufacture.*

---

*Limits for the assembly L and U are to be met in a $\pm 3\sigma$ normal curve sense, so that $\sigma = T/6$. Then in general we have (14.3) for independent components:

$$\sigma = \sqrt{\sigma_1^2 + \sigma_2^2 + \cdots + \sigma_k^2}.$$

Multiplying both sides by 6 gives

$$6\sigma = \sqrt{(6\sigma_1)^2 + (6\sigma_2)^2 + \cdots + (6\sigma_k)^2}$$

Defining

$$6\sigma_i = T_i$$

we have (14.23). (14.24)

4. Set desired component means μ_i so that, following (14.1), we have (14.2)

$$\mu_y = \mu_1 \pm \mu_2 \pm \cdots \pm \mu_k \tag{14.2}$$

5. Thus for the i'th component we have the desired nominal μ_i and tolerance T_i. These are then used to set up process controls or acceptance sampling plans as below (instead of being used directly as limits in the conventional sense). By following these process controls or sampling plans thus determined for the components, all three of the sources of variation given above will be adequately controlled, so that the production system will yield assemblies meeting $\mu_y \pm 3\sigma_y$ with quite normal distributions.
6. If production or processing of the component is done nearby, it is recommended that one of the two process controls be used (which are much like $\bar{x}$ and R charts). But if production is done at a distance, as for example, for purchased parts, then the acceptance sampling program is recommended.

Acceptance Sampling Program for Components

Program

1. Initially on an order, use Plan A, until five consecutive lots have been passed, then use Plan B. Continue using Plan B until two lots out of any five consecutive lots have been rejected for screening, when Plan A is reverted to for qualification.
2. If a producer has had a satisfactory history of production on this or similar parts, may start with Plan B.

A. Acceptance Sampling Plan A (Sample Size 30)

1. From the lot draw at random three samples, each of 10 parts, and measure the characteristic x_i for each of the 30 parts. Find the average $\bar{x}$ of the 30. Also find three ranges, R, one for each sample of 10.

2. The tests for this, the i'th part are determined from the desired mean μ_i and tolerance T_i, as follows:

   a. Accept the lot if <u>both</u> of the following are met:

      i. $\mu_i - .13T_i \leq \bar{x} \leq \mu_i + .13T_i$ (14.25)

      ii. $R_1 + R_2 + R_3 \leq 1.45\ T_i$ (14.26)

   b. Reject the lot if either or both requirements in Section a are not met, and screen the lot 100% to limits of $\mu_i \pm (3T_i/8)$. (Some parts may be reworkable.)

B. Acceptance Sampling Plan B (Sample Size 10)

1. From the lot, draw at random a sample of 10 parts, and measure the characteristic x_i for each of the 10. Find the average $\bar{x}$ and range R.

2. The tests for this the i'th part are determined from the desired mean μ_i and tolerance T_i, as follows:

   a. Accept the lot if <u>both</u> of the following are met:

      i. $\mu_i - .174T_i \leq \bar{x} \leq \mu_i + .174T_i$ (14.27)

      ii. $R \leq .521T_i$ (14.28)

   b. Reject the lot if either or both requirements in Section a are not met, and screen the lot 100% to limits of $\mu_i \pm (3T_i/8)$. (Some parts may be reworkable.)

This sampling acceptance program will adequately control the <u>distribution</u> of outgoing product (including accepted and screened product) regardless of incoming distributions. If used on each of the parts in an assembly, the requirements for the assembly characteristic will be met.

Process Controls - Introduction

The process control procedures are provided to assure that the variability of outgoing parts is not too great relative to T_i, and that the average is relatively near the desired μ_i. Commonly the control takes the form of adjusting the process average as indicated, provided that the natural variability in the process is not excessive. Two classes of controls are for the cases of the production process (1) subject to the average level consistently changing in one direction, as for example, as a tool or grinder wears, and (2) subject to erratic jumps in average level. Also two methods of adjusting the process level are provided for (1) automatic control and (2) hand control.

Process Control Procedures

Given for the i'th component part characteristic a tolerance T_i from (14.23) and a nominal mean μ_i from (14.2). Then use the following rules to control the distribution of the i'th part:

A. Take samples of n = 5 parts periodically from the most recent production, measure and find mean $\bar{x}$ and range R for the x's.

B. Approve the process and let it continue if <u>both</u> of the following are met:

1. $R \leq .55T_i$ (14.29)

2. $\mu_i - .16T_i \leq \bar{x} \leq \mu_i + .16T_i$ (14.30)

C. If $R > .55\ T_i$ sort the product produced since the previous sample of five, 100%, to the limits $\mu_i \pm (T_i/2)$, if feasible, and in any case take steps to reduce the process variability.

D. If $\bar{x}$ lies outside the limits $\mu_i \pm .16T_i$, the process level is to be reset according to the following cases:

1. Consistent process change, as in tool wear.

   a. Resetting fixed amount, as in automatic control:

      i. downward by $.3T_i$ if dimension increases, or

      ii. upward by $.3T_i$ if dimension decreases.

   b. Resetting to fixed level, as in hand control:

      i. to $\mu_i - (T_i/8)$ if dimension increases, or

      ii. to $\mu_i + (T_i/8)$ if dimension decreases.

   c. 100% inspection is not necessary in this case of Section 1.

2. Erratic changes in process level

   a. Resetting by amount depending upon $\bar{x}$, as in automatic control:

      i. downward by $.8(\bar{x}-\mu_i)$ if $\bar{x} > \mu_i + .16T_i$

      ii. upward by $.8(\mu_i-\bar{x}_i)$ if $\bar{x} < \mu_i - .16T_i$

   b. Resetting to desired level, as in hand control: to μ_i as nearly as possible.

   c. Product should be sorted 100% to limits $\mu_i \pm (T_i/2)$ in this case of Section 2.

Summary of Approach

By use of the foregoing acceptance sampling and process control program, the design engineer can safely use (14.23) $T = \sqrt{T_1^2 + T_2^2 + \ldots + T_k^2}$ instead of the very conservative $T = T_1 + T_2 + \ldots + T_k$, because he can have full confidence that the component means will be sufficiently well held that the assembly characteristic will meet the specified limits L, U very well. This will in general mean that with such assurance, he can greatly loosen up on the variability, anywhere from 40% to 200% or more. This can bring great savings in inspection, rework and scrap costs, and make relations between Production and Engineering much more realistic.

14.5 Products, Quotients, and Other Functions. Although functions like (14.1) and (14.5) are extremely common in industry one might have

$$y = x_1 \cdot x_2 \tag{14.31}$$

$$y = x_1/x_2 \tag{14.32}$$

Or, there may be several variables multiplied and divided, or any functions of almost unlimited variety. We now describe ways of handling non-linear functions for characteristic y.

As we shall see later on, the important point in functions such as (14.31) and (14.32) is the relative variability within each component, not the absolute variability, as was the case for additive and subtractive functions. We define as is usual the coefficient of relative variation by

$$V_x = \sigma_x/\mu_x = \text{coefficient of variation} \tag{14.33}$$

where V_x may be expressed decimally or in percent.

We may show that for two independent random variables x_1 and x_2 which are always positive we have the approximate relations

$$V^2_{x_1x_2} \doteq V^2_{x_1} + V^2_{x_2} \tag{14.34}$$

$$V^2_{x_1/x_2} = V^2_{x_1} + V^2_{x_2} \tag{14.35}$$

Thus if $V_{x_1} = .001$ and $V_{x_2} = .030$, which mean respectively that typical percentage variations from the mean are .1% and 3%, we have approximately:

$$V^2_{x_1x_2} \doteq .001^2 + .030^2 = .000,901 \text{ or}$$

$$V_{x_1x_2} \doteq .03001$$

Therefore the product is subject to the same relative variation as is x_2 alone. Decreasing σ_{x_1} or V_{x_1} on down to zero will have negligible effect: x_2 will still impose its 3% variation upon the product, x_1x_2. Further decreasing of V_{x_1} will not help at all. Precisely the same thing can be said for x_1/x_2.

Next we may inquire about the means of products and quotients. We have the following

$$\mu_{x_1x_2} = \mu_{x_1}\mu_{x_2} \qquad x_1, x_2 \text{ independent} \tag{14.36}$$

$$\mu_{x_1/x_2} \doteq \mu_{x_1}/\mu_{x_2} \qquad x_1, x_2 \text{ independent} \tag{14.37}$$

Note that the first is exact, the second approximate. As a consequence of (14.34) to (14.37) we have, for example, for always positive independent components

$$y = \frac{x_1x_2}{x_3x_4} \tag{14.38}$$

$$\mu_y \doteq \frac{\mu_1\mu_2}{\mu_3\mu_4} \tag{14.39}$$

$$V_y^2 \doteq V_{x_1}^2 + V_{x_2}^2 + V_{x_3}^2 + V_{x_4}^2 \tag{14.40}$$

Example 1. Consider the volume y of steel plates. Suppose we have the following

x_1: $\mu_1 = 300$ cm. $V_1 = .001$ $\therefore \sigma_1 = .3$ cm.

x_2: $\mu_2 = 120$ cm. $V_2 = .001$ $\therefore \sigma_2 = .12$ cm.

x_3: $\mu_3 = 1$ cm. $V_3 = .020$ $\therefore \sigma_3 = .020$ cm.

Then for the volume $y = x_1x_2x_3$ we have approximately

$$\mu_y \doteq 36{,}000 \text{ cc}$$

$$V_y^2 \doteq .001^2 + .001^2 + .020^2 = .000402 \qquad V_y \doteq .02005$$

which shows that the small σ_3, contributes nearly all of the variation to y because μ_3 is so small that V_3 is far greater than V_1 or V_2. If volume is of basic importance, further refinement of control of length and width will have negligible effect on the volume. Effort on further control of thickness is where real improvement can be made. As to distribution, the component with the largest V has the most influence upon the distribution of y. Here, if x_3 is "normal," so will y be, or if of "rectangular" distribution so will y be. When the V's are more nearly equal the distribution tends toward normality, much like the Central Limit theorem.

Example 2. How can we best describe the result of the calculation with approximate numbers given below?

$$y = \frac{37.2(462)}{.0402(45{,}700)}$$

Now just what does 37.2 mean? If a measurement result, it should mean that the result is accurate to the nearest .1 of a unit, that is, that we believe the true value to lie somewhere between 37.15 and 37.25, for otherwise some other tenth unit would be more appropriate. Moreover, it seems reasonable to suppose that the true value is equally likely to lie anywhere between the limits, that is, to be rectangularly distributed between. If such is not the case, it should be so stated!

Assuming independence, which is usually good, we use the model of (14.38), and assume x_1 averages 37.2 and so on. Then for the mean result we have

$$\mu_y = \frac{37.2(462)}{.0402(45{,}700)} = 9.355$$

Now to how much error is this subject? We need the four coefficients of variation to answer the question. We assume a rectangular distribution of the true x_i's, and thus for the 37.2 measurement having $L_1 = 37.15$, $U_1 = 37.25$, $U-L = .1$, which yields by (14.21) $\sigma_1 = .1/2\sqrt{3}$, and thus $V_1 = .1/[2\sqrt{3}(37.2)]$. Hence

$$V_1 = \frac{.1}{2\sqrt{3}(37.2)} = .000{,}776$$

$$V_2 = \frac{1}{2\sqrt{3}(462)} = .000{,}625$$

$$V_3 = \frac{.0001}{2\sqrt{3}(.0402)} = .000{,}718$$

$$V_4 = \frac{100}{2\sqrt{3}(45{,}700)} = .000{,}632.$$

Therefore we use (14.40) finding

$$V_y \doteq .001382 = \sigma_y/\mu_y.$$

Substituting $\mu_y = 9.355$ gives $\sigma_y = .0129$. Now what of the distribution? With four factors all more or less equal in relative variability we can count on the distribution being quite normal. Hence the best way to express the result y is to treat it as a distribution giving

$$\mu_y = 9.355, \quad \sigma_y = .0129, \text{ approximately normal distribution.}$$

Examining this we see that the probability of a true value lying inside $9.355 \pm .013 = 9.342, 9.368$ is about .68 and that the deviation from 9.355 can quite possibly be as much as .03. Thus to write the value of y as 9.35 with the implication that it is accurate to the nearest .01, just because the four factors were each good for "three significant figures" is thoroughly misleading.

Significant Figures

Example 2 gives us reason to doubt the usefulness of such descriptions as "three significant figure accuracy." Let us consider this a bit further. Compare

$$x_1 = 10.2 \ (\mu_1 = 10.2) \ \sigma_1 = .1/2\sqrt{3} = .0289 \quad V_1 = .00283$$

$$x_2 = 99.7 \ (\mu_2 = 99.7) \ \sigma_2 = .1/2\sqrt{3} = .0289 \quad V_2 = .000290$$

and we see that although both x_1 and x_2 carry "three significant figure accuracy," one has about ten times the <u>relative</u> accuracy of the other. They are like a foot rule which may be anywhere from 12 in. to 120 in. long! Such a description of accuracy should be forever laid to rest. But statistical sophistication being by no means universal, the practice will doubtless linger on.

Example 3. In order to give a little light on the distribution of products and quotients of rectangularly distributed components, consider the distributions of $y = x_1x_2$ and $z = x_1/x_2$ where x_1 has limits 8 to 12 and x_2 17 to 23. Figure 14.1 shows the <u>exact</u> distributions of the two functions, found by integration over appropriate regions. The two distributions are surprisingly similar in shape, with their single straight sections and two curved sections.

14.5.1.* The Delta Technique. This is the name given to a technique for finding moments of the distribution of a function of random variables. Define the deviation

$$\Delta x = x-\mu_x \tag{14.41}$$

then the expections (E) are as follows

$$E(\Delta_x) = E(x-\mu_x) = 0 \tag{14.42}$$

$$E[\Delta_x^{\ 2}] = E[(x-\mu_x)^2] = \sigma_x^2. \tag{14.43}$$

Moreover by (14.12) using $\Delta y = y-\mu_y$

$$E(\Delta x\Delta y) = \text{covar}\ (x,y) \tag{14.44}$$

so that if x and y are independent (or at least uncorrelated) (14.14) gives

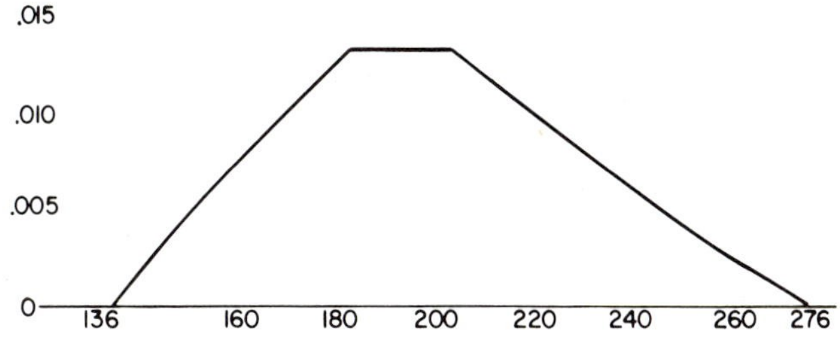

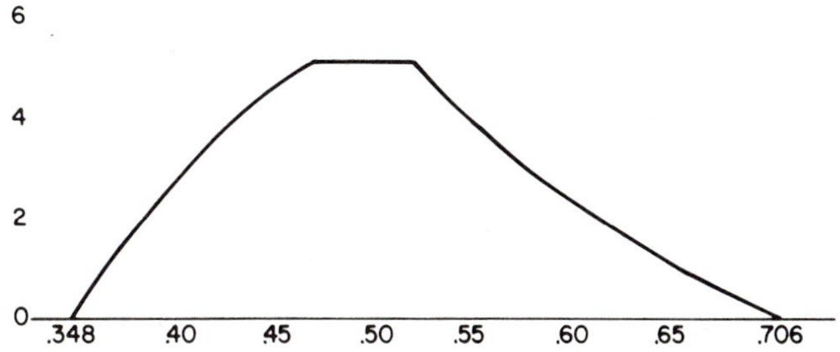

FIG. 14.1. Distributions of $y = x_1x_2$ and $z = x_1/x_2$ where x_1 is rectangularly distributed 8 to 12, and x_2 likewise, but 17 to 23, with x_1, x_2 being independent.

$$E(\Delta x \Delta y) = 0 \quad x,y \text{ independent.} \tag{14.45}$$

We now consider x,y independent and each ≥ 0

$$z = xy \qquad \Delta z = z - \mu_z. \tag{14.46}$$

Then using (14.13)

$$\mu_z = E(z) = E(xy) = \int_0^\infty \int_0^\infty xy\ f(x,y)\,dx\,dy = \int_0^\infty \int_0^\infty xy\ f_1(x) f_2(y)\ dx\ dy$$

$$= \int_0^\infty x\ f_1(x)\ dx \int_0^\infty y\ f_2(y)\ dy = \mu_x\ \mu_y, \tag{14.47}$$

And so using (14.41)

$$\Delta z = xy - \mu_x\mu_y = (\mu_x + \Delta x)\cdot(\mu_y + \Delta y) - \mu_x\mu_y$$

$$= \mu_y \Delta x + \mu_x \Delta y + \Delta x \Delta y.$$

Squaring yields

$$(\Delta z)^2 = \mu_y^2(\Delta x)^2 + \mu_x^2(\Delta y)^2 + (\Delta x)^2(\Delta y)^2 + 2\mu_x\mu_y\Delta x\Delta y + 2\mu_y(\Delta x)^2\Delta y + + 2\mu_x\Delta x(\Delta y)^2.$$

Taking the expectation of each side gives σ_z^2 on the left. The first three terms on the right have expectations $\mu_y^2\sigma_x^2$, $\mu_x^2\sigma_y^2$, $\sigma_x^2\sigma_y^2$ by independence, which also gives a zero expectation for each of the last three terms proceeding as for (14.47). Thus

$$\sigma_z^2 = \sigma_{xy}^2 = \mu_y^2\sigma_x^2 + \mu_x^2\sigma_y^2 + \sigma_x^2\sigma_y^2 \tag{14.48}$$

Now dividing through by the respective sides of (14.47) squared and using (14.33) gives the exact version of (14.34)

$$V_{xy}^2 = V_x^2 + V_y^2 + V_x^2V_y^2. \tag{14.49}$$

The last term on the right is nearly always negligible.

For z = x/y we may again use the Δ technique, but now things do not go quite as smoothly. For example

$$E(z) = E(x/y) = E(x)\cdot E(1/y)$$

if x and y are independent. Then

$$1/y = 1/(\mu_y + \Delta y) = 1/[\mu_y(1 + \Delta y/\mu_y)]$$

$$= \frac{1}{\mu_y}[1 - \frac{\Delta y}{\mu_y} + \frac{(\Delta y)^2}{\mu_y^2} - \frac{(\Delta y)^3}{\mu_y^3} + \ldots].$$

Taking the expectation of both sides we find

$$E(1/y) = (1/\mu_y)[1 + V_y^2 - V_y^3\,\alpha_{3:y} + \ldots]$$

where $\alpha_3 = E[(z-\mu_y)^3]/\sigma_y^3$ is the skewness of y.

Thus for E(1/y) to even exist we need a type of convergence. Since V_y is likely to be very small in industry we usually assume

$$E(1/y) \doteq (1 + V_y^2)/\mu_y \tag{14.50}$$

or even merely $1/\mu_y$.

An extension of the foregoing enables us to handle functions of more general type, if the functions are "well behaved." For example, for

$$z = g(x,y) \tag{14.51}$$

we may expand into a Taylor series around μ_x, μ_y:

$$z = g(\mu_x, \mu_y) + \left[\frac{\partial g}{\partial x}\right]_{\mu_x \mu_y} \cdot \Delta x + \left[\frac{\partial g}{\partial y}\right]_{\mu_x \mu_y} \cdot \Delta y + \dots .$$

From this, subject to rapidity of convergence, we have

$$\mu_z \doteq g(\mu_x, \mu_y) \tag{14.52}$$

$$\sigma_z^2 \doteq \left\{\left[\frac{\partial g}{\partial x}\right]_{\mu_x \mu_y}\right\}^2 \sigma_x^2 + \left\{\left[\frac{\partial g}{\partial y}\right]_{\mu_x \mu_y}\right\}^2 \sigma_y^2. \tag{14.53}$$

Actually these approximations are surprisingly good in most practical situations.

Example 4. Consider $y = x_1^3\sqrt{x_2}/x_3^2$. Using extensions of (14.51)-(14.53) we have

$$\mu_y \doteq \mu_1^3\sqrt{\mu_2}/\mu_3^2$$

$$\sigma_y^2 \doteq \left\{\frac{3\mu_1^2\sqrt{\mu_2}}{\mu_3^2}\right\}^2 \cdot \sigma_1^2 + \left\{\frac{\mu_1^3}{2\sqrt{\mu_2}\,\mu_3^2}\right\}^2 \cdot \sigma_2^2 + \left\{\frac{-2\mu_1^3\sqrt{\mu_2}}{\mu_3^3}\right\}^2 \cdot \sigma_3^2$$

which may be greatly simplified by use of V's to

$$V_y^2 = 9V_1^2 + V_2^2/4 + 4V_3^2.$$

Note how the exponents whether plus or minus enter into equation squared. In particular we have

$$V(x^p) \doteq pV_x \text{ and } V(\sqrt[r]{x}) \doteq \frac{1}{r} V_x \tag{14.54}$$

A final example of a function to which we may apply (14.52) and (14.53) is the resultant eccentricity (distance between two center points) coming from eccentricity x_1: A to B, eccentricity x_2: B to C, with random angle $\theta = \angle ABC$. The distribution of θ is usually taken as rectangular 0° to 180°. But the distributions of x_1 and x_2 might be more or less normal or else skewed toward the few higher values, because of the impossibility of negative eccentricities. Thus we seek information on the distribution of

$$y = \sqrt{x_1^2 + x_2^2 - 2x_1x_2\cos\theta}.$$

This can be handled by (14.52) and (14.53) but the shape of y's distribution is more difficult to determine. The distribution of y may be approximated by random drawings of x_1, x_2, θ.

14.6. Summary. In this chapter we have been concerned with the distribution of a function of random variables, given the distributions of these component variables. In particular we had the simple case of additive and subtractive functions, the mean and standard deviation of which are easily obtained. Such functions are extremely common in industry wherever there are assemblies of component parts. By taking advantage of the compensating errors or variation in the part dimensions it is possible to considerably loosen up tolerances of the parts, provided assurance is obtained of reasonable control of the mean dimensions. A system to ensure this was given in Section 14.4.

While we have been illustrating with more or less mechanical or electrical cases, the methods apply quite broadly, for example, to chemical mixtures.

The less frequently encountered non-linear functions can also be handled, though not quite as easily. Pure products and quotients, perhaps involving powers or roots give little difficulty through use of approximation methods and the coefficients of variation. Applications were made to calculations with approximate measurement data .

PROBLEMS

14.1. A brass washer and a mica washer are to be assembled together, one on top of the other. For the former μ_1 = .1155 in., σ_1 = .00045 in., while for the latter μ_2 = .0832 in. and σ_2 = .00180 in. (a). Find μ,σ for the total thickness. (b). Make what assumptions you need, and stating them, find the percentage of pairs in random assembly which will not meet specifications of .200 $\pm$ .005 in. (c). Would it help much if σ_1 were to be cut in half? (d). How in practice would you determine the μ's and σ's?

14.2. A 1000 ohm resistor and two 200 ohm resistors are to be connected in series so that their resistances add. The former is from a process yielding $\bar{\bar{x}}$ = 998, $\bar{R}$ = 12 ohms for samples of 5 each, while the latter two are drawn at random from a process with $\bar{\bar{x}}$ = 202, $\bar{R}$ = 3.1 ohms for samples of 4 each. (a). What percent can be expected to meet specifications of 1400 $\pm$ 15 ohms? (b). What did you assume in your solution?

14.3. The outside diameter of a pin showed good control with $\bar{\bar{x}}_1$ = .53600, $\bar{R}_1$ = .00080 cm, and the inside diameter of a mating collar also, with $\bar{\bar{x}}_2$ = .53733, $\bar{R}_2$ = .00100 cm, sample size being five in each case. In choosing for random assembly a pin and a collar, what percentage of pairs will have less diametral clearance than .00025 cm?

14.4. For electric light bulbs the distance from the rim at the base of the thread to the glass had $\bar{x}_1$ = 2.6667, s_1 = .1087 cm. For the inner liner of a socket the distance from rim to top had $\bar{x}_2$ = 2.3061, s_2 = .0305 cm. Both samples were substantial. What can you say about the distribution of x_1-x_2, stating any assumptions used?

14.5. A problem from a letter was the following: Three parts were assembled as shown. A and C are flush (that is the same point), B and F flush. Knowing the distributions of AB, CD and EF, how can one find the mean and standard deviation of the distance DE?

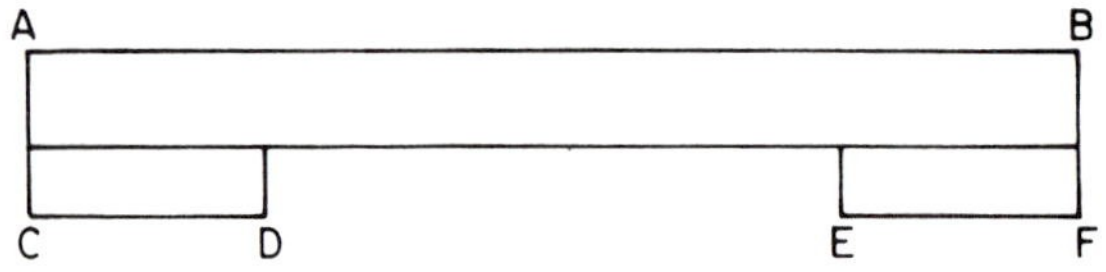

14.6. The true dimension x has added to it a measurement error e, the total being the observed dimension y. (a). Suppose that in production $\sigma_x = .05$ mm and $\sigma_e = .01$ mm, what is the standard deviation of the observed dimension y? (b). Would it help to measure each piece twice and average the two? (c). Same as (a) and (b) except $\sigma_x = .02$, $\sigma_e = .04$ mm.

14.7. Eight different parts are assembled along a shaft, additively. Each is given specification limits of $\pm.020$ cm. If the resulting minimum and maximum limits are merely added, what would the specification limits be on the sum? (a). If each part is produced to just meet its limits in a $\pm3\sigma$ normal curve sense, what specification limits could be set for the sum? (b). If each part is produced to just meet its limits with a rectangular distribution, what specifications can the sum meet with $\pm3\sigma$ assurance?

14.8. In a plant manufacturing soaps and cosmetics, a question arose as to whether to use a lot of caps for cologne bottles, which was on hand. Accordingly tests were made on the distribution of cap strength (that is, torque which would break the cap), and the distribution of the torque which the capping machine would apply to a cap. They found in inch-pounds respectively

$$\bar{x}_1 = 11.1,\ s_1 = 2.80;\ \bar{x}_2 = 8.0,\ s_2 = 1.51.$$

Assuming independence (very safe here) and normality, estimate the mean and standard deviation of x_1-x_2 (margin of safety), and estimate the percentage of caps which can be expected to be broken, by negative x_1-x_2.

14.9. Two parts are to fit one inside the other. The clearance between them is to meet specifications of $.0015 \pm .0005$ in., that is, difference of diameters. (a). If the parts can be produced under conditions of good statistical control, but you want to give 25% more tolerance to the inside diameter of the outer piece than to the outside diameter of the inside piece, what $\pm$ specifications would you recommend for each? (b). What problems are presented if the $\pm$ specifications so set are only $\pm 2\sigma$ of the processes? (c). If $\pm 4\sigma$ of the processes? (d). One $\pm 2\sigma$, one $\pm 4\sigma$?

14.10. A bearing and shaft assembly is as shown:

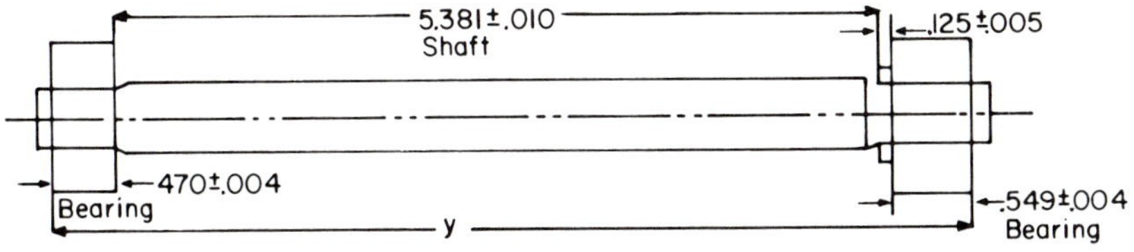

Now first suppose that each part meets its specifications precisely, in a normal curve $\pm 3\sigma$ sense. (a). Then what $\pm 3\sigma$ limits is y able to meet (give $\mu \pm 3\sigma$)? (b). Suppose on the other hand that each part is only meeting its specifications with a rectangular distribution between the limits. Then what limits can be met?

14.11. In manufacturing rectangular copper wire, the width has $\mu_1 = .365$ and $\sigma_1 = .0100$ in., and thickness $\mu_2 = .152$, $\sigma_2 = .0048$ in. (a). Assuming independence, find the mean and standard deviation of the cross-sectional area. (b). Is independence a very safe assumption?

14.12. Suppose that for sheets of metal, nominally 14 x 20 inches, the respective dimensions x_1 and x_2 have $\mu_1 = 14.01$, $\sigma_1 = .030$ in., $\mu_2 = 20.02$, $\sigma_2 = .011$ in., and x_1 and x_2 are independent. What can you say about the distribution of the area $y = x_1x_2$? (Assume the sheets are always perfectly rectangular.)

14.13. For Example 3 of Section 14.5, find for the product y: μ, V and σ. Do the same for the quotient z.

14.14. In a certain packaging operation, the net weight is estimated from the total weight w, cover weight c, and weight of coverless empty container e. Thus if x is the net weight, then $x = w - c - e$. Suppose examination yields $\bar{w} = 107.5$, $s_w = .4$, $\bar{c} = 2.0$, $s_c = .1$, $\bar{e} = 5.0$, $s_e = .2$ in pounds, for substantial samples. What can you say about the distribution of x? (Careful with this one!)

14.15. Three independent variables x_1, x_2 and x_3 are to be combined to give $y = x_1 + (x_2 x_3)$. If $\mu_1 = 200$, $\sigma_1 = 2$, $\mu_2 = 12$, $\sigma_2 = .05$, $\mu_3 = 20$, $\sigma_3 = .1$. (a). Find μ_y and σ_y approximately. (Hint: first find $\mu_{x_2 x_3}$, $\sigma_{x_2 x_3}$.) (b). If the three x_i's follow approximately normal distributions what would you expect for y?

14.16. Derive (14.54).

14.17. For the combined resistance for two resistors connected in parallel we have $y = \dfrac{1}{1/x_1 + 1/x_2}$, where x_1 and x_2 are the separate resistances. Knowing μ_1, σ_1, μ_2, σ_2, what can you say about the distribution of y?

14.18. If x has $\mu = 10$, $\sigma = .1$, what would you expect to have for (a). $y = x^2$ (b). $y = 1/x$ (c). $y = \sqrt{x}$?

14.19. Assuming rectangular distributions to the precisions indicated, approximate μ_y and σ_y for

$$y = \frac{25.2 \times 10.2}{87.9 \times .933}$$

14.20. Same as 14.19 but $y = 12.8 \times 93.77 \times .146$.

14.21. Same as 14.19 but $y = 36.7/35.8$.

14.22. Bottles are, we shall suppose, manufactured perfectly cylindrical inside. Let the radius be r, with $\mu_r = 1.00$, $\sigma_r = .02$ in., and normal distribution. The volume v of fill of liquid is also normally distributed, with $\mu_v = 15.10$, $\sigma_v = .05$ cu. in. Find the mean height of liquid in a bottle μ_y and its standard deviation σ_y, approximately.

14.23. Five pieces on a shaft have been carrying specifications as follows:

Two with 2.0140 ± .0040 in.

Two with 3.2060 ± .0035 in.

One with 3.8400 ± .0060 in.

They have been set to meet specifications of 14.2800 ± .0210 in., for the sum but difficulty has been encountered meeting the component limits. Preserving the relative ratios of the tolerances, set T_i's by $T = T_1^2 + T_2^2 + \ldots + T_5^2$. Then using Section 14.4, set up (a) an acceptance sampling program for the first kind of part, and (b) set up a machine-reset process control of the first kind of part if the average tends to rise, as is commonly the case on outer dimensions of machined parts.

References

1. I. W. Burr, Some theoretical and practical aspects of tolerances for mating parts. Indust. Quality Control 15 (No.3), 18-22 (1958).
2. I. W. Burr, On the distribution of products and quotients of random variables. Indust. Quality Control 18 (No.3), 16-18 (1961).
3. I. W. Burr, Unpublished research done while on sabbatical leave at University of Washington, 1966-1967, supported by the National Science Foundation.
4. I. W. Burr, Specifying the desired distribution rather than maximum and minimum limits. Indust. Quality Control 24 (No.2), 94-101 (1967).
5. R. E. James, Statistical process controls and acceptance sampling methods to control the distribution of combined output characteristics. Ph.D. Thesis, Purdue Univ., Lafayette, Indiana, 1970.

6. I. W. Burr and R. E. James, Specifying the desired distribution in lieu of tolerance limits - a systems approach. Paper presented to the Design Engineering Division of the Amer. Soc. of Mech. Engrs., Conference, Chicago, Apr. 2, 1974.

7. G. Bennett, The application of probability theory to the allocation of engineering tolerances. Ph.D. Thesis, Univ. of New South Wales, 1964.

CHAPTER 15

SOME OTHER FREQUENCY DISTRIBUTIONS

15.1 Introduction. It is desirable for the quality control worker in industry to know something about distributions other than those we have been working with so much, namely, the binomial, Poisson and hypergeometric discrete distributions, and the normal curve continuous one. We have also used the rectangular distribution in Chapter 14 as one type of extreme alternative to the normal curve.

In this chapter we first take up moments of frequency distributions for descriptive purposes, and then present several more frequency distributions, both discrete and continuous. Then finally a standard test for comparing observed and calculated frequencies is presented.

15.2. Moments of Distributions. Population and sample means and variances are examples of moments. The mean is the <u>first</u> moment figured from the origin of the x axis. The variance is a second moment around the mean, the square root being the standard deviation. Thus the student already has some background in moments. Let us now add to this. We first consider population moments. Discrete and continuous populations lead to different formulas, which are basically the same. We define the expected or theoretical average values of powers of x by

$$\mu_k' = E(x^k). \tag{15.1}$$

The weights applied to the x^k values are determined by the theoretical distribution in question. The two cases are

$$\mu_k' = \sum_x p(x)x^k, \text{ discrete} \qquad \int_{-\infty}^{\infty} f(x)x^k\,dx, \text{ continuous.} \tag{15.2}$$

These are quite analogous to moments in mechanics. We note that

$$\mu_0' = 1 \quad \text{and} \quad \mu_1' = \mu = \text{population mean} \tag{15.3}$$

In the foregoing the prime to the μ's meant "moments about the origin." We next take up central moments, that is, moments around the mean instead of around the origin. Thus by definition

$$\mu_k = E[(x-\mu)^k] \tag{15.4}$$

and so

$$\mu_k = \sum_x p(x)(x-\mu)^k \qquad \int_{-\infty}^{\infty} f(x)(x-\mu)^k dx. \tag{15.5}$$

We note that as in (14.10) and (14.11)

$$\mu_0 = 1, \qquad \mu_1 = 0, \qquad \mu_2 = \sigma_x^2. \tag{15.6}$$

The μ's may be found from the μ''s as follows, by expanding out the binomials in (15.5), distributing the summation or integral symbols along, and using the definitions (15.2) and (15.3):

$$\mu_2 = \mu_2' - \mu^2 = \sigma^2 \tag{15.7}$$

$$\mu_3 = \mu_3' - 3\mu_2'\mu + 2\mu^3 \tag{15.8}$$

$$\mu_4 = \mu_4' - 4\mu_3'\mu + 6\mu_2'\mu^2 - 3\mu^4. \tag{15.9}$$

Now what good are the various moments? They aid in characterizing or describing a theoretical distribution. The central moments μ_3 and μ_4 give major importance to the numerically large deviations $x - \mu$, since we are raising them to the third or fourth power. In μ_3 the third power of negative deviations are still negative, so that μ_3 is an algebraic sum and there is a contest as to which sign will win! But in μ_4 all fourth powers are positive. These two moments are related to the curve shape of the distribution. However, they really reach full value in description only when "standardized," that is, when the variability is taken into account. This is done as follows:

$$\alpha_3 = \frac{\mu_3}{\mu_2^{3/2}} = \text{skewness} \tag{15.10}$$

$$\alpha_4 = \frac{\mu_4}{\mu_2^{2}} = \text{kurtosis or contact.} \tag{15.11}$$

Such measures are free of units and of the variability, and thus comparable from one distribution to another.

In all of these population moments considered, Greek letters were used. The sample counterparts or equivalents use Roman letters corresponding, except that for the sample mean we use $\bar{x}$ rather than m. Thus for observed frequency distributions with mid-values x_i and corresponding frequencies f_i we define

$$\bar{x} = \frac{\Sigma f_i x_i}{\Sigma f_i} = m_1' \tag{15.12}$$

$$m_k' = \frac{\Sigma f_i x_i^k}{\Sigma f_i} \tag{15.13}$$

$$m_k = \frac{\Sigma f_i (x_i - \bar{x})^k}{\Sigma f_i}. \tag{15.14}$$

Some workers define m_2 as the "sample variance." But if we use (2.4) we have

$$m_2 = s_x^2(n-1)/n. \tag{15.15}$$

Also parallel to (15.7) - (15.9) are

$$m_2 = m_2' - \bar{x}^2 \tag{15.16}$$

$$m_3 = m_3' - 3m_2'\bar{x} + 2\bar{x}^3 \tag{15.17}$$

$$m_4 = m_4' - 4m_3'\bar{x} + 6m_2'\bar{x}^2 - 3\bar{x}^4. \tag{15.18}$$

The measures of curve shape are

$$a_3 = \frac{m_3}{m_2^{3/2}} \tag{15.19}$$

$$a_4 = \frac{m_4}{m_2^2}. \tag{15.20}$$

Most of the time, when one has an observed frequency distribution, the calculations will be done through coding via, say

$$u = \frac{x-a}{c}, \text{ or } x = a + cu, \quad c > 0 \tag{15.21}$$

It can readily be shown that the values of a_3 and a_4 are unchanged by such a general linear transformation, that is, that

$$a_{3:x} = a_{3:u} \qquad a_{4:x} = a_{4:u}. \tag{15.22}$$

The objective in fitting an observed distribution by a theoretical one is to try to match up α_3 with the observed a_3, or α_3 and α_4 with a_3 and a_4. Thus we try to match the curve shapes of the two distributions, observed and theoretical.

As a rough rule of thumb, if $|\alpha_3| \leq .3$ the distribution will appear quite symmetrical, $\alpha_3 = 0$ indicates perfect symmetry by this criterion. $|\alpha_3|$ .3 to .7 is moderate asymmetry. Skewness of $\pm$ 1.0 is substantial asymmetry. As pointed out, α_4 coming from μ_4 is always positive, in fact at least one. For the normal curve $\alpha_4 = 3.0$, and this is a good basis for judging contact, that is, the relative rapidity with which the frequency or the curve approaches zero. For the rectangular distribution $\alpha_4 = 1.8$.

It can be shown that if one takes random independent samples of n x's from the same population having α_3 and α_4 finite, then we have for the distribution of $\bar{x}$'s

$$\alpha_{3:\bar{x}} = \frac{\alpha_{3:x}}{\sqrt{n}} \tag{15.23}$$

$$\alpha_{4:\bar{x}} = 3 + \frac{\alpha_{4:x} - 3}{n}. \tag{15.24}$$

Note what happens in (15.23) and (15.24) when n increases, namely, α_3 for the $\bar{x}$'s approaches zero and α_4 for the $\bar{x}$'s approaches three, the respective normal curve values. This is in line with the Central Limit theorem which states that, under very general conditions, the distribution of the $\bar{x}$'s becomes normal as n increases. Also we may remark that (15.23) and (15.24) hold for sums of x's as well as $\bar{x}$'s, since the former are only a multiple of the latter, using (15.22).

15.3. Some Frequency Distribution Models. Let us first give moments for the distributions already included in this book:

$$\alpha_3 = 0, \quad \alpha_4 = 3 \quad \text{normal distribution} \tag{15.25}$$

$$\alpha_3 = 0, \quad \alpha_4 = 1.8 \quad \text{rectangular distribution.} \tag{15.26}$$

These are constant for all examples for each distribution, since there are no "shape parameters" and all examples of each distribution are merely linear transformations of each other. This follows from the fact that α_3 and α_4 are unaffected by any linear transformation (as are a_3 and a_4 for observed data).

On the other hand for the Poisson, binomial and hypergeometric distributions the curve-shape moments, α_3 and α_4, vary according to the various parameters. We have for the Poisson probability as defined in (2.35)

$$\alpha_3 = 1/\sqrt{c'}, \qquad \alpha_4 = 3 + 1/c', \qquad \text{Poisson distribution.} \tag{15.27}$$

As c' becomes very small, the skewness α_3 becomes very large, with the great bulk of probability concentrated at c = 0, namely, $e^{-c'}$. Also α_4 is large. On the other hand as c' increases, α_3 approaches zero and α_4 approaches 3. In fact by the time c' reaches 16, α_3 is already down to .25 and α_4 3.06. α_3 is always positive for the Poisson.

For the binomial (2.30) with parameters p' and n, we have

$$\alpha_3 = \frac{1-2p'}{\sqrt{np'(1-p')}}, \qquad \alpha_4 = 3 + \frac{1-6p'(1-p')}{np'(1-p')} \quad \text{binomial distribution} \tag{15.28}$$

Here we see that if $p' = .5$, $\alpha_3 = 0$. For all n's such binomials are perfectly symmetrical around n/2. If we use the conditions for a Poisson as the limit of the binomial as $n\to\infty$, $p'\to 0$, and $np'\to c'$, it is easily seen that (15.28) approaches (15.27) as a limit, as is of course expected. On the other hand if p' is fixed and $n\to\infty$, then the limits of (15.28) are $\alpha_3 = 0$ and $\alpha_4 = 3$, namely the normal curve values. This is in line with the Central Limit theorem which would in the binomial case say that if we draw n from the "binomial" for n = 1:

$$\begin{aligned} p(0) &= 1-p' \\ p(1) &= p' \end{aligned} \tag{15.29}$$

we approach a normal distribution function as $n\to\infty$. It is also worth mentioning that if in (15.29) we use $p' = .5$, we have a distribution with $\alpha_4 = 1$, which is the absolute minimum for the moment.

The formulas for the hypergeometric distribution (9.29) with N, D and n are the following complicated ones, in which we define D/N as p'[1]:

$$\alpha_3 = \frac{1-2p'}{\sqrt{np'(1-p')}} \sqrt{\frac{N-1}{N-n}}\,\frac{N-2n}{N-2} \qquad \text{hypergeometric} \tag{15.30}$$

$$\alpha_4 = \frac{(N-1)[N(N+1)-6n(N-n)+3p'(1-p')\{N^2(n-2)-Nn^2+6n(N-n)\}]}{np'(1-p')(N-n)(N-2)(N-3)}. \tag{15.31}$$

If N increases with D/N = p' held constant and n constant these moments approach those for the binomial (15.28), as we would expect.

15.3.1. The Geometric Distribution. This discrete distribution occurs frequently in applications, expecially in life testing and reliability. It is concerned with consecutive trials of an event each of which may be either a "success" or a "failure." The random variable is the number of trials i taken until the <u>first</u> "failure." Thus i can be 1, 2,... Let p' be the probability of a failure, q' the probability of a success. Then if the first failure occurs on the ith trial or test, there must have been i-1 successes followed by a failure. Therefore if we assume independence of trials and constant p', as in the binomial distribution, we have

$$p(i)=P(SS\cdots SF) = P(S)P(S)\cdots P(S)P(F) = q'^{\,i-1}p' \quad \text{geometric distribution.} \tag{15.32}$$

From the form of (15.32), we see that the probabilities p(i) start at a maximum of p' and gradually decrease, giving rise to what is often called a "J-shaped" distribution. By use of the geometric progression sum we may show that

$$\sum_{i=1}^{\infty} p(i) = 1 \tag{15.33}$$

as indeed it must be. For the various desired moments we need to use techniques for summing infinite series. For example

$$\mu = \sum_{i=1}^{\infty} i(q')^{i-1}p'$$
$$\mu = p' + p'2q' + p'3q'^2 + \ldots = p'(1+2q' + 3q'^2 + \ldots)$$

Call

$$S = 1 + 2q' + 3q'^2 + \ldots + jq'^{\,j-1} + \ldots$$

then multiply both sides by q' giving

$$q'S = q' + 2q'^2 + \ldots + (j-1)q'^{\,j-1} + \ldots$$

and subtract yielding

$$(1-q')S = 1 + q' + q'^2 + \ldots + q'^{\,j-1} + \ldots$$

Again multiply by q' obtaining

$$q'(1-q')S = q' + q'^2 + \ldots + q'^{\,j-1} + \ldots$$

and subtract giving

$$(1-q')^2 S = 1.$$

But since $1 - q' = p'$ we have

$$S = 1/p'^2.$$

Substituting this S into the expression for μ gives

$$\mu = 1/p' \qquad \text{geometric distribution.} \tag{15.34}$$

This formula for the average number of trials is just what one might expect, since, for example, if $p' = .10$, it seems reasonable to suppose that we should take 10 trials on the average before observing a one-chance-in-ten occurrence.

We may also derive

$$\sigma = \sqrt{1-p'}/p' \qquad \text{geometric distribution} \tag{15.35}$$

$$\alpha_3 = (2-p')/\sqrt{1-p'} \tag{15.36}$$

$$\alpha_4 = (9 - 9p' + p'^2)/(1-p'). \tag{15.37}$$

A typical application is to subject a part or an assembly to a series of severe tests and require that the first failure shall not occur before some minimum, for example, 1000 tests. Or one may test each of a series of parts once each, and prescribe that the first failure observed be not before the 200th part tested.

15.3.2. The Negative Binomial Distribution. This distribution is concerned with the same kind of drawing as are the binomial and the geometric distributions, namely independent, with a constant probability at each drawing, for the two kinds of objects or outcomes.

The problem solved by the negative binomial is that of asking for the probability that the <u>k</u>th failure occurs at the <u>i</u>th trial. (Note that if $k = 1$, then this specializes into the geometric distribution.)

The problem appears difficult at first, but a little thought soon indicates that the probability $p(i)$ is made up from two separate independent events whose probabilities are easily found. For the <u>k</u>th failure to occur on the ith trial, we must have had precisely k-1 failures on the first i-1 trials, and then this must be followed by a failure on the ith trial. The first is an ordinary binomial probability, and the latter is merely p'. Thus

$$p(i) = C(i-1,k-1)p'^{k-1}(1-p')^{i-k}.p'$$

or

$$p(i) = C(i-1,k-1)p'^{k}(1-p')^{i-k} \quad \text{negative binomial distribution.} \qquad (15.38)$$

Just as in the geometric distribution, it is the number of trials i that is the random variable. That is, (15.38) gives the probability that the kth failure occurred at the ith trial, where i = k, k+1,....

The negative binomial is in reality a logical extension of the geometric distribution. For suppose k = 2. Then let $i = i_1 + i_2$, where i_1 is the number of trials till the first failure occurs, and i_2 the number of additional trials till the second failure occurs. Or in general for k, $i = i_1 + i_2 + \ldots + i_k$.

Then, using (14.5) through (14.7) with the a_i's all +1, we find

$$\mu = k/p' \qquad \text{negative binomial for k} \qquad (15.39)$$

$$\sigma = (1/p')\sqrt{k(1-p')} \qquad (15.40)$$

Moreover, using (15.23), (15.24) which are true of sums of x's as well as $\bar{x}$'s we find from (15.36) and (15.37)

$$\alpha_3 = (2-p')/\sqrt{k(1-p')} \qquad (15.41)$$

$$\alpha_4 = 3 + \frac{6(1-p')+p'^2}{k(1-p')}. \qquad (15.42)$$

In general for k > 1, the probabilities p(i) increase for a while and then decrease.*

---

* The maximum p(i) has i satisfying

$$\frac{k-1}{p'} \le i \le \frac{k-1}{p'} + 1, \quad \text{or } \mu - \frac{1}{p'} \le i \le \mu - \frac{1}{p'} + 1.$$

There may be two modes, within one, of each other.

Individual probabilities and cumulative probabilities for an interval for the negative binomial may be found from appropriate tables of the binomial. This rests upon the equivalence of two events:

1. The *k*th failure occurs at the nth trial or earlier.

2. There are *k* or more failures in n trials.

This is because in 1, there must be at least k failures in the n trials, if we complete the n trials. But if so, then surely in the n trials there are k, k+1,...,n failures. But this is 2.

Now the probability of event 1 is the sum of negative binomial terms from i = k to i = n, whereas the probability of event 2 is the sum of binomial terms from d = k to n. Thus

$$\sum_{i=k}^{n} p(i|k,p')_{\text{neg. binom.}} = \sum_{d=k}^{n} p(d|n,p')_{\text{binomial.}} \tag{15.43}$$

15.3.3. The Discrete Uniform Distribution. This is a simple distribution, where we have consecutive whole numbers for the random variable i, say, 0, 1, 2,...,M-1, the probability of each being the same. Since there are M of them and the sum of the probabilities is 1 we have

$$p(i) = 1/M \qquad i = 0,1,\ldots,M-1 \qquad \text{uniform distribution} \tag{15.44}$$
$$= 0 \qquad \text{otherwise.}$$

One can use formulas for sums of powers of consecutive integers, along with (15.2), (15.3), (15.7) — (15.9) to prove

$$\mu = (M-1)/2 \quad \text{uniform distribution } 0, 1,\ldots,M-1 \tag{15.45}$$

$$\sigma = \sqrt{(M^2-1)/12} \tag{15.46}$$

$$\alpha_3 = 0 \tag{15.47}$$

$$\alpha_4 = .6(3M^2-7)/(M^2-1). \tag{15.48}$$

This is the discrete analog of the continuous rectangular distribution of (14.19) — (14.21).

Probably the most important application of the uniform distribution is in random digits, especially where M = 10, that is, 0, 1, 2,...,9.

These are to be drawn with constant probability .1, and independently. For them $\mu = 4.5$, $\sigma = \sqrt{33/4} = 2.87$, and $\alpha_4 = 1.78$. Two-digit random numbers 00 to 99 have M = 100 and thus $\mu = 49.5$, $\sigma = 28.87$ and $\alpha_4 = 1.80$.

We finally note that whenever we have discrete values of a random variable, j, of the form a, a+d, a+2d,...,a + nd, we can always transform them, to use (15.44) by letting n = M-1 and (j-a)/d = i from which we may find moments of j from (15.45) — (15.48), the last two being unaffected by the linear transformation.

15.3.4. The Exponential Distribution. This is a continuous distribution of wide use in testing for length of life to failure, with the time to failure being a continuous variable. In fact in life testing, the exponential distribution is about as standard a model as the normal distribution is in other fields of application. The idea is that there may well be early failures, but that the probability of failure occurring in uniform lengths of time decreases as time goes on.

The density function is defined as follows:

$$f(x) = \begin{cases} \mu^{-1}e^{-x/\mu} & 0 \leq x \\ 0 & x > 0 \end{cases} \tag{15.49}$$

We may easily show that μ in (15.49) is in fact the population mean E(x), as indeed it should be if we are to use the letter μ for the constant. Thus we have by (15.2)*

$$\mu_1' = \int_0^\infty x\mu^{-1}e^{-x/\mu}\,dx = \mu \quad \text{exponential distribution.} \tag{15.50}$$

---

*Using the transformation $x/\mu = w$ on the integral in (15.50) gives

$$\mu\int_0^\infty we^{-w}\,dw$$

which by the definition of the gamma function (5.28) becomes

$$\mu\Gamma(2) = \mu\cdot 1\cdot\Gamma(1) = \mu$$

Using properties of the gamma function (5.28) we may find

$$\sigma = \mu, \ \alpha_3 = 2, \ \alpha_4 = 9 \ \text{exponential distribution.} \tag{15.51}$$

Here we see that the mean and standard deviation have both the same value μ. The curve-shape characteristics indicate quite extreme asymmetry and long tailing out toward the x axis. Figure 15.1 shows the curve shape for the exponential distribution for two values of μ, one and two. All exponential distributions are merely changes of scale from each other, and therefore α_3 and α_4 are constant.

Since μ governs both mean and variance, there is little freedom of action. It is possible to use

$$f(x) = \begin{cases} \mu^{-1}e^{-(x-\theta)/\mu} & \theta \le x \\ 0 & x < \theta \end{cases} \tag{15.52}$$

So that the mean is $\theta + \mu$ and variance μ, which gives more freedom. But, at least for length of life, it makes little sense, because it

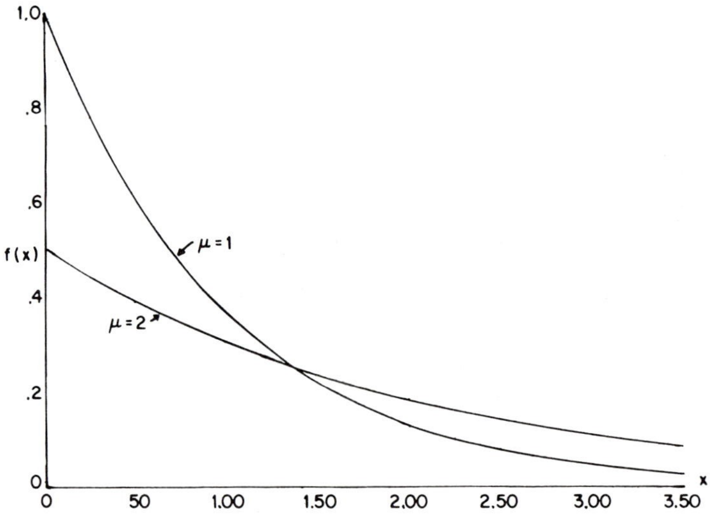

FIG. 15.1. Exponential distributions with means $\mu = 1$ and $\mu = 2$.

implies no failures till $x = \theta$, at which point, suddenly maximum incidence of failures occurs.

This distribution is the continuous analogy of the discrete geometric distribution (15.32).

15.3.5. The Gamma Distribution. This distribution is also called the Type III distribution in the system of distributions developed by Karl Pearson. A definition is

$$f(x;p) = \begin{cases} \dfrac{x^{p-1}e^{-x}}{\Gamma(p)} & 0 < x, \quad p > 0 \\ 0 & x \leq 0, \end{cases} \quad \text{gamma distribution} \tag{15.53}$$

where we have defined the gamma function in (5.28). One can then obtain the following moments by use of definitions (15.2), (15.3) and formulas (15.7) through (15.11), by use of the properties of the gamma function (5.29) - (5.32).

$$\mu_x = p \qquad \text{gamma distribution} \tag{15.54}$$

$$\sigma_x = \sqrt{p} \tag{15.55}$$

$$\alpha_3 = 2/\sqrt{p} \tag{15.56}$$

$$\alpha_4 = 3 + (6/p) = 3 + 1.5\alpha_3^2. \tag{15.57}$$

As defined in (15.53) we have but a one-parameter family of distributions, which seems quite restricted for applications, since mean, standard deviation and curve shape are all tied together being functions of p. But we can generalize to make a three-parameter family if we first standardize x by

$$z = (x-p)/\sqrt{p} \tag{15.58}$$

whence $\mu_z = 0$, $\sigma_z = 1$.

Then if our desired mean is μ and standard deviation σ for a variable, say, y, we can let

$$z = (y-\mu)/\sigma. \tag{15.59}$$

We could equate the two expressions for z, and solve for x in terms of y. But in practice in fitting data by the gamma distribution it is easiest to use tables by L. R. Salvosa [2], [3], if they are available.

First one finds $\bar{y}$, s_y and a_3, as usual, the last by (15.19). Then the first two are called μ and σ if n is, say, at least 100. Next for desired limits, such as class boundaries of y, we convert y to z by (15.59). We are then in a position to find the area or probability below y, enter Salvosa's tables with the corresponding z values (called t there) and a_3 (called "skewness" there). Interpolation is usually necessary on z and a_3, but the intervals are small enough to permit ordinary interpolation.

It is not necessary that p in (15.53) be integral, as it can be any positive real number. But if p is a positive integer then there is an intimate relation with the exponential distribution. For, if we let μ be 1 in (15.49), and then let $x = x_1 + x_2 + \cdots + x_p$ with the x_i's independently distributed by (15.49), their sum x will follow (15.53). Hence the gamma distribution bears the same relation to the exponential distribution (15.49) that the negative binomial distribution (15.38) does to the geometric distribution (15.32). Thus we may think of the gamma and negative binomial as being analogous to each other.

Figure 15.2 shows several gamma distributions for comparison. They have p = 1, 2, 4 which by (15.54) are also the means. We also have respectively by (15.55) and (15.56), σ = 1, 1.41, 2 and α_3 = 2, 1.41, 1.00. Another comparison could have been drawn with the horizontal scale z in (15.59) while arranging the vertical scales to make the total area unity. This would compare the curve shapes on a uniform basis.

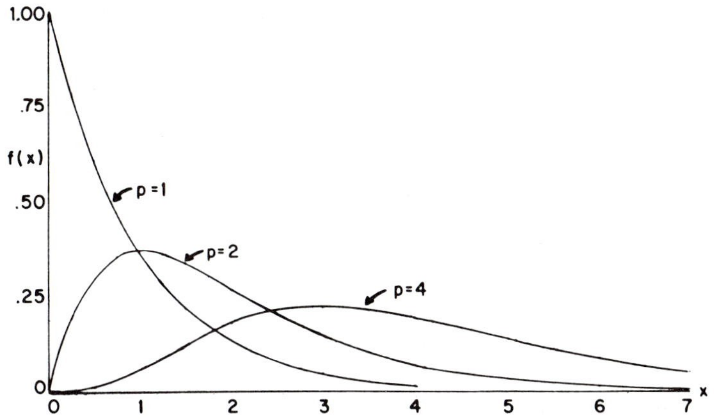

FIG. 15.2. Three gamma distributions with parameters p = 1, 2, 4, showing varying degrees of skewness. For p=1, this is an exponential distribution.

As p increases the curve approaches normality. It is quite normal at $p = 100$ with $\alpha_3 = .2$, $\alpha_4 = 3.06$. This of course follows from the Central Limit theorem since for example, x for p = 100, is $x_1 + x_2 + \dots + x_{100}$ where x_i follows an exponential distribution.

Regarding the normal curve as a "first approximation" to continuous data, we can think of the gamma distribution as a "second approximation" to continuous data.

We may add that the chi-square distribution is a gamma distribution.

15.3.6*. A General System of Distributions. A flexible system of frequency distributions may be defined as follows. If a distribution has a density function f(y), then we can let the <u>cumulative probability</u> F(y) be defined by the probability that the random variable Y be less than or equal to y. That is,

$$F(y) = P(Y \leq y) = \int_{-\infty}^{y} f(v)\,dv. \qquad (15.60)$$

Then we have the probability of Y lying in an interval a to b by

$$P(a < Y \leq b) = \int_{a}^{b} f(v)\,dv = \int_{-\infty}^{b} f(v)\,dv - \int_{-\infty}^{a} f(v)\,dv = F(b) - F(a). \qquad (15.61)$$

We now define the cumulative probability for the system by: [4]

$$F(y) = \begin{cases} 1-(1+y^c)^{-k} & y > 0 \quad c,k > 0 \\ 0 & y < 0. \end{cases} \tag{15.62}$$

At first this looks like a two-parameter family, but in reality it can be used to fit four moments, typically the observed $\bar{x}, s_x$, a_3 and a_4. In fitting observed data a_3 and a_4 are used to determine c and k, which are in essence curve-shape parameters, for they yield α_3, α_4 combinations. They also yield μ_y, σ_y. Then we convert observed x values to y's by

$$\frac{x-\bar{x}}{s_x} = \frac{y-\mu_y}{\sigma_y}$$

The x values are often class boundaries, from which we want the probabilities between. Are you still with me? Probably not. So we need an example.

Consider Table 15.1, which is actually of discrete data but is fitted just the same as though continuous data. The y's in column (1) are treated as mid-values with corresponding frequencies in column (2). For calculational purposes we code by $v = y-12$, and cumulate the sums shown at the bottom of the table. These yield the m_j' moments for v's as shown next. Then the central moments m_j's for the v's are found next, from (15.16) - (15.18), and finally the curve shape characteristics a_3 and a_4.

The closest entry in Table XV to $a_3 = .3804$, $a_4 = 3.240$ is $\alpha_3 = .40$, $\alpha_4 = 3.30$, which has $c = 3.341$, $k = 5.369$. Also given in Table XV for this c,k are $\mu = .5638$, $\sigma = .2039$.

The class boundaries for the v scale, column (4), are next converted to x's for (15.62) by equating standardized variables for x and v as shown.

TABLE 15.1

Total Number of Letters in Alphabet in People's First and Last Names, Fitted by (15.62)

(1) No. of Letters in names,y	(2) Observed Freq.,f	Code v	(4) Boundary,v	(5) x in (15.62)	(6) $1+x^c=$ $1+x^{3.341}$	(7) $1-F(x)=$ $(1+x^c)^{-5.369}$	(8) Class Probab.	(9) Calc. Freq., F
8	3	-4					.01935	2.1
			-3.5	.18626	1.00364	.98065		
9	5	-3					.06105	6.6
			-2.5	.28866	1.01575	.91960		
10	11	-2					.12364	13.2
			-1.5	.39106	1.04342	.79596		
11	23	-1					.17995	19.3
			- .5	.49346	1.09444	.61601		
12	22	0					.19973	21.4
			+ .5	.59586	1.17731	.41628		
13	17	+1					.17303	18.5
			+1.5	.69826	1.30121	.24325		
14	13	+2					.11955	12.8
			+2.5	.80066	1.47581	.12370		
15	7	+3					.06780	7.3
			+3.5	.90306	1.71132	.05590		
16	4	+4					.03286	3.5
			+4.5	1.00546	2.01840	.02304		
17	0	+5					.01411	1.5
			+5.5	1.10786	2.40813	.00893		
18	2	+6					.00893	1.0
	107							107.1

$\Sigma vf = +20 \quad \Sigma v^2 f = 428 \quad \Sigma v^3 f = 560 \quad \Sigma v^4 f = 5780$

For v's $m_1' = 20/107 = .186{,}916 \quad m_2' = 4.000{,}000 \quad m_3' = 5.233{,}645 \quad m_4' = 54.018{,}692$

$m_2 = m_2' - (m_1')^2 = 3.965{,}062 \quad m_3 = m_3' - 3m_2'm_1' + 2(m_1')^3 = 3.003{,}713$

$m_4 = m_4' - 4m_3'm_1' + 6m_2'(m_1')^2 - 3(m_1')^4 = 50.940{,}533.$

$a_3 = m_3/(m_2)^{3/2} = .3804 \quad a_4 = m_4/(m_2)^2 = 3.240$

Take $\alpha_3=.4 \quad \alpha_4=3.3 \quad c=3.341 \quad k=5.369 \quad \mu_x=.5638 \quad \sigma_x=.2039$ Table XV

$\bar{v} = .18692, \sqrt{m_2} = 1.9912.$ Use $\frac{v-.18692}{1.9912} = \frac{x-.5638}{.2039} \quad x = .1024v + .54466$

Columns (6) and (7) are steps on the way to finding probabilities for classes by use of differences of $1 - F(x_i)$. Thus $P(x_{i-1} < x < x_i) = [1-F(x_{i-1})] - [1-F(x_i)]$, $1-F(x_i)$ being $(1+x_i^c)^{-k}$ from (15.62). These differences are given in column (8). Multiplying column (8) by n = 107 gives the calculated class frequencies F in column (9). Note the good agreement between columns (2) and (9). Also note that the first entry in column (8) is for x = 0 to x = .18626 and the last entry is for x = 1.10786 to ∞.

Much more complete tables are given in [4], with α_3 = .0(.05)1.0(.1)2.0, and a good variety of α_4 values by intervals of .1 or .2 to 9.0 maximum. But since the a_3, a_4 estimates of α_3, α_4 are subject to considerable error unless n is large, use of the nearest entry in Table XV is at least reasonable. If a_3 is negative one can pick the entry with α_3 nearest to $-a_3$, then reverse the scale direction by

$$\frac{v-m_1'}{\sqrt{m_2}} = -\frac{x-\mu_x}{\sigma_x} \qquad (15.63)$$

This causes our negative a_3 to be positive in relation to the x variable of (15.62).

15.4. A Goodness of Fit Test for Frequency Data. It is sometimes desirable to compare a set of observed frequencies with those from some theoretical distribution, for example normal or Poisson. Could the observed data reasonably have come from the proposed theoretical distribution? In general we have numerical class intervals, within each of which there is an observed frequency, f_i, out of a sample of n observations. The theoretical distribution being compared provides theoretical probabilities, p_i, for each such class, and thus expected frequencies, $np_i = F_i$. Now we must expect that the observed f_i's will differ from their corresponding F_i's even when we draw all the sample results at random from the proposed theoretical distribution. We therefore need a measure of the departures, taken as a whole. This is supplied by the statistic named "chi-square." Now let us be more specific:

1. The hypothesis being tested is that the population is of some given type, for example, normal, binomial or Poisson.
2. The parameters (for example μ, σ, p' or c') may be specified, in which case the parameters are part of the hypothesis. Or they may have to be estimated from the data at hand, in which case the hypothesis is only on the type of population.
3. A series of n observations of measurements x, or counts d or c, is tabulated into numerical classes giving a frequency table. Let f_i be the observed frequency in the ith class.
4. Class probabilities p_i's are calculated from the more or less completely specified population. Then the expected class frequencies are $F_i = np_i$. Comparison is now to be made between the f_i and F_i frequencies to see whether or not the discrepancies are readily explainable by chance variation. This comparison is made on the set of discrepancies taken as a whole.
5. The test criterion is called chi-square (the same as that used for variances from a normal curve population). It is

$$\chi_\nu^2 = \sum_{i=1}^{k} \frac{(f_i - F_i)^2}{F_i} \tag{15.64}$$

   where the subscript, nu, is the "degrees of freedom," and there are k class frequencies to compare.
6. If any "tail" frequency F_i is less than 1.0, combine it with the next adjacent F_i so as to make the lumped F_i, 1.0 at least.

This could occur at either or at each tail.* Then k is the number of classes after any such lumping of frequencies is done.

7. Further, we let p stand for the number of independent parameters estimated from the data, such as, μ by $\bar{x}$, σ by s, c' by $\bar{c}$ or p' by d/n. Then:

   Degrees of freedom = ν = k-1-p. (15.65)

8. The larger the observed chi-square (15.64) is, the relatively poorer the fit. Chi-squares values of ν are about average, by chance variation. But if our observed chi-square is above the 1-α point the fit is judged unsatisfactory, where α is the risk of erroneously rejecting an hypothesis when true. That is:

$$\begin{aligned} &\text{Observed } \chi^2 > \chi^2_{\nu,1-\alpha \text{ below}} \quad \text{fit unsatisfactory} \\ &\text{Observed } \chi^2 \leq \chi^2_{\nu,1-\alpha \text{ below}} \quad \text{fit satisfactory.} \end{aligned} \qquad (15.66)$$

   In the former case we may try some other theoretical distribution. In the latter case, there is no proof that the sample did come from the theoretical distribution, but only that it perfectly well could have come from it.

15.4.1. A Normal Curve Example. In Table 15.2 are shown observed frequencies for an electrical contact, showing a "normal" looking distribution. We may then wish to test whether the data could readily have arisen from some normal population. We therefore find frequencies F_i to compare with the observed f_i. The first step is to find the mean $\bar{x}$ and standard deviation s.

---

*This rule is given in [5]. It is a more liberal rule than those most often given in textbooks.

TABLE 15.2

Observed Distribution of Dimensions of 1000 Electrical Contacts. Class Frequencies Calculated Under Hypothesis of Normality, and Goodness of Fit Tested by Chi-square.

(1)	(2)	(3)	(4)	(5)	(6)	(7)	(8)
Mid-value, x in.	Observed freq., f_i	Class bounds, x	Standard var. z	Area below z in (4)	Area between	Calcul. freq. F_i	Contr. to χ^2
.412	6				.0152	15.2	5.5
		.4145	-2.166	.0152			
.417	34				.0409	40.8	1.1
		.4195	-1.588	.0561			
.422	132				.1001	100.1	10.2
		.4245	-1.010	.1562			
.427	179				.1767	176.7	.0
		.4295	- .432	.3329			
.432	218				.2251	225.1	.2
		.4345	+.146	.5580			
.437	183				.2074	207.4	2.9
		.4395	+.724	.7654			
.442	146				.1381	138.1	.5
		.4445	+1.302	.9035			
.447	69				.0664	66.4	.1
		.4495	+1.880	.9699			
.452	30				.0232	23.2	2.0
		.4545	+2.458	.9931			
.457	3				.0069	6.9	2.2
	1000					χ^2	=24.7

Proceeding by usual methods for coding in a frequency distribution we find

$$\bar{x} = .43324 \text{ in.}, \qquad s = .008651 \text{ in.}$$

Next we convert the class boundaries in column (3) to the standard variable z:

$$z = (x-\bar{x})/s = (x-.43324)/.008651 = 115.594x - 50.0798$$

(retaining more places than strictly justified to avoid calculational errors on the differences). This gives column (4). The area or cumulative probabilities below the class boundaries are then found from Table I.

interpolating, to yield column (5). The area or probabilities between class boundaries are found by subtraction of entries in (5) giving (6).

Note that the first class includes all x's from $-\infty$ (pure fiction in practice) to .4145 in., while the upper class is .4545 in. to ∞(1.0000 − .9931). The class probabilities are then multiplied by n to give the expected frequencies F_i of column (7).

To test the goodness of fit we form column (8). For example, in the first class the contribution to χ^2 is

$$(6-15.2)^2/15.2 = 5.5$$

The total is $\chi^2 = 24.7$. Is this indicative of a good or poor fit? It depends upon α and ν. Suppose we take $\alpha = .01$, and $\nu = 10 - 1 - 2 = 7$, the "two" being for using $\bar{x}$ for μ and s for σ. Then looking up the required entry in Table II we find $\chi^2_{7,.99\text{ below}} = 18.475$. Since the observed chi-square value of 24.7 is above this we regard the fit as unsatisfactory, and conclude that the population is not normal. Not once in 100 times would we get as poor a fit by chance, from a normal population. The large contributions to χ^2 show the trouble spots.

15.4.2. A Poisson Example. For an example of the Poisson distribution consider Table 15.3, wherein counts c_i of a defect called "plug holes" were tabulated for 54 large aircraft assemblies, yielding frequencies f_i as in column (2). Do these frequencies reasonably well follow a Poisson distribution? In order to check this, we must estimate c' for (2.35), that is, just the one parameter which fully determines the Poisson distribution. The usual estimate is

$$\bar{c} = \Sigma f_i c_i / n = 308/54 = 5.70.$$

TABLE 15.3

Defects Called "Plug Holes" in Large Aircraft Assemblies. Fitting of F_i Done by Poisson Distribution.

(1) Plug Holes, c_i	(2) Frequency, f_i	(3) f_ic_i	(4) $p_i=P(c\|5.70)$	(5) $F_i=p_in$	(6) Contr. to χ^2
12	2 } 3	24	.0141	.8 } 1.7	1.0
11	1	11	.0172	.9	
10	1	10	.0334	1.8	.4
9	5	45	.0586	3.2	1.0
8	4	32	.0925	5.0	.2
7	7	49	.1298	7.0	.0
6	8	48	.1594	8.6	.0
5	9	45	.1678	9.1	.0
4	5	20	.1472	7.9	1.1
3	5	15	.1033	5.6	.1
2	3	6	.0543	2.9	.0
1	3 } 4	3	.0191	1.0 } 1.2	2.7
0	1	0	.0033	.2	
	54	308	1.0000		$\chi_9^2 = 6.9$

$\bar{c} = 308/54 = 5.70$

Then we can either use (2.35) for $P(c|c'=5.70)$, or use a table [6] or [7]. Then multiplying column (4) entries by n = 54 gives the expected frequencies of column (5):

$$F_i = p_i n.$$

To study the discrepancies between the f_i and F_i, we work out the contributions to χ^2:

$$(f_i - F_i)^2/F_i$$

as given in column (6). The sum is $\chi^2 = 6.9$. Note that we have combined the frequencies f_i and F_i for c = 11 and 12, and also for c = 0 and 1, in line with instruction 6 of Section 15.4 so as to have all F_i's at least 1.0.

To interpret this observed χ^2 we need the α risk and degrees of freedom. Suppose we set $\alpha = .05$. Then for ν we use (15.65)

$$\nu = k - 1 - p = 11 - 1 - 1 = 9$$

p being one for estimating c' from $\bar{c}$, and k being 11 classes after lumping. Then by Table II,

$$\chi^2_{9,.95 \text{ below}} = 16.919$$

so that the discrepancies between the f_i and F_i are readily explainable by chance. The true distribution of defects could perfectly well be a Poisson distribution.

15.5. Summary. In this chapter we have presented some methods of describing the curve-shape characteristics of frequency distributions, theoretical and observed, by use of moments. The curve-shape characteristics were given for distributions previously studied. Then we presented discrete distributions: geometric, negative binomial and uniform, and continuous distributions: exponential and gamma and a general system. Interrelations between them were pointed out. Finally a test of goodness of fit by use of chi-square was described and illustrated. Many other distributions could have been included such as the beta and Weibull distributions.

PROBLEMS

15.1. Choose a random starting spot in the random digits of Table IV. Then tabulate this digit and continue down the column, then the next column to the right and so on for 200 random digits, to find observed frequencies f_0, $f_1,\ldots,f_9$. The expected frequencies F_i are each 20. Test goodness of agreement by χ^2 with $\alpha = .05$ and interpret your result.

15.2. Random digits one to six may be generated by rolling a single perfectly balanced die. For this uniform distribution find μ, σ, α_4. What do these parameters become for the total on a roll of two balanced dice? For three dice?

15.3. To illustrate the Central Limit theorem, one may delete the face cards from two 52-card decks then draw five cards randomly from the 80 cards without replacement (with replacement is better, but five does not unduly deplete the 80). These five form a sample. Regard aces as x=1, to 10's as x = 10. Find $\mu_{\bar{x}}$, $\sigma_{\bar{x}}$, α_4, for the distribution. Would it be quite normal appearing?

15.4. Sampling as in 15.3, each class member draws 10 samples, and records his 10 $\bar{x}$'s. These are pooled, tabulating into appropriate classes, and fitted by a normal curve, using μ,σ from 15.3 and tested for goodness of fit by χ^2.

15.5. Derive (15.7) and (15.8) for the continuous case.

15.6. Derive (15.7) and (15.8) for the discrete case.

15.7. Show that (15.19) is unchanged under any linear transformation for which the coefficient, a, of the variable is positive, in y = ax + b.

15.8. Using analogous definitions to (15.10), what are α_0, α_1, α_2?

15.9. Data on the number of "Mislocations" on each of 54 large aircraft assemblies follow:

No. defects, c	0	1	2	3	4	5	6	7	8	9
Frequency, f	12	9	11	8	7	0	4	0	1	2

Fit by a Poisson distribution, using Table VII, or [6] or [7], and test goodness of fit. (The discrepancies are of a form indicating a "contagious" tendency, that is, non-independence of defects.) Use $\alpha = .01$.

15.10. Test for normality the following data on ladle carbon in open-hearth heats of steel using $\alpha = .05$.

%C	.69	.70	.71	.72	.73	.74	.75	.76	.77	.78	.79	.80
f_i	2	21	32	63	89	106	97	51	33	4	1	1

15.11. Test for normality the following data on density of glass at 20°C [8], using $\alpha = .05$.

Density g/cc	2.5012	2.5022	2.5032	2.5042	2.5052	2.5062	2.5072
Freq.f_i	2	6	25	33	19	10	4

15.12. Test the goodness of fit for Table 15.1, that is, columns (2) vs. (9), using $\alpha = .05$. (Note that here p=4.)

15.13. From a box having two colors of beads, with a small proportion of one, say, red colored, scoop up a sample of n=50. Count the red ones, return all beads, mix and scoop another 50. Continue for 100 samples. If the number N in the box is large relative to n=50, the counts should be close to being binomially distributed. Use the known p' and a binomial distribution for finding the theoretical probabilities F_i, and test for goodness of fit.

15.14. Prove (15.51), that $\sigma = \mu$ by using properties of the gamma function and (15.7).

15.15. Consider the following data on the "eccentricity" distance between a conical point and the pitch diameter, of a needle value in .0001 in.

Fit a gamma distribution to the data and test the goodness of fit, with $\alpha = .05$.

Eccentricity	4.5	14.5	24.5	34.5	44.5	54.5	64.5	74.5	84.5	94.5
Frequency, f_i	75	195	180	164	119	77	52	19	10	4

104.5	114.5	124.5	134.5
2	1	1	1

15.16. Fit the data of Problem 15.15 by (15.62) and test the goodness of fit, with $\alpha = .05$. (From [4], if $\alpha_3 = .95$, $\alpha_4 = 4.4$, then c = 1.993,536, k = 8.293,100, $\mu = .321,572$, $\sigma = .181,784$.)

References

1. M. G. Kendall and A. Stuart, "The Advanced Theory of Statistics." Charles Griffin, London, 1969.

2. L. R. Salvosa, Tables of Pearson's Type III function. Ann. Math. Statist. 1, 191-198 (1930).

3. H. C. Carver, "Statistical Tables." Edwards, Ann Arbor, Michigan, 1970.

4. I. W. Burr, Parameters for a general system of distributions to match a grid of α_3 and α_4. Commun. in Stat. 2, 1-21 (1973).

5. W. G. Cochran, Some methods for strengthening the common χ^2 tests. Biometrics 10, 417-451 (1954).

6. T. Kitagawa, "Tables of Poisson Distribution." Baifukan, Tokyo, 1952.

7. E. C. Molina, "Poisson's Exponential Binomial Limit." Van Nostrand-Reinhold, Princeton, New Jersey, 1947.

8. L. G. Ghering, Refined method of control of cordiness and workability of glass during production. J. Amer. Ceram. Soc. 27, 373-387 (1944).

ANSWERS TO ODD-NUMBERED PROBLEMS

Numerical answers rounded to about slide-rule accuracy. Discussions could well be made more complete than given here.

4.3 (a) $\bar{x}$ control lines: 46.74, 48.38, 50.02 R control lines: 0, 2.25, 5.13

(b) Both charts show lack of control. Erratic control of process.

(c) Not justified in estimating either μ nor σ_x.

4.5 (a) $\bar{x}$ control lines: 822.6, 836.8, 851.0 R control lines: 0, 13.9, 35.7

(b) R Chart is in perfect control

(c) Can estimate σ_x by $\bar{R}/d_2$ at 8.19, because R chart in control.

(d) No $\bar{x}$ points within control limits. Not suprising, since samples can be expected to give differing sets of repeated measurements.

4.7 From $\bar{x}$, R $\bar{x}$ control lines: 505.8, 546.2, 586.6 R control lines: 0, 83.7, 167.7

From $\bar{x}$, s $\bar{x}$ control lines: 504.0, 546.2, 588.4 s control lines: 1.0, 32.8, 64.6

Can regard within-group variabilities as homogeneous.

Cannot assume all data are from one homogeneous lot. μ's vary.

4.9 Early data $\bar{x}$ control lines: -.833, .333, 1.499 R control lines: 0, 2.02, 4.27. Both charts in control, need a fundamental change in process to meet specifications. Estimate, $\hat{\sigma}_x = .868$ by $\bar{R}/d_2$. Later data $\bar{x}$ control lines: 1.100, 1.328, 1.556 R control lines: 0, .395, .835. Four $\bar{x}$'s outside, all R's inside. Estimate $\hat{\sigma}_x = .170$ by $\bar{R}/d_2$. For early data: tolerance/$\hat{\sigma}_x$ = 2.30 which is bad. Can estimate percent out by (limit - $\bar{\bar{x}}$)/$\hat{\sigma}_x$ to check percents given. For later data: tolerance/$\hat{\sigma}_x$ = 11.8, leaving room for μ to vary.

4.11 $\bar{x}$ control lines: 29.4, 30.8, 32.2 s control lines: .41, 1.44, 2.47 Variabilities in control, but Bismuth and Selenium significantly low.

4.13 From $\bar{\bar{x}}$, $\bar{R}$ $\bar{x}$ control lines: .09, 2.03, 3.97 R control lines: 0, 3.36, 7.11. From μ, σ $\bar{x}$ control lines: -.30, 2.00, 4.30 R control lines: 0, 3.99, 8.43. Control is perfect with respect to either set of lines.

4.15 $\bar{x}$ points closer to center line. R or s points running low (often below center line)

CHAPTER 5

5.1 Need to use $\Gamma(4) = 3!$, $\Gamma(3.5) = (2.5)(1.5)(.5)\sqrt{\pi}$

5.3 Continuing, since s's vary $\sigma_s^2 > 0$ and $\sigma_s^2 = \sigma^2 - (c_4\sigma)^2$.

CHAPTER 6

6.1 Concern with defectives. np chart simplest. $n\bar{p} = 1.72$ UCL = 5.57. Might use (6.20). Three consecutive points out. Probably to provide 36 good articles nearly always. P = .096.

6.3 p chart. $\bar{p} = .00687$. Use $.00687 \pm .2477/\sqrt{n}$. Groupings could be $\bar{n}$: 15,575, 7316, 2056, 1122. Five points out. Use weighted average of p's, not $\sum p/k$. Use (6.12).

6.5 Defectives, p chart. $\bar{p} = .0909$. Limits$_p$: $.0909 \pm .8623/\sqrt{\bar{n}}$ with $\bar{n}$ = 2416, 1229, 783. Three high at beginning and end, seven low between. Out of control. Use (6.12) for weighted average of p's.

6.7 Assignable causes might be associated with (a) time of shift A, B or C, (b) foreman and his crew, (c) day of week (d) first versus second sample, or (e) the week. Use np chart, lines 0, 3.75, 9.5. Two points out, for which the only thing in common is (a). High p's during the day shift. Reason: excess light and eye strain. (Could make a three-point or six-point chart by shifts.)

6.9 Control lines for c: 6.1, 19.3, 32.5. One above UCL, four very close.

6.11 Control lines for u: 2.64, 3.69, 4.74 One high.

6.13 Control lines for u: 2.00, 2.94, 3.88 Two out.

6.15 Set up formula for UCL_p with $\bar{p}$ = .150 and set to .180 and solve for n. Gives 1275.

CHAPTER 7

7.1 $\bar{x}$ = 32.0, $\bar{R}$ = 6.94, UCL_R = 22.7, $limits_x$ = 19.7, 44.3. R's in control, six x's outside. One low x for 3s limits.

7.3 $\bar{x}$ = 50.8, $\bar{R}$ = 1.06, UCL_R = 3.45, $limits_x$ = 48.9, 52.7. Variability steady, but process average drifts.

7.5 $\bar{x}$ = 23.2, $\bar{R}$ = 4.64, UCL_R = 15.2, $limits_x$ = 15.0, 31.4. Some assignable cause for high initial point. Ask the bug!

7.7 $\bar{R}$ = 3.02, UCL_R = 6.4. Two R points out. Cause: before resetting and after resetting pieces included in same sample. Can catch this. Hence revise $\bar{R}$ to 2.76. Slanting control limits at ± 1.59 from trend line. The 3.2 should probably be called out of control depending on sketch of trend.

7.9 Weighted averages $\bar{\bar{x}}$ = .245, $\bar{R}$ = 2.02, $Limits_s$ = 1.63, 2.42, using $\bar{n}$ = 117.5, badly out of control. $Limits_{\bar{x}}$ = -.316, +.805, all within.

7.11 $Limits_{\bar{\bar{x}}}$ = .22500, .22520 in. $Limits_{\bar{R}}$ = .00051, .00091 in. Using $\bar{\bar{x}}$ $\hat{\sigma}_x$, estimate about .002 of pieces will be out of specifications. Assume good control and normality of x's.

CHAPTER 9

	Suggested increment, p'	Typical value of p'	Pa	AOQ	ASN	ATI	approx. AOQL
9.1	.004	.008	.525	.0042	200	2005	.0043
9.3	.008	.024	.651	.0156	200	1526	.016
9.5	.020	.060	.651	.0391	80	401	.040
9.7	.050	.200	.583	.116	50	238	.13
9.9	.020	.040	.518	.0207	65	516	.021
9.11	.004	.008	.503	.0040	171	2068	.0042
9.13	.008	.024	.639	.0153	174	1551	.016
9.15	.020	.060	.639	.0384	70	404	.04
9.17	.050	.200	.553	.111	45	247	.13
9.19	.020	.040	.653	.0261	75	381	.02

9.21 $p'_{95} = .00657$ $p'_{50} = 0.62$ $p'_{10} = .0267$

9.23 Ac = 5, n = 232. Might use Ac = 4, n = 200, giving α a bit over .05.

9.25 Ac = 6, n = 150. Might use Ac = 5, n = 132, giving α a bit over .05.

9.27 Find p' values, one giving Pa < .10, and the other Pa > .10, and interpolate.

9.29 $Pa = q'^2(1+2p')$. Then p' = .25 Pa = .84; p' = .50, Pa = .50; p' = .75, Pa = .16.

9.31 It cuts off second sampling in the more or less "hopeless" cases.

9.33 Arbitrary rules of thumb: (1) $p' \leq .05$ and $n \geq 20$, (2) $N \geq 8n$, (3) both (1) and (2) met.

9.37 .3929

9.39 .1095

9.41 $P(d + 1 \text{ in } n) = \frac{(n-d)p'}{(d+1)q'} P(d \text{ in } n)$ $P(c+1) = \frac{c'}{(c+1)} P(c)$

CHAPTER 10

10.1 $n_1 = 85$ $c_1 = 0$ $n_2 = 220$ $c_2 = 5$ $p_t = 3.3\%$

10.3 $n = 135$ $c = 2$ $p_t = 3.9\%$

10.5 $n_1 = 110$ $c_1 = 1$ $n_2 = 175$ $c_2 = 6$ AOQL = 1.3%

10.7 Try neighborhood of $p' = .02$. AOQL about 1.01%

10.9 Pa = .108 by Poisson

10.11 p_t decreases as $\bar{p}$ increases. Latter requires increasing samples which give more discriminating OC curves which drop to the p_t point for lower p' values. (The AOQL occurs at p''s for which Pa is about .55.) Also p_t decreases as N increases for same reason.

10.13 We might well still be using sampling inspection when p' is at or just below the AOQL, but would have discontinued sampling if Pa were to drop to anywhere near .10.

10.15 n = 80 Ac = 1 Re = 2

10.17 n = 32 Ac = 0 Re = 2

10.19 $n_i = 20$

Ac	#	#	0	0	1	1	2
Re	2	2	2	3	3	3	3

10.21 n = 20 Ac = 0 Re = 1

10.23 n = 125 Ac = 0 Re = 1

10.25 $n_1 = 20$ $Ac_1 = 11$ $Re_1 = 16$ $\sum n = 40$ $Ac_2 = 26$ $Re_2 = 27$

10.27 $n_1 = 3$ $Ac_1 = 0$ $Re_1 = 3$ $\sum n = 6$ $Ac_2 = 3$ $Re_2 = 4$

10.29 Use large enough scale on p' to enable reduced to show a curve.

10.31 AOQL = .0067. So that when on tightened inspection we can be sure that the average quality is at the acceptable level.

10.33 Because of the bolstering conditions which make receipt of bad quality lots quite unlikely.

10.35 Yes. Useful in analyzing producer's process and in dealing with him.

10.37 Up to 10.0 defects or defectives per 100 units, the two are very nearly equal. Above 10.0 they become less so.

10.39 Single: .525, .171. Double: .503, .172.

10.41 $q' = P(\text{no defect on unit} | c' = .1) = .905$ $p' = .095$ $np' = 100(.095) = 9.5$ vs. 10 $q' = P(\text{no defect on unit} | c' = .4) = .670$ $p' = .330$ $np' = 33$ vs. 40

CHAPTER 11

11.3 n = 11. Preserving α, K = 100,016 psi. Accept if $\bar{x} \geq K$. Sketch, using points at, say, 98,000, 100,016, and 102,000 psi.

11.5 n = 24. Preserving α, k = .00028 in. Accept if $\bar{x}$ is between $12.7435 \pm k$ in. $\beta = .098$.

11.7 n = 7. Accept if $1.1761 \leq \bar{x} \leq 1.1764$ in. Use one-way tests, because $U-L = 12.5\sigma_x$. Two-way tests unnecessarily tight.

11.9 (a) n = 31. Accept if $\bar{x} - 2.052s \geq 427$g

(b) n = 10. Accept if $\bar{x} \geq 444.88$g

(c) n = 107, Ac = 2

11.11 Test on homogeneous material, n = 23. Accept if $s^2 \leq 3.46(\text{ppm})^2$

11.13 n = 22. Accept if $s^2 \leq .0564(^{\circ}C)^2$. Assume homogeneity, normality.

11.15 T = .0050 in. $\sigma_1 = T/10$, $\sigma_2 = T/6$, n = 23. Accept if $s^2 \leq 3.85(10^{-7})$sq.in. From the same lot close together in time of production.

CHAPTER 12

12.1 $h_1 = 1.239$ $h_2 = 1.591$ s = .0139 at p' = s, Pa = .562

p'	0	.005	.0139	.030	1
ASN	89	122	143	82	2

n	2-29	30-88	89-100	101-159	160-172
Acc.	*	*	0	0	1
Rej.	2	3	3	4	4

12.3 $h_1 = .958$ $h_2 = 1.231$ s = .0197 at p' = s, Pa = .562

p'	0	.005	.0197	.050	1
ASN	49	58	61	33	2

12.5 Most convenient is to work from 100,000 psi as zero, in, say, 1000 psi units. Then $h_1 = 11.78$ $h_2 = 11.78$ s = 0, $\mu_1 = -2$, $\mu_2 = +2$ Acc. no. = +11.78, rej. no. = -11.78 for all n's. Pa(at 0) = .5.

μ	-2	0	+2
ASN	5.3	8.7	5.3

12.7 Take $\mu_1 = 441.31$, $\mu_2 = 449.40$, $\sigma = 8.7$. $h_1 = 27.04$, $h_2 = 21.06$, s = 445.355

$Pa(\mu_1) = .10$ Pa(s) = .562 $Pa(\mu_2) = .95$

$ASN(\mu_1) = 5.50$ ASN(s) = 7.52 $ASN(\mu_2) = 4.61$

12.9 For lower specification μ_1 = 1.17608", μ_2 = 1.17612". Code in .0001" units from 1.17610". Then s = 0, $h_1 = h_2 = .879$ from $\sigma = .4$

$Pa(\mu_1) = .1$ Pa(s) = .5 $Pa(\mu_2) = .9$

$ASN(\mu_1) = 3.52$ ASN(s) = 4.83 $ASN(\mu_2) = 3.52$

12.11 For coded variable z, $\mu_0 = 0$, $\mu_0 \pm d = \pm 5$, $\sigma = 8.34$. $h_1 = 20.93$, $h_2 = 40.21$ s = 2.5. Acc. intercept = 12.2. $Pa(\mu_0) = .90$ $Pa(\mu_0 \pm d) = .10$

12.13 $h_1 = 20.70 = h_2$ s = 3.592 Pa(1.5) = .9 $Pa(\sqrt{s}) = .5$ Pa(2.5) = .1
ASN(1.5) = 14.9 $ASN(\sqrt{s}) = 17.6$ ASN(2.5) = 8.0

12.15 $h_1 = 3.16 = h_2$, s = .0584 $\sigma_1 = .2$ $\sigma_2 = .3$ $Pa(\sigma_1) = .90$ $Pa(\sqrt{s}) = .5$
$Pa(\sigma_2) = .10$ For μ known: ASN(0) = 6 $ASN(\sigma_1) = 13.8$ $ASN(\sqrt{s}) = 14.7$
$ASN(\sigma_2) = 8.0$

12.17 Coding in .0001": $\sigma_1 = 5$, $\sigma_2 = 8.33$ $h_1 = 230.0 = h_2$ s = 39.91
$Pa(\sigma_1) = .95$ $Pa(\sqrt{s}) = .50$ $Pa(\sigma_2) = .05$
μ known: ASN(0) = 6 $ASN(\sigma_1) = 13.9$ $ASN(\sqrt{s}) = 16.6$ $ASN(\sigma_2) = 7.0$
For μ unknown add 1 to above ASN's.

12.19 $\sigma_1 = T/10 = 1.7$ $\sigma_2 = T/6 = 2.83$

No. of R's of 8	1	2	3	
Acc. for $\sum R$	3.2	9.3	15.6	etc.
Rej. for $\sum R$	9.2	15.3	21.4	

12.21 $h_1 = 4.41$ $h_2 = 5.66$ s = 7.83 Pa(0) = 1 Pa(6) = .95 Pa(s) = .562
Pa(10) = .10 ASN(0) = 1 ASN(6) = 2.13 ASN(s) = 3.18 ASN(10) = 2.14

12.23 The sequential plan has a lower ASN curve than the single plan having comparable OC curves.

12.25 If $\alpha + \beta > 1$ then $Pa(q_1) = 1 - \alpha < \beta = Pa(q_2)$, that is, the Pa for acceptable quality would be less than for rejectable quality.

CHAPTER 13

13.1 i = 37, $p_t = 2.2\%$; f = 1%, $p_t = 20.6\%$ by (13.1); i = 92, f = 3%;
AOQL = 1%, $p_t = .8\%$

13.3 $i_1 = 72$, $i_2 = 96$

13.5 Increasing either i or f gives greater protection, that is, lower AOQL. So for fixed AOQL, if one increases, the other must decrease.

13.7 Use of same f permits use of the same sample of pieces for inspection, after qualifying on both classes of defects. Use of same i does not help at all.

13.9 93%, 87% compared to 81%. 48%, 41% compared to 40.5%.

13.11 Gives a higher Pa for lots under relatively good p' process levels.

13.13 For procedure 1: i = 22, i = 29

CHAPTER 14

14.1 (a) $\mu = .1987$, $\sigma = .00186$ in. (b) Assuming normality and independence, 2.28% will be outside. (c) Very little. (d) From control charts in control.

14.3 $\mu \doteq .00133$, $\sigma \doteq .00055$ cm. 2.5% below.

14.5 $\mu = \mu_1 - \mu_2 - \mu_3$, $\sigma = \sqrt{\sigma_1^2 + \sigma_2^2 + \sigma_3^2}$, distribution approximately normal.

14.7 If purely additive approach, assembly meeting $\pm$ .160 cm. (a) $\pm$ .0566 cm. (b) $\pm$ .0980 cm.

14.9 (a) Specifications $\pm$ .00039, $\pm$ .00031 in. or in practice probably $\pm$ .0004, $\pm$ .0003 in. (b) Need better process, or ask relaxation of tolerance or sort 100% (c) Use different allocation of tolerances.

14.11 (a) $\mu = .0555$, $\sigma = .00232$ sq. in. (b) Could well be related, for example inversely.

14.13 $\mu = 200$, $\sigma = 28.9$; $\mu = .50$, $\sigma = .0722$

14.15 $\mu \doteq 440$, $\sigma_y \doteq 2.54$. Expect approximate normality.

14.17 $\mu_y \doteq (1/\mu_1 + 1/\mu_2)^{-1}$, $\sigma_y \doteq \mu_y^4[\sigma_1^2/\mu_1^4 + \sigma_2^2/\mu_2^4]$ Cannot tell much about shape of distribution.

14.19 $\mu \doteq 3.134$, $\sigma \doteq .00967$

14.21 $\mu \doteq 1.025$, $\sigma \doteq .0115$

14.23 $T_1 = .0175$, $T_2 = .0153$, $T_3 = .0262$ in.
Acc. Plan A: Acc. if $\bar{x}_{30}$ between 2.01172, 2.01628 in. and $R_1+R_2+R_3 \leq .0254$ in.
Process control for rising by tool wear: acc. if $R_5 \leq .00963$ and $\bar{x}_5$ between 2.0112 and 2.0168 in. Reset by D1ai or D1bi.

CHAPTER 15

15.3 $\mu_{\bar{x}} = 5.5$, $\sigma_{\bar{x}} = 1.28$, $\alpha_4 = 2.76$, yes.

15.5 Simply expand $(x-\mu)^j$ under the integral and break into separate integrals.

15.7 The cube of the coefficient a is in both numerator and denominator, thus cancelling out.

15.9 $\bar{c} = 2.4630$ $F_c = 4.6, 11.3, 14.0, 11.5, 7.1, 3.5, 1.4, .5, .2, .1.$
$\chi_5^2 = 28.1$, not a Poisson distribution.

15.11 $F_i = 2.2, 8.6, 21.4, 30.0, 23.5, 10.4, 3.0$ $\chi_4^2 = 2.9$ Excellent fit by normal.

15.15 Using $u = (x-34.5)/10$, $\bar{u} = -.19444$, $s_u = 1.9835$, $a_3 = .956$, $a_4 = 4.396$. Use class boundaries for u variable, convert to z by $(u-\bar{u})/s_u$. Then use [2] or [3], interpolating on z to .001 for $\alpha_3 = .9, 1.0$, then on α_3 for .96, giving cumulative probabilities. Then find differences and multiply by n = 900 for F_i. Gives F_i 82.9, 170.4, 198.0, 168.0, 118.6, 74.1, 42.6, 22.8, 11.7, 5.8, 2.8, 1.3, .6, .5. Gives $\chi_9^2 = 10.7$. Satisfactory fit.

APPENDIX

TABLE I

Cumulative Probability, -∞ to z for the Standardized Normal Distribution.

$\Phi(z) = \int_{-\infty}^{z} \phi(w)\,dw = P(Z \leq z)$ in Body of Table.

z	.00	.01	.02	.03	.04	.05	.06	.07	.08	.09
−3.5	.0002	.0002	.0002	.0002	.0002	.0002	.0002	.0002	.0002	.0002
−3.4	.0003	.0003	.0003	.0003	.0003	.0003	.0003	.0003	.0003	.0002
−3.3	.0005	.0005	.0005	.0004	.0004	.0004	.0004	.0004	.0004	.0003
−3.2	.0007	.0007	.0006	.0006	.0006	.0006	.0006	.0005	.0005	.0005
−3.1	.0010	.0009	.0009	.0009	.0008	.0008	.0008	.0008	.0007	.0007
−3.0	.0013	.0013	.0013	.0012	.0012	.0011	.0011	.0011	.0010	.0010
−2.9	.0019	.0018	.0018	.0017	.0016	.0016	.0015	.0015	.0014	.0014
−2.8	.0026	.0025	.0024	.0023	.0023	.0022	.0021	.0021	.0020	.0019
−2.7	.0035	.0034	.0033	.0032	.0031	.0030	.0029	.0028	.0027	.0026
−2.6	.0047	.0045	.0044	.0043	.0041	.0040	.0039	.0038	.0037	.0036
−2.5	.0062	.0060	.0059	.0057	.0055	.0054	.0052	.0051	.0049	.0048
−2.4	.0082	.0080	.0078	.0075	.0073	.0071	.0069	.0068	.0066	.0064
−2.3	.0107	.0104	.0102	.0099	.0096	.0094	.0091	.0089	.0087	.0084
−2.2	.0139	.0136	.0132	.0129	.0125	.0122	.0119	.0116	.0113	.0110
−2.1	.0179	.0174	.0170	.0166	.0162	.0158	.0154	.0150	.0146	.0143
−2.0	.0228	.0222	.0217	.0212	.0207	.0202	.0197	.0192	.0188	.0183
−1.9	.0287	.0281	.0274	.0268	.0262	.0256	.0250	.0244	.0239	.0233
−1.8	.0359	.0351	.0344	.0336	.0329	.0322	.0314	.0307	.0301	.0294
−1.7	.0446	.0436	.0427	.0418	.0409	.0401	.0392	.0384	.0375	.0367
−1.6	.0548	.0537	.0526	.0516	.0505	.0495	.0485	.0475	.0465	.0455
−1.5	.0668	.0655	.0643	.0630	.0618	.0606	.0594	.0582	.0571	.0559
−1.4	.0808	.0793	.0778	.0764	.0749	.0735	.0721	.0708	.0694	.0681
−1.3	.0968	.0951	.0934	.0918	.0901	.0885	.0869	.0853	.0838	.0823
−1.2	.1151	.1131	.1112	.1093	.1075	.1056	.1038	.1020	.1003	.0985
−1.1	.1357	.1335	.1314	.1292	.1271	.1251	.1230	.1210	.1190	.1170
−1.0	.1587	.1562	.1539	.1515	.1492	.1469	.1446	.1423	.1401	.1379
−0.9	.1841	.1814	.1788	.1762	.1736	.1711	.1685	.1660	.1635	.1611
−0.8	.2119	.2090	.2061	.2033	.2005	.1977	.1949	.1922	.1894	.1867
−0.7	.2420	.2389	.2358	.2327	.2296	.2266	.2236	.2206	.2177	.2148
−0.6	.2743	.2709	.2676	.2643	.2611	.2578	.2546	.2514	.2483	.2451
−0.5	.3085	.3050	.3015	.2981	.2946	.2912	.2877	.2843	.2810	.2776
−0.4	.3446	.3409	.3372	.3336	.3300	.3264	.3228	.3192	.3156	.3121
−0.3	.3821	.3783	.3745	.3707	.3669	.3632	.3594	.3557	.3520	.3483
−0.2	.4207	.4168	.4129	.4090	.4052	.4013	.3974	.3936	.3897	.3859
−0.1	.4602	.4562	.4522	.4483	.4443	.4404	.4364	.4325	.4286	.4247
−0.0	.5000	.4960	.4920	.4880	.4840	.4801	.4761	.4721	.4681	.4641

TABLE I (continued)

z	.00	.01	.02	.03	.04	.05	.06	.07	.08	.09
+0.0	.5000	.5040	.5080	.5120	.5160	.5199	.5239	.5279	.5319	.5359
+0.1	.5398	.5438	.5478	.5517	.5557	.5596	.5636	.5675	.5714	.5753
+0.2	.5793	.5832	.5871	.5910	.5948	.5987	.6026	.6064	.6103	.6141
+0.3	.6179	.6217	.6255	.6293	.6331	.6368	.6406	.6443	.6480	.6517
+0.4	.6554	.6591	.6628	.6664	.6700	.6736	.6772	.6808	.6844	.6879
+0.5	.6915	.6950	.6985	.7019	.7054	.7088	.7123	.7157	.7190	.7224
+0.6	.7257	.7291	.7324	.7357	.7389	.7422	.7454	.7486	.7517	.7549
+0.7	.7580	.7611	.7642	.7673	.7704	.7734	.7764	.7794	.7823	.7852
+0.8	.7881	.7910	.7939	.7967	.7995	.8023	.8051	.8078	.8106	.8133
+0.9	.8159	.8186	.8212	.8238	.8264	.8289	.8315	.8340	.8365	.8389
+1.0	.8413	.8438	.8461	.8485	.8508	.8531	.8554	.8577	.8599	.8621
+1.1	.8643	.8665	.8686	.8708	.8729	.8749	.8770	.8790	.8810	.8830
+1.2	.8849	.8869	.8888	.8907	.8925	.8944	.8962	.8980	.8997	.9015
+1.3	.9032	.9049	.9066	.9082	.9099	.9115	.9131	.9147	.9162	.9177
+1.4	.9192	.9207	.9222	.9236	.9251	.9265	.9279	.9292	.9306	.9319
+1.5	.9332	.9345	.9357	.9370	.9382	.9394	.9406	.9418	.9429	.9441
+1.6	.9452	.9463	.9474	.9484	.9495	.9505	.9515	.9525	.9535	.9545
+1.7	.9554	.9564	.9573	.9582	.9591	.9599	.9608	.9616	.9625	.9633
+1.8	.9641	.9649	.9656	.9664	.9671	.9678	.9686	.9693	.9699	.9706
+1.9	.9713	.9719	.9726	.9732	.9738	.9744	.9750	.9756	.9761	.9767
+2.0	.9772	.9778	.9783	.9788	.9793	.9798	.9803	.9808	.9812	.9817
+2.1	.9821	.9826	.9830	.9834	.9838	.9842	.9846	.9850	.9854	.9857
+2.2	.9861	.9864	.9868	.9871	.9875	.9878	.9881	.9884	.9887	.9890
+2.3	.9893	.9896	.9898	.9901	.9904	.9906	.9909	.9911	.9913	.9916
+2.4	.9918	.9920	.9922	.9925	.9927	.9929	.9931	.9932	.9934	.9936
+2.5	.9938	.9940	.9941	.9943	.9945	.9946	.9948	.9949	.9951	.9952
+2.6	.9953	.9955	.9956	.9957	.9959	.9960	.9961	.9962	.9963	.9964
+2.7	.9965	.9966	.9967	.9968	.9969	.9970	.9971	.9972	.9973	.9974
+2.8	.9974	.9975	.9976	.9977	.9977	.9978	.9979	.9979	.9980	.9981
+2.9	.9981	.9982	.9982	.9983	.9984	.9984	.9985	.9985	.9986	.9986
+3.0	.9987	.9987	.9987	.9988	.9988	.9989	.9989	.9989	.9990	.9990
+3.1	.9990	.9991	.9991	.9991	.9992	.9992	.9992	.9992	.9993	.9993
+3.2	.9993	.9993	.9994	.9994	.9994	.9994	.9994	.9995	.9995	.9995
+3.3	.9995	.9995	.9995	.9996	.9996	.9996	.9996	.9996	.9996	.9997
+3.4	.9997	.9997	.9997	.9997	.9997	.9997	.9997	.9997	.9997	.9998
+3.5	.9998	.9998	.9998	.9998	.9998	.9998	.9998	.9998	.9998	.9998

TABLE II

Cumulative Probability Points for the Chi-Square Distribution, Zero to χ^2 for Degrees of Freedom ν in Row Heading. P = P(random variable $\chi^2 \leq$ tabled value).

	P					
ν	0.005	0.01	0.025	0.05	0.10	0.25
1	-	-	0.001	0.004	0.016	0.102
2	0.010	0.020	0.051	0.103	0.211	0.575
3	0.072	0.115	0.216	0.352	0.584	1.213
4	0.207	0.297	0.484	0.711	1.064	1.923
5	0.412	0.554	0.831	1.145	1.610	2.675
6	0.676	0.872	1.237	1.635	2.204	3.455
7	0.989	1.239	1.690	2.167	2.833	4.255
8	1.344	1.646	2.180	2.733	3.490	5.071
9	1.735	2.088	2.700	3.325	4.168	5.899
10	2.156	2.558	3.247	3.940	4.865	6.737
11	2.603	3.053	3.816	4.575	5.578	7.584
12	3.074	3.571	4.404	5.226	6.304	8.438
13	3.565	4.107	5.009	5.892	7.042	9.299
14	4.075	4.660	5.629	6.571	7.790	10.165
15	4.601	5.229	6.262	7.261	8.547	11.037
16	5.142	5.812	6.908	7.962	9.312	11.912
17	5.697	6.408	7.564	8.672	10.085	12.792
18	6.265	7.015	8.231	9.390	10.865	13.675
19	6.844	7.633	8.907	10.117	11.651	14.562
20	7.434	8.260	9.591	10.851	12.443	15.452
21	8.034	8.897	10.283	11.591	13.240	16.344
22	8.643	9.542	10.982	12.338	14.042	17.240
23	9.260	10.196	11.689	13.091	14.848	18.137
24	9.886	10.856	12.401	13.848	15.659	19.037
25	10.520	11.524	13.120	14.611	16.473	19.939
26	11.160	12.198	13.844	15.379	17.292	20.843
27	11.808	12.879	14.573	16.151	18.114	21.749
28	12.461	13.565	15.308	16.928	18.939	22.657
29	13.121	14.257	16.047	17.708	19.768	23.567
30	13.787	14.954	16.791	18.493	20.599	24.478
31	14.458	15.655	17.539	19.281	21.434	25.390
32	15.134	16.362	18.291	20.072	22.271	26.304
33	15.815	17.074	19.047	20.867	23.110	27.219
34	16.501	17.789	19.806	21.664	23.952	28.136
35	17.192	18.509	20.569	22.465	24.797	29.054
36	17.887	19.233	21.336	23.269	25.643	29.973
37	18.586	19.960	22.106	24.075	26.492	30.893
38	19.289	20.691	22.878	24.884	27.343	31.815
39	19.996	21.426	23.654	25.695	28.196	32.737
40	20.707	22.164	24.433	26.509	29.051	33.660
41	21.421	22.906	25.215	27.326	29.907	34.585
42	22.138	23.650	25.999	28.144	30.765	35.510
43	22.859	24.398	26.785	28.965	31.625	36.436
44	23.584	25.148	27.575	29.787	32.487	37.363
45	24.311	25.901	28.366	30.612	33.350	38.291

TABLE II (continued)

ν	0.75	0.90	0.95	0.975	0.99	0.995
1	1.323	2.706	3.841	5.024	6.635	7.879
2	2.773	4.605	5.991	7.378	9.210	10.597
3	4.108	6.251	7.815	9.348	11.345	12.838
4	5.385	7.779	9.488	11.143	13.277	14.860
5	6.626	9.236	11.071	12.833	15.086	16.750
6	7.841	10.645	12.592	14.449	16.812	18.548
7	9.037	12.017	14.067	16.013	18.475	20.278
8	10.219	13.362	15.507	17.535	20.090	21.955
9	11.389	14.684	16.919	19.023	21.666	23.589
10	12.549	15.987	18.307	20.483	23.209	25.188
11	13.701	17.275	19.675	21.920	24.725	26.757
12	14.845	18.549	21.026	23.337	26.217	28.299
13	15.984	19.812	22.362	24.736	27.688	29.819
14	17.117	21.064	23.685	26.119	29.141	31.319
15	18.245	22.307	24.996	27.488	30.578	32.801
16	19.369	23.542	26.296	28.845	32.000	34.267
17	20.489	24.769	27.587	30.191	33.409	35.718
18	21.605	25.989	28.869	31.526	34.805	37.156
19	22.718	27.204	30.144	32.852	36.191	38.582
20	23.828	28.412	31.410	34.170	37.566	39.997
21	24.935	29.615	32.671	35.479	38.932	41.401
22	26.039	30.813	33.924	36.781	40.289	42.796
23	27.141	32.007	35.172	38.076	41.638	44.181
24	28.241	33.196	36.415	39.364	42.980	45.559
25	29.339	34.382	37.652	40.646	44.314	46.928
26	30.435	35.563	38.885	41.923	45.642	48.290
27	31.528	36.741	40.113	43.194	46.963	49.645
28	32.620	37.916	41.337	44.461	48.278	50.993
29	33.711	39.087	42.557	45.722	49.588	52.336
30	34.800	40.256	43.773	46.979	50.892	53.672
31	35.887	41.422	44.985	48.232	52.191	55.003
32	36.973	42.585	46.194	49.480	53.486	56.328
33	38.058	43.745	47.400	50.725	54.776	57.648
34	39.141	44.903	48.602	51.966	56.061	58.964
35	40.223	46.059	49.802	53.203	57.342	60.275
36	41.304	47.212	50.998	54.437	58.619	61.581
37	42.383	48.363	52.192	55.668	59.892	62.883
38	43.462	49.513	53.384	56.896	61.162	64.181
39	44.539	50.660	54.572	58.120	62.428	65.476
40	45.616	51.805	55.758	59.342	63.691	66.766
41	46.692	52.949	56.942	60.561	64.950	68.053
42	47.766	54.090	58.124	61.777	66.206	69.336
43	48.840	55.230	59.304	62.990	67.459	70.616
44	49.913	56.369	60.481	64.201	68.710	71.893
45	50.985	57.505	61.656	65.410	69.957	73.166

TABLE II (continued)

ν	P 0.005	0.01	0.025	0.05	0.10	0.25
46	25.041	26.657	29.160	31.439	34.215	39.220
47	25.775	27.416	29.956	32.268	35.081	40.149
48	26.511	28.177	30.755	33.098	35.949	41.079
49	27.249	28.941	31.555	33.930	36.818	42.010
50	27.991	29.707	32.357	34.764	37.689	42.942
51	28.735	30.475	33.162	35.600	38.560	43.874
52	29.481	31.246	33.968	36.437	39.433	44.808
53	30.230	32.018	34.776	37.276	40.308	45.741
54	30.981	32.793	35.586	38.116	41.183	46.676
55	31.735	33.570	36.398	38.958	42.060	47.610
56	32.490	34.350	37.212	39.801	42.937	48.546
57	33.248	35.131	38.027	40.646	43.816	49.482
58	34.008	35.913	38.844	41.492	44.696	50.419
59	34.770	36.698	39.662	42.339	45.577	51.356
60	35.534	37.485	40.482	43.188	46.459	52.294
61	36.300	38.273	41.303	44.038	47.342	53.232
62	37.068	39.063	42.126	44.889	48.226	54.171
63	37.838	39.855	42.950	45.741	49.111	55.110
64	38.610	40.649	43.776	46.595	49.996	56.050
65	39.383	41.444	44.603	47.450	50.883	56.990
66	40.158	42.240	45.431	48.305	51.770	57.931
67	40.935	43.038	46.261	49.162	52.659	58.872
68	41.713	43.838	47.092	50.020	53.548	59.814
69	42.494	44.639	47.924	50.879	54.438	60.756
70	43.275	45.442	48.758	51.739	55.329	61.698
71	44.058	46.246	49.592	52.600	56.221	62.641
72	44.843	47.051	50.428	53.462	57.113	63.585
73	45.629	47.858	51.265	54.325	58.006	64.528
74	46.417	48.666	52.103	55.189	58.900	65.472
75	47.206	49.475	52.942	56.054	59.795	66.417
76	47.997	50.286	53.782	56.920	60.690	67.362
77	48.788	51.097	54.623	57.786	61.586	68.307
78	49.582	51.910	55.466	58.654	62.483	69.252
79	50.376	52.725	56.309	59.522	63.380	70.198
80	51.172	53.540	57.153	60.391	64.278	71.145
81	51.969	54.357	57.998	61.261	65.176	72.091
82	52.767	55.174	58.845	62.132	66.076	73.038
83	53.567	55.993	59.692	63.004	66.976	73.985
84	54.368	56.813	60.540	63.876	67.876	74.933
85	55.170	57.634	61.389	64.749	68.777	75.881
86	55.973	58.456	62.239	65.623	69.679	76.829
87	56.777	59.279	63.089	66.498	70.581	77.777
88	57.582	60.103	63.941	67.373	71.484	78.726
89	58.389	60.928	64.793	68.249	72.387	79.675
90	59.196	61.754	65.647	69.126	73.291	80.625

TABLE II (continued)

ν	0.75	0.90	0.95	0.975	0.99	0.995
46	52.056	58.641	62.830	66.617	71.201	74.437
47	53.127	59.774	64.001	67.821	72.443	75.704
48	54.196	60.907	65.171	69.023	73.683	76.969
49	55.265	62.038	66.339	70.222	74.919	78.231
50	56.334	63.167	67.505	71.420	76.154	79.490
51	57.401	64.295	68.669	72.616	77.386	80.747
52	58.468	65.422	69.832	73.810	78.616	82.001
53	59.534	66.548	70.993	75.002	79.843	83.253
54	60.600	67.673	72.153	76.192	81.069	84.502
55	61.665	68.796	73.311	77.380	82.292	85.749
56	62.729	69.919	74.468	78.567	83.513	86.994
57	63.793	71.040	75.624	79.752	84.733	88.236
58	64.857	72.160	76.778	80.936	85.950	89.477
59	65.919	73.279	77.931	82.117	87.166	90.715
60	66.981	74.397	79.082	83.298	88.379	91.952
61	68.043	75.514	80.232	84.476	89.591	93.186
62	69.104	76.630	81.381	85.654	90.802	94.419
63	70.165	77.745	82.529	86.830	92.010	95.649
64	71.225	78.860	83.675	88.004	93.217	96.878
65	72.285	79.973	84.821	89.177	94.422	98.105
66	73.344	81.085	85.965	90.349	95.626	99.330
67	74.403	82.197	87.108	91.519	96.828	100.554
68	75.461	83.308	88.250	92.689	98.028	101.776
69	76.519	84.418	89.391	93.856	99.228	102.996
70	77.577	85.527	90.531	95.023	100.425	104.215
71	78.634	86.635	91.670	96.189	101.621	105.432
72	79.690	87.743	92.808	97.353	102.816	106.648
73	80.747	88.850	93.945	98.516	104.010	107.862
74	81.803	89.956	95.081	99.678	105.202	109.074
75	82.858	91.061	96.217	100.839	106.393	110.286
76	83.913	92.166	97.351	101.999	107.583	111.495
77	84.968	93.270	98.484	103.158	108.771	112.704
78	86.022	94.374	99.617	104.316	109.958	113.911
79	87.077	95.476	100.749	105.473	111.144	115.117
80	88.130	96.578	101.879	106.629	112.329	116.321
81	89.184	97.680	103.010	107.783	113.512	117.524
82	90.237	98.780	104.139	108.937	114.695	118.726
83	91.289	99.880	105.267	110.090	115.876	119.927
84	92.342	100.980	106.395	111.242	117.057	121.126
85	93.394	102.079	107.522	112.393	118.236	122.325
86	94.446	103.177	108.648	113.544	119.414	123.522
87	95.497	104.275	109.773	114.693	120.591	124.718
88	96.548	105.372	110.898	115.841	121.767	125.913
89	97.599	106.469	112.022	116.989	122.942	127.106
90	98.650	107.565	113.145	118.136	124.116	128.299

TABLE III

Cumulative Probabilities of Students t Distribution. P = P(t $\leq$ tabled value) for Given Degrees of Freedom ν in Row.

ν	P = 0.75	0.90	0.95	0.975	0.99	0.995
1	1.0000	3.0777	6.3138	12.7062	31.8205	63.6567
2	0.8165	1.8856	2.9200	4.3027	6.9646	9.9248
3	0.7649	1.6377	2.3534	3.1824	4.5407	5.8409
4	0.7407	1.5332	2.1318	2.7764	3.7469	4.6041
5	0.7267	1.4759	2.0150	2.5706	3.3649	4.0322
6	0.7176	1.4398	1.9432	2.4469	3.1427	3.7074
7	0.7111	1.4149	1.8946	2.3646	2.9980	3.4995
8	0.7064	1.3968	1.8595	2.3060	2.8965	3.3554
9	0.7027	1.3830	1.8331	2.2622	2.8214	3.2498
10	0.6998	1.3722	1.8125	2.2281	2.7638	3.1693
11	0.6974	1.3634	1.7959	2.2010	2.7181	3.1058
12	0.6955	1.3562	1.7823	2.1788	2.6810	3.0545
13	0.6938	1.3502	1.7709	2.1604	2.6503	3.0123
14	0.6924	1.3450	1.7613	2.1448	2.6245	2.9768
15	0.6912	1.3406	1.7531	2.1315	2.6025	2.9467
16	0.6901	1.3368	1.7459	2.1199	2.5835	2.9208
17	0.6892	1.3334	1.7396	2.1098	2.5669	2.8982
18	0.6884	1.3304	1.7341	2.1009	2.5524	2.8784
19	0.6876	1.3277	1.7291	2.0930	2.5395	2.8609
20	0.6870	1.3253	1.7247	2.0860	2.5280	2.8453
21	0.6864	1.3232	1.7207	2.0796	2.5177	2.8314
22	0.6858	1.3212	1.7171	2.0739	2.5083	2.8188
23	0.6853	1.3195	1.7139	2.0687	2.4999	2.8073
24	0.6848	1.3178	1.7109	2.0639	2.4922	2.7969
25	0.6844	1.3163	1.7081	2.0595	2.4851	2.7874
26	0.6840	1.3150	1.7056	2.0555	2.4786	2.7787
27	0.6837	1.3137	1.7033	2.0518	2.4727	2.7707
28	0.6834	1.3125	1.7011	2.0484	2.4671	2.7633
29	0.6830	1.3114	1.6991	2.0452	2.4620	2.7564
30	0.6828	1.3104	1.6973	2.0423	2.4573	2.7500
31	0.6825	1.3095	1.6955	2.0395	2.4528	2.7440
32	0.6822	1.3086	1.6939	2.0369	2.4487	2.7385
33	0.6820	1.3077	1.6924	2.0345	2.4448	2.7333
34	0.6818	1.3070	1.6909	2.0322	2.4411	2.7284
35	0.6816	1.3062	1.6896	2.0301	2.4377	2.7238
36	0.6814	1.3055	1.6883	2.0281	2.4345	2.7195
37	0.6812	1.3049	1.6871	2.0262	2.4314	2.7154
38	0.6810	1.3042	1.6860	2.0244	2.4286	2.7116
39	0.6808	1.3036	1.6849	2.0227	2.4258	2.7079
40	0.6807	1.3031	1.6839	2.0211	2.4233	2.7045
41	0.6805	1.3025	1.6829	2.0195	2.4208	2.7012
42	0.6804	1.3020	1.6820	2.0181	2.4185	2.6981
43	0.6802	1.3016	1.6811	2.0167	2.4163	2.6951
44	0.6801	1.3011	1.6802	2.0154	2.4141	2.6923
45	0.6800	1.3006	1.6794	2.0141	2.4121	2.6896

Reproduced with permission from D. B. Owen, "Handbook of Statistical Tables," Addison-Wesley, Reading, Massachusetts, 1962, pp. 28, 29.

TABLE III (continued)

ν	P 0.75	0.90	0.95	0.975	0.99	0.995
46	0.6799	1.3002	1.6787	2.0129	2.4102	2.6870
47	0.6797	1.2998	1.6779	2.0117	2.4083	2.6846
48	0.6796	1.2994	1.6772	2.0106	2.4066	2.6822
49	0.6795	1.2991	1.6766	2.0096	2.4049	2.6800
50	0.6794	1.2987	1.6759	2.0086	2.4033	2.6778
51	0.6793	1.2984	1.6753	2.0076	2.4017	2.6757
52	0.6792	1.2980	1.6747	2.0066	2.4002	2.6737
53	0.6791	1.2977	1.6741	2.0057	2.3988	2.6718
54	0.6791	1.2974	1.6736	2.0049	2.3974	2.6700
55	0.6790	1.2971	1.6730	2.0040	2.3961	2.6682
56	0.6789	1.2969	1.6725	2.0032	2.3948	2.6665
57	0.6788	1.2966	1.6720	2.0025	2.3936	2.6649
58	0.6787	1.2963	1.6716	2.0017	2.3924	2.6633
59	0.6787	1.2961	1.6711	2.0010	2.3912	2.6618
60	0.6786	1.2958	1.6706	2.0003	2.3901	2.6603
61	0.6785	1.2956	1.6702	1.9996	2.3890	2.6589
62	0.6785	1.2954	1.6698	1.9990	2.3880	2.6575
63	0.6784	1.2951	1.6694	1.9983	2.3870	2.6561
64	0.6783	1.2949	1.6690	1.9977	2.3860	2.6549
65	0.6783	1.2947	1.6686	1.9971	2.3851	2.6536
66	0.6782	1.2945	1.6683	1.9966	2.3842	2.6524
67	0.6782	1.2943	1.6679	1.9960	2.3833	2.6512
68	0.6781	1.2941	1.6676	1.9955	2.3824	2.6501
69	0.6781	1.2939	1.6672	1.9949	2.3816	2.6490
70	0.6780	1.2938	1.6669	1.9944	2.3808	2.6479
71	0.6780	1.2936	1.6666	1.9939	2.3800	2.6469
72	0.6779	1.2934	1.6663	1.9935	2.3793	2.6459
73	0.6779	1.2933	1.6660	1.9930	2.3785	2.6449
74	0.6778	1.2931	1.6657	1.9925	2.3778	2.6439
75	0.6778	1.2929	1.6654	1.9921	2.3771	2.6430
76	0.6777	1.2928	1.6652	1.9917	2.3764	2.6421
77	0.6777	1.2926	1.6649	1.9913	2.3758	2.6412
78	0.6776	1.2925	1.6646	1.9908	2.3751	2.6403
79	0.6776	1.2924	1.6644	1.9905	2.3745	2.6395
80	0.6776	1.2922	1.6641	1.9901	2.3739	2.6387
81	0.6775	1.2921	1.6639	1.9897	2.3733	2.6379
82	0.6775	1.2920	1.6636	1.9893	2.3727	2.6371
83	0.6775	1.2918	1.6634	1.9890	2.3721	2.6364
84	0.6774	1.2917	1.6632	1.9886	2.3716	2.6356
85	0.6774	1.2916	1.6630	1.9883	2.3710	2.6349
86	0.6774	1.2915	1.6628	1.9879	2.3705	2.6342
87	0.6773	1.2914	1.6626	1.9876	2.3700	2.6335
88	0.6773	1.2912	1.6624	1.9873	2.3695	2.6329
89	0.6773	1.2911	1.6622	1.9870	2.3690	2.6322
90	0.6772	1.2910	1.6620	1.9867	2.3685	2.6316

TABLE IV

Random Numbers

1368	9621	9151	2066	1208	2664	9822	6599	6911	5112
5953	5936	2541	4011	0408	3593	3679	1378	5936	2651
7226	9466	9553	7671	8599	2119	5337	5953	6355	6889
8883	3454	6773	8207	5576	6386	7487	0190	0867	1298
7022	5281	1168	4099	8069	8721	8353	9952	8006	9045
4576	1853	7884	2451	3488	1286	4842	7719	5795	3953
8715	1416	7028	4616	3470	9938	5703	0196	3465	0034
4011	0408	2224	7626	0643	1149	8834	6429	8691	0143
1400	3694	4482	3608	1238	8221	5129	6105	5314	8385
6370	1884	0820	4854	9161	6509	7123	4070	6759	6113
4522	5749	8084	3932	7678	3549	0051	6761	6952	7041
7195	6234	6426	7148	9945	0358	3242	0519	6550	1327
0054	0810	2937	2040	2299	4198	0846	3937	3986	1019
5166	5433	0381	9686	5670	5129	2103	1125	3404	8785
1247	3793	7415	7819	1783	0506	4878	7673	9840	6629
8529	7842	7203	1844	8619	7404	4215	9969	6948	5643
8973	3440	4366	9242	2151	0244	0922	5887	4883	1177
9307	2959	5904	9012	4951	3695	4529	7197	7179	3239
2923	4276	9467	9868	2257	1925	3382	7244	1781	8037
6372	2808	1238	8098	5509	4617	4099	6705	2386	2830
6922	1807	4900	5306	0411	1828	8634	2331	7247	3230
9862	8336	6453	0545	6127	2741	5967	8447	3017	5709
3371	1530	5104	3076	5506	3101	4143	5845	2095	6127
6712	9402	9588	7019	9248	9192	4223	6555	7947	2474
3071	8782	7157	5941	8830	8563	2252	8109	5880	9912
4022	9734	7852	9096	0051	7387	7056	9331	1317	7833
9682	8892	3577	0326	5306	0050	8517	4376	0788	5443
6705	2175	9904	3743	1902	5393	3032	8432	0612	7972
1872	8292	2366	8603	4288	6809	4357	1072	6822	5611
2559	7534	2281	7351	2064	0611	9613	2000	0327	6145
4399	3751	9783	5399	5175	8894	0296	9483	0400	2272
6074	8827	2195	2532	7680	4288	6807	3101	6850	6410
5155	7186	4722	6721	0838	3632	5355	9369	2006	7681
3193	2800	6184	7891	9838	6123	9397	4019	8389	9508
8610	1880	7423	3384	4625	6653	2900	6290	9286	2396
4778	8818	2992	6300	4239	9595	4384	0611	7687	2088
3987	1619	4164	2542	4042	7799	9084	0278	8422	4330
2977	0248	2793	3351	4922	8878	5703	7421	2054	4391
1312	2919	8220	7285	5902	7882	1403	5354	9913	7109
3890	7193	7799	9190	3275	7840	1872	6232	5295	3148
0793	3468	8762	2492	5854	8430	8472	2264	9279	2128
2139	4552	3444	6462	2524	8601	3372	1848	1472	9667
8277	9153	2880	9053	6880	4284	5044	8931	0861	1517
2236	4778	6639	0862	9509	2141	0208	1450	1222	5281
8837	7686	1771	3374	2894	7314	6856	0440	3766	6047
6605	6380	4599	3333	0713	8401	7146	8940	2629	2006
8399	8175	3525	1646	4019	8390	4344	8975	4489	3423
8053	3046	9102	4515	2944	9763	3003	3408	1199	2791
9837	9378	3237	7016	7593	5958	0068	3114	0456	6840
2557	6395	9496	1884	0612	8102	4402	5498	0422	3335

Reproduced with permission from D. B. Owen, "Handbook of Statistical Tables," Addison-Wesley, Reading, Massachusetts, 1962, pp. 519, 520.

TABLE IV (continued)

2671	4690	1550	2262	2597	8034	0785	2978	4409	0237
9111	0250	3275	7519	9740	4577	2064	0286	3398	1348
0391	6035	9230	4999	3332	0608	6113	0391	5789	9926
2475	2144	1886	2079	3004	9686	5669	4367	9306	2595
5336	5845	2095	6446	5694	3641	1085	8705	5416	9066
6808	0423	0155	1652	7897	4335	3567	7109	9690	3739
8525	0577	8940	9451	6726	0876	3818	7607	8854	3566
0398	0741	8787	3043	5063	0617	1770	5048	7721	7032
3623	9636	3638	1406	5731	3978	8068	7238	9715	3363
0739	2644	4917	8866	3632	5399	5175	7422	2476	2607
6713	3041	8133	8749	8835	6745	3597	3476	3816	3455
7775	9315	0432	8327	0861	1515	2297	3375	3713	9174
8599	2122	6842	9202	0810	2936	1514	2090	3067	3574
7955	3759	5254	1126	5553	4713	9605	7909	1658	5490
4766	0070	7260	6033	7997	0109	5993	7592	5436	1727
5165	1670	2534	8811	8231	3721	7947	5719	2640	1394
9111	0513	2751	8256	2931	7783	1281	6531	7259	6993
1667	1084	7889	8963	7018	8617	6381	0723	4926	4551
2145	4587	8585	2412	5431	4667	1942	7238	9613	2212
2739	5528	1481	7528	9368	1823	6979	2547	7268	2467
8769	5480	9160	5354	9700	1362	2774	7980	9157	8788
6531	9435	3422	2474	1475	0159	3414	5224	8399	5820
2937	4134	7120	2206	5084	9473	3958	7320	9878	8609
1581	3285	3727	8924	6204	0797	0882	5945	9375	9153
6268	1045	7076	1436	4165	0143	0293	4190	7171	7932
4293	0523	8625	1961	1039	2856	4889	4358	1492	3804
6936	4213	3212	7229	1230	0019	5998	9206	6753	3762
5334	7641	3258	3769	1362	2771	6124	9813	7915	8960
9373	1158	4418	8826	5665	5896	0358	4717	8232	4859
6968	9428	8950	5346	1741	2348	8143	5377	7695	0685
4229	0587	8794	4009	9691	4579	3302	7673	9629	5246
3807	7785	7097	5701	6639	0723	4819	0900	2713	7650
4891	8829	1642	2155	0796	0466	2946	2970	9143	6590
1055	2968	7911	7479	8199	9735	8271	5339	7058	2964
2983	2345	0568	4125	0894	8302	0506	6761	7706	4310
4026	3129	2968	8053	2797	4022	9838	9611	0975	2437
4075	0260	4256	0337	2355	9371	2954	6021	5783	2827
8488	5450	1327	7358	2034	8060	1788	6913	6123	9405
1976	1749	5742	4098	5887	4567	6064	2777	7830	5668
2793	4701	9466	9554	8294	2160	7486	1557	4769	2781
0916	6272	6825	7188	9611	1181	2301	5516	5451	6832
5961	1149	7946	1950	2010	0600	5655	0796	0569	4365
3222	4189	1891	8172	8731	4769	2782	1325	4238	9279
1176	7834	4600	9992	9449	5824	5344	1008	6678	1921
2369	8971	2314	4806	5071	8908	8274	4936	3357	4441
0041	4329	9265	0352	4764	9070	7527	7791	1094	2008
0803	8302	6814	2422	6351	0637	0514	0246	1845	8594
9965	7804	3930	8803	0268	1426	3130	3613	3947	8086
0011	2387	3148	7559	4216	2946	2865	6333	1916	2259
1767	9871	3914	5790	5287	7915	8959	1346	5482	9251

TABLE V

Control Chart Constants for Averages $\bar{x}$, Standard Deviations s and Ranges R; from Normal Populations. Factors for Computing Central Lines and Three Sigma Control Limits

Sample Size n	Factors for Control limits for $\bar{x}$			Factors for Standard deviations, s		Factors for control limits for s				Factors for ranges R		Factors for control limits for R			
n	A	A_2	A_3	c_4	c_5	B_3	B_4	B_5	B_6	d_2	d_3	D_1	D_2	D_3	D_4
2	2.121	1.880	2.659	.798	.603	0	3.267	0	2.606	1.128	.853	0	3.686	0	3.267
3	1.732	1.023	1.954	.886	.463	0	2.568	0	2.276	1.693	.888	0	4.358	0	2.575
4	1.500	.729	1.628	.921	.389	0	2.266	0	2.088	2.059	.880	0	4.698	0	2.282
5	1.342	.577	1.427	.940	.341	0	2.089	0	1.964	2.326	.864	0	4.918	0	2.115
6	1.225	.483	1.287	.952	.308	.030	1.970	.029	1.874	2.534	.848	0	5.078	0	2.004
7	1.134	.419	1.182	.959	.282	.118	1.882	.113	1.806	2.704	.833	.205	5.203	.076	1.924
8	1.061	.373	1.099	.965	.262	.185	1.815	.179	1.751	2.847	.820	.387	5.307	.136	1.864
9	1.000	.337	1.032	.969	.246	.239	1.761	.232	1.707	2.970	.808	.546	5.394	.184	1.816
10	.949	.308	.975	.973	.232	.284	1.716	.276	1.669	3.078	.797	.687	5.469	.223	1.777
11	.905	.285	.927	.975	.221	.321	1.679	.313	1.637	3.173	.787	.812	5.534	.256	1.744
12	.866	.266	.886	.978	.211	.354	1.646	.346	1.610	3.258	.778	.924	5.592	.284	1.716
13	.832	.249	.850	.979	.202	.382	1.618	.374	1.585	3.336	.770	1.026	5.646	.308	1.692
14	.802	.235	.817	.981	.194	.406	1.594	.399	1.563	3.407	.762	1.121	5.693	.329	1.671
15	.775	.223	.789	.982	.187	.428	1.572	.421	1.544	3.472	.755	1.207	5.737	.348	1.652
16	.750	.212	.763	.983	.181	.448	1.552	.440	1.526	3.532	.749	1.285	5.779	.364	1.636
17	.728	.203	.739	.985	.175	.466	1.534	.458	1.511	3.588	.743	1.359	5.817	.379	1.621
18	.707	.194	.718	.985	.170	.482	1.518	.475	1.496	3.640	.738	1.426	5.854	.392	1.608
19	.688	.187	.698	.986	.165	.497	1.503	.490	1.483	3.689	.733	1.490	5.888	.404	1.596
20	.671	.180	.680	.987	.161	.510	1.490	.504	1.470	3.735	.729	1.548	5.922	.414	1.586
21	.655	.173	.663	.988	.157	.523	1.477	.516	1.459	3.778	.724	1.606	5.950	.425	1.575
22	.640	.167	.647	.988	.153	.534	1.466	.528	1.448	3.819	.720	1.659	5.979	.434	1.566
23	.626	.162	.633	.989	.150	.545	1.455	.539	1.438	3.858	.716	1.710	6.006	.443	1.557
24	.612	.157	.619	.989	.147	.555	1.445	.549	1.429	3.895	.712	1.759	6.031	.452	1.548
25	.600	.153	.606	.990	.144	.565	1.435	.559	1.420	3.931	.709	1.804	6.058	.459	1.541

Formulas for Control Charts for Variables, $\bar{x}$, s, R

Purpose of Chart	Chart for	Central Line	3-sigma Control Limits
No Standard Given - used for analyzing past data for control. ($\bar{\bar{x}}$, $\bar{R}$, $\bar{s}$ are average values for data being analyzed.)	Averages, $\bar{x}$	$\bar{\bar{x}}$	$\bar{\bar{x}} \pm A_2\bar{R}$, or $\bar{\bar{x}} \pm A_3\bar{s}$
	Ranges, R	$\bar{R}$	$D_3\bar{R}$, $D_4\bar{R}$
	Std. Devs., s	$\bar{s}$	$B_3\bar{s}$, $B_4\bar{s}$
Standards Given - Used for controlling quality with respect to standards given μ,σ. $R'_n = d_2\sigma$ for n.	Averages, $\bar{x}$	μ	$\mu \pm A\sigma$
	Ranges, R	$d_2\sigma$ or R'_n	$D_1\sigma$, $D_2\sigma$; $D_3R'_n$, $D_4R'_n$
	Std. devs., s	$c_4\sigma$	$B_5\sigma$, $B_6\sigma$

$E(s) = c_4\sigma$, $\sigma_s = c_5\sigma$, $E(R) = d_2\sigma$, $\sigma_R = d_3\sigma$.

TABLE VI

Moment Constants for Distribution of Sample Standard Deviation s*

n	c_4	c_5	$\alpha_{3:s}$	n	c_4	c_5	$\alpha_{3:s}$
2	.7979	.6028	.995	32	.9920	.1265	.130
3	.8862	.4633	.631	34	.9925	.1226	.125
4	.9213	.3888	.486	36	.9929	.1191	.122
5	.9400	.3412	.406	38	.9933	.1158	.118
6	.9515	.3075	.354	40	.9936	.1129	.115
7	.9594	.2822	.318	42	.9939	.1101	.112
8	.9650	.2621	.291	44	.9942	.1075	.109
9	.9693	.2458	.269	46	.9945	.1051	.107
10	.9727	.2322	.252	48	.9947	.1029	.105
11	.9754	.2207	.237	50	.9949	.1008	.102
12	.9776	.2107	.225	52	.9951	.0988	.100
13	.9794	.2019	.215	54	.9953	.0969	.098
14	.9810	.1942	.205	56	.9955	.0951	.096
15	.9823	.1872	.197	58	.9956	.0935	.095
16	.9835	.1810	.190	60	.9958	.0919	.093
17	.9845	.1753	.184	62	.9959	.0903	.091
18	.9854	.1702	.178	64	.9960	.0889	.090
19	.9862	.1655	.172	66	.9962	.0875	.089
20	.9869	.1611	.168	68	.9963	.0862	.087
21	.9876	.1571	.163	70	.9964	.0850	.086
22	.9882	.1534	.159	72	.9965	.0838	.085
23	.9887	.1499	.155	74	.9966	.0826	.083
24	.9892	.1466	.151	76	.9967	.0815	.082
25	.9896	.1436	.148	78	.9968	.0805	.081
26	.9901	.1407	.145	80	.9968	.0794	.080
27	.9904	.1380	.142	84	.9970	.0775	.078
28	.9908	.1354	.139	88	.9971	.0757	.076
29	.9911	.1330	.137	92	.9973	.0740	.075
30	.9914	.1307	.134	96	.9974	.0725	.073
31	.9917	.1286	.132	100	.9975	.0710	.072

[a] $E(s) = c_4\sigma$; $\sigma_s = c_5\sigma$; $\alpha_{3:s}$.

Reproduced with permission from I. W. Burr, "Applied Statistical Methods," Academic Press, New York, 1974, p. 437.

TABLE VII

Poisson Distribution. Probabilities of c or less, given c', appears in body of table multiplied by 1000.

c' or np' \ c	0	1	2	3	4	5	6	7	8	9
0.02	980	1,000								
0.04	961	999	1,000							
0.06	942	998	1,000							
0.08	923	997	1,000							
0.10	905	995	1,000							
0.15	861	990	999	1,000						
0.20	819	982	999	1,000						
0.25	779	974	998	1,000						
0.30	741	963	996	1,000						
0.35	705	951	994	1,000						
0.40	670	938	992	999	1,000					
0.45	638	925	989	999	1,000					
0.50	607	910	986	998	1,000					
0.55	577	894	982	998	1,000					
0.60	549	878	977	997	1,000					
0.65	522	861	972	996	999	1,000				
0.70	497	844	966	994	999	1,000				
0.75	472	827	959	993	999	1,000				
0.80	449	809	953	991	999	1,000				
0.85	427	791	945	989	998	1,000				
0.90	407	772	937	987	998	1,000				
0.95	387	754	929	984	997	1,000				
1.00	368	736	920	981	996	999	1,000			
1.1	333	699	900	974	995	999	1,000			
1.2	301	663	879	966	992	998	1,000			
1.3	273	627	857	957	989	998	1,000			
1.4	247	592	833	946	986	997	999	1,000		
1.5	223	558	809	934	981	996	999	1,000		
1.6	202	525	783	921	976	994	999	1,000		
1.7	183	493	757	907	970	992	998	1,000		
1.8	165	463	731	891	964	990	997	999	1,000	
1.9	150	434	704	875	956	987	997	999	1,000	
2.0	135	406	677	857	947	983	995	999	1,000	

TABLE VII (continued)

c' or np' \ c	0	1	2	3	4	5	6	7	8	9
2.2	111	355	623	819	928	975	993	998	1,000	
2.4	091	308	570	779	904	964	988	997	999	1,000
2.6	074	267	518	736	877	951	983	995	999	1,000
2.8	061	231	469	692	848	935	976	992	998	999
3.0	050	199	423	647	815	916	966	988	996	999
3.2	041	171	380	603	781	895	955	983	994	998
3.4	033	147	340	558	744	871	942	977	992	997
3.6	027	126	303	515	706	844	927	969	988	996
3.8	022	107	269	473	668	816	909	960	984	994
4.0	018	092	238	433	629	785	889	949	979	992
4.2	015	078	210	395	590	753	867	936	972	989
4.4	012	066	185	359	551	720	844	921	964	985
4.6	010	056	163	326	513	686	818	905	955	980
4.8	008	048	143	294	476	651	791	887	944	975
5.0	007	040	125	265	440	616	762	867	932	968
5.2	006	034	109	238	406	581	732	845	918	960
5.4	005	029	095	213	373	546	702	822	903	951
5.6	004	024	082	191	342	512	670	797	886	941
5.8	003	021	072	170	313	478	638	771	867	929
6.0	002	017	062	151	285	446	606	744	847	916

	10	11	12	13	14	15	16
2.8	1,000						
3.0	1,000						
3.2	1,000						
3.4	999	1,000					
3.6	999	1,000					
3.8	998	999	1,000				
4.0	997	999	1,000				
4.2	996	999	1,000				
4.4	994	998	999	1,000			
4.6	992	997	999	1,000			
4.8	990	996	999	1,000			
5.0	986	995	998	999	1,000		
5.2	982	993	997	999	1,000		
5.4	977	990	996	999	1,000		
5.6	972	988	995	998	999	1,000	
5.8	965	984	993	997	999	1,000	
6.0	957	980	991	996	999	999	1,000

TABLE VII (continued)

c' or np' \ c	0	1	2	3	4	5	6	7	8	9
6.2	002	015	054	134	259	414	574	716	826	902
6.4	002	012	046	119	235	384	542	687	803	886
6.6	001	010	040	105	213	355	511	658	780	869
6.8	001	009	034	093	192	327	480	628	755	850
7.0	001	007	030	082	173	301	450	599	729	830
7.2	001	006	025	072	156	276	420	569	703	810
7.4	001	005	022	063	140	253	392	539	676	788
7.6	001	004	019	055	125	231	365	510	648	765
7.8	000	004	016	048	112	210	338	481	620	741
8.0	000	003	014	042	100	191	313	453	593	717
8.5	000	002	009	030	074	150	256	386	523	653
9.0	000	001	006	021	055	116	207	324	456	587
9.5	000	001	004	015	040	089	165	269	392	522
10.0	000	000	003	010	029	067	130	220	333	458
	10	11	12	13	14	15	16	17	18	19
6.2	949	975	989	995	998	999	1,000			
6.4	939	969	986	994	997	999	1,000			
6.6	927	963	982	992	997	999	999	1,000		
6.8	915	955	978	990	996	998	999	1,000		
7.0	901	947	973	987	994	998	999	1,000		
7.2	887	937	967	984	993	997	999	999	1,000	
7.4	871	926	961	980	991	996	998	999	1,000	
7.6	854	915	954	976	989	995	998	999	1,000	
7.8	835	902	945	971	986	993	997	999	1,000	
8.0	816	888	936	966	983	992	996	998	999	1,000
8.5	763	849	909	949	973	986	993	997	999	999
9.0	706	803	876	926	959	978	989	995	998	999
9.5	645	752	836	898	940	967	982	991	996	998
10.0	583	697	792	864	917	951	973	986	993	997
	20	21	22							
8.5	1,000									
9.0	1,000									
9.5	999	1,000								
10.0	998	999	1,000							

TABLE VII (continued)

c' or np' \ c	0	1	2	3	4	5	6	7	8	9
10.5	000	000	002	007	021	050	102	179	279	397
11.0	000	000	001	005	015	038	079	143	232	341
11.5	000	000	001	003	011	028	060	114	191	289
12.0	000	000	001	002	008	020	046	090	155	242
12.5	000	000	000	002	005	015	035	070	125	201
13.0	000	000	000	001	004	011	026	054	100	166
13.5	000	000	000	001	003	008	019	041	079	135
14.0	000	000	000	000	002	006	014	032	062	109
14.5	000	000	000	000	001	004	010	024	048	088
15.0	000	000	000	000	001	003	008	018	037	070
	10	11	12	13	14	15	16	17	18	19
10.5	521	639	742	825	888	932	960	978	988	994
11.0	460	579	689	781	854	907	944	968	982	991
11.5	402	520	633	733	815	878	924	954	974	986
12.0	347	462	576	682	772	844	899	937	963	979
12.5	297	406	519	628	725	806	869	916	948	969
13.0	252	353	463	573	675	764	835	890	930	957
13.5	211	304	409	518	623	718	798	861	908	942
14.0	176	260	358	464	570	669	756	827	883	923
14.5	145	220	311	413	518	619	711	790	853	901
15.0	118	185	268	363	466	568	664	749	819	875
	20	21	22	23	24	25	26	27	28	29
10.5	997	999	999	1,000						
11.0	995	998	999	1,000						
11.5	992	996	998	999	1,000					
12.0	988	994	997	999	999	1,000				
12.5	983	991	995	998	999	999	1,000			
13.0	975	986	992	996	998	999	1,000			
13.5	965	980	989	994	997	998	999	1,000		
14.0	952	971	983	991	995	997	999	999	1,000	
14.5	936	960	976	986	992	996	998	999	999	1,000
15.0	917	947	967	981	989	994	997	998	999	1,000

TABLE VII (continued)

c' or np' \ c	4	5	6	7	8	9	10	11	12	13
16	000	001	004	010	022	043	077	127	193	275
17	000	001	002	005	013	026	049	085	135	201
18	000	000	001	003	007	015	030	055	092	143
19	000	000	001	002	004	009	018	035	061	098
20	000	000	000	001	002	005	011	021	039	066
21	000	000	000	000	001	003	006	013	025	043
22	000	000	000	000	001	002	004	008	015	028
23	000	000	000	000	000	001	002	004	009	017
24	000	000	000	000	000	000	001	003	005	011
25	000	000	000	000	000	000	001	001	003	006
	14	15	16	17	18	19	20	21	22	23
16	368	467	566	659	742	812	868	911	942	963
17	281	371	468	564	655	736	805	861	905	937
18	208	287	375	469	562	651	731	799	855	899
19	150	215	292	378	469	561	647	725	793	849
20	105	157	221	297	381	470	559	644	721	787
21	072	111	163	227	302	384	471	558	640	716
22	048	077	117	169	232	306	387	472	556	637
23	031	052	082	123	175	238	310	389	472	555
24	020	034	056	087	128	180	243	314	392	473
25	012	022	038	060	092	134	185	247	318	394
	24	25	26	27	28	29	30	31	32	33
16	978	987	993	996	998	999	999	1,000		
17	959	975	985	991	995	997	999	999	1,000	
18	932	955	972	983	990	994	997	998	999	1,000
19	893	927	951	969	980	988	993	996	998	999
20	843	888	922	948	966	978	987	992	995	997
21	782	838	883	917	944	963	976	985	991	994
22	712	777	832	877	913	940	959	973	983	989
23	635	708	772	827	873	908	936	956	971	981
24	554	632	704	768	823	868	904	932	953	969
25	473	553	629	700	763	818	863	900	929	950
	34	35	36	37	38	39	40	41	42	43
19	999	1,000								
20	999	999	1,000							
21	997	998	999	999	1,000					
22	994	996	998	999	999	1,000				
23	988	993	996	997	999	999	1,000			
24	979	987	992	995	997	998	999	999	1,000	
25	966	978	985	991	994	997	998	999	999	1,000

TABLE VIII

Single Sample Tests for $\sigma = \sigma_1$ versus $\sigma = \sigma_2 > \sigma_1$, with Risks $\alpha = \beta$.

Sample size	(1) Ratio of σ_2/σ_1 for $\alpha = \beta$: .10	(2) .05	(3) .02	(4) .01	(5) Multiplier for σ_1^2 to get K. $\alpha = \beta$: .10	(6) .05	(7) .02	(8) .01 (9)
2	13.1	31.3	92.6	206.	2.71	3.84	5.41	6.64
3	4.67	7.63	13.9	21.4	2.30	3.00	3.91	4.60
4	3.27	4.71	7.29	9.93	2.08	2.60	3.28	3.78
5	2.70	3.65	5.22	6.69	1.94	2.37	2.92	3.32
6	2.40	3.11	4.22	5.22	1.85	2.21	2.68	3.02
7	2.20	2.76	3.64	4.39	1.77	2.10	2.51	2.80
8	2.06	2.55	3.26	3.86	1.72	2.01	2.37	2.64
9	1.96	2.38	2.99	3.49	1.67	1.94	2.27	2.51
10	1.88	2.26	2.79	3.22	1.63	1.88	2.19	2.41
11	1.81	2.16	2.63	3.01	1.60	1.83	2.12	2.32
12	1.76	2.07	2.50	2.85	1.57	1.79	2.06	2.25
13	1.72	2.01	2.40	2.71	1.55	1.75	2.00	2.18
14	1.68	1.95	2.31	2.60	1.52	1.72	1.96	2.13
15	1.64	1.90	2.24	2.50	1.50	1.69	1.92	2.08
16	1.61	1.86	2.17	2.42	1.49	1.67	1.88	2.04
17	1.59	1.82	2.12	2.35	1.47	1.64	1.85	2.00
18	1.57	1.78	2.07	2.28	1.46	1.62	1.82	1.97
19	1.55	1.75	2.02	2.23	1.44	1.60	1.80	1.93
20	1.53	1.73	1.98	2.18	1.43	1.59	1.77	1.90
21	1.51	1.70	1.95	2.13	1.42	1.57	1.75	1.88
22	1.50	1.68	1.91	2.09	1.41	1.56	1.73	1.85
23	1.48	1.66	1.88	2.05	1.40	1.54	1.71	1.83
24	1.47	1.64	1.86	2.02	1.39	1.53	1.69	1.81
25	1.46	1.62	1.83	1.99	1.38	1.52	1.68	1.79
26	1.44	1.61	1.81	1.96	1.38	1.51	1.66	1.77
27	1.43	1.59	1.79	1.93	1.37	1.50	1.65	1.76
28	1.42	1.58	1.77	1.91	1.36	1.49	1.63	1.74
29	1.41	1.56	1.75	1.89	1.35	1.48	1.62	1.72
30	1.41	1.55	1.73	1.87	1.35	1.47	1.61	1.71
31	1.40	1.54	1.72	1.84	1.34	1.46	1.60	1.70
40	1.34	1.46	1.60	1.71	1.30	1.40	1.52	1.60
50	1.30	1.40	1.52	1.61	1.27	1.35	1.46	1.53
60	1.27	1.36	1.46	1.54	1.24	1.32	1.41	1.48
70	1.25	1.33	1.42	1.49	1.22	1.30	1.38	1.44
80	1.23	1.30	1.39	1.45	1.21	1.28	1.36	1.41
90	1.21	1.28	1.36	1.42	1.20	1.26	1.33	1.38
100	1.20	1.26	1.34	1.39	1.19	1.24	1.31	1.36

For n, seek entry in columns (2) - (5), $\leq \sigma_2/\sigma_1$, giving n. Then for this n and $\alpha = \beta$, find in columns (6) - (9), the multiplier for σ_1^2 to give K. Then $s^2 < K$, accept $\sigma = \sigma_1$, $s^2 > K$, reject $\sigma = \sigma_1$, and conclude $\sigma > \sigma_1$.

Reproduced with permission from I. W. Burr, "Applied Statistical Methods," Academic Press, New York, 1974, p. 448.

TABLE IX

Logarithms of Factorials-Tens to Left, Units in Columns

	0	1	2	3	4	5	6	7	8	9
00	0.0000	0.0000	0.3010	0.7782	1.3802	2.0792	2.8573	3.7024	4.6055	5.5598
10	6.5598	7.6012	8.6803	9.7943	10.9404	12.1165	13.3206	14.5511	15.8063	17.0851
20	18.3861	19.7083	21.0508	22.4125	23.7927	25.1906	26.6056	28.0370	29.4841	30.9465
30	32.4237	33.9150	35.4202	36.9387	38.4702	40.0142	41.5705	43.1387	44.7185	46.3096
40	47.9116	49.5244	51.1477	52.7811	54.4246	56.0778	57.7406	59.4127	61.0939	62.7841
50	64.4831	66.1906	67.9066	69.6309	71.3633	73.1037	74.8519	76.6077	78.3712	80.1420
60	81.9202	83.7055	85.4979	87.2972	89.1034	90.9163	92.7359	94.5619	96.3945	98.2333
70	100.0784	101.9297	103.7870	105.6503	107.5196	109.3946	111.2754	113.1619	115.0540	116.9516
80	118.8547	120.7632	122.6770	124.5961	126.5204	128.4498	130.3843	132.3238	134.2683	136.2177
90	138.1719	140.1310	142.0948	144.0632	146.0364	148.0141	149.9964	151.9831	153.9744	155.9700
100	157.9700	159.9743	161.9829	163.9958	166.0128	168.0340	170.0593	172.0887	174.1221	176.1595
110	178.2009	180.2462	182.2955	184.3485	186.4054	188.4661	190.5306	192.5988	194.6707	196.7462
120	198.8254	200.9082	202.9945	205.0844	207.1779	209.2748	211.3751	213.4790	215.5862	217.6967
130	219.8107	221.9280	224.0485	226.1724	228.2995	230.4298	232.5634	234.7001	236.8400	238.9830
140	241.1291	243.2783	245.4306	247.5860	249.7443	251.9057	254.0700	256.2374	258.4076	260.5808
150	262.7569	264.9359	267.1177	269.3024	271.4899	273.6803	275.8734	278.0693	280.2679	282.4693
160	284.6735	286.8803	289.0898	291.3020	293.5168	295.7343	297.9544	300.1771	302.4024	304.6303
170	306.8608	309.0938	311.3293	313.5674	315.8079	318.0509	320.2965	322.5444	324.7948	327.0477
180	329.3030	331.5606	333.8207	336.0832	338.3480	340.6152	342.8847	345.1565	347.4307	349.7071
190	351.9859	354.2669	356.5502	358.8358	361.1236	363.4136	365.7059	368.0003	370.2970	372.5959

Reproduced with permission from I. W. Burr, "Engineering Statistics and Quality Control," McGraw-Hill, New York, 1953, pp. 412-416.

	0	1	2	3	4	5	6	7	8	9
200	374.8969	377.2001	379.5054	381.8129	384.1226	386.4343	388.7482	391.0642	393.3822	395.7024
210	398.0246	400.3489	402.6752	405.0036	407.3340	409.6664	412.0009	414.3373	416.6758	419.0162
220	421.3587	423.7031	426.0494	428.3977	430.7480	433.1002	435.4543	437.8103	440.1682	442.5281
230	444.8898	447.2534	449.6189	451.9862	454.3555	456.7265	459.0994	461.4742	463.8508	466.2292
240	468.6094	470.9914	473.3752	475.7608	478.1482	480.5374	482.9283	485.3210	487.7154	490.1116
250	492.5096	494.9093	497.3107	499.7138	502.1186	504.5252	506.9334	509.3433	511.7549	514.1682
260	516.5832	518.9999	521.4182	523.8381	526.2597	528.6830	531.1078	533.5344	535.9625	538.3922
270	540.8236	543.2566	545.6912	548.1273	550.5651	553.0044	555.4453	557.8878	560.3318	562.7774
280	565.2246	567.6733	570.1235	572.5753	575.0287	577.4835	579.9399	582.3977	584.8571	587.3180
290	589.7804	592.2443	594.7097	597.1766	599.6449	602.1147	604.5860	607.0588	609.5330	612.0087
300	614.4858	616.9644	619.4444	621.9258	624.4087	626.8930	629.3787	631.8659	634.3544	636.8444
310	639.3357	641.8285	644.3226	646.8182	649.3151	651.8134	654.3131	656.8142	659.3166	661.8204
320	664.3255	666.8320	669.3399	671.8491	674.3596	676.8715	679.3847	681.8993	684.4152	686.9324
330	689.4509	691.9707	694.4918	697.0143	699.5380	702.0631	704.5894	707.1170	709.6460	712.1762
340	714.7076	717.2404	719.7744	722.3097	724.8463	727.3841	729.9232	732.4635	735.0051	737.5479
350	740.0920	742.6373	745.1838	747.7316	750.2806	752.8308	755.3823	757.9349	760.4888	763.0439
360	765.6002	768.1577	770.7164	773.2764	775.8375	778.3997	780.9632	783.5279	786.0937	788.6608
370	791.2290	793.7983	796.3689	798.9406	801.5135	804.0875	806.6627	809.2390	811.8165	814.3952
380	816.9749	819.5559	822.1379	824.7211	827.3055	829.8909	832.4775	835.0652	837.6540	840.2440
390	842.8351	845.4272	848.0205	850.6149	853.2104	855.8070	858.4047	861.0035	863.6034	866.2044

TABLE IX (continued)

	0	1	2	3	4	5	6	7	8	9
400	868.8064	871.4096	874.0138	876.6191	879.2255	881.8329	884.4415	887.0510	889.6617	892.2734
410	894.8862	897.5001	900.1150	902.7309	905.3479	907.9660	910.5850	913.2052	915.8264	918.4486
420	921.0718	923.6961	926.3214	928.9478	931.5751	934.2035	936.8329	939.4633	942.0948	944.7272
430	947.3607	949.9952	952.6307	955.2672	957.9047	960.5431	963.1826	965.8231	968.4646	971.1071
440	973.7505	976.3949	979.0404	981.6868	984.3342	986.9825	989.6318	992.2822	994.9334	997.5857
450	1000.2389	1002.8931	1005.5482	1008.2043	1010.8614	1013.5194	1016.1783	1018.8383	1021.4991	1024.1609
460	1026.8237	1029.4874	1032.1520	1034.8176	1037.4841	1040.1516	1042.8200	1045.4893	1048.1595	1050.8307
470	1053.5028	1056.1758	1058.8498	1061.5246	1064.2004	1066.8771	1069.5547	1072.2332	1074.9127	1077.5930
480	1080.2742	1082.9564	1085.6394	1088.3234	1091.0082	1093.6940	1096.3806	1099.0681	1101.7565	1104.4458
490	1107.1360	1109.8271	1112.5191	1115.2119	1117.9057	1120.6003	1123.2958	1125.9921	1128.6893	1131.3874
500	1134.0864	1136.7862	1139.4869	1142.1885	1144.8909	1147.5942	1150.2984	1153.0034	1155.7093	1158.4160
510	1161.1236	1163.8320	1166.5412	1169.2514	1171.9623	1174.6741	1177.3868	1180.1003	1182.8146	1185.5298
520	1188.2458	1190.9626	1193.6803	1196.3988	1199.1181	1201.8383	1204.5593	1207.2811	1210.0037	1212.7272
530	1215.4514	1218.1765	1220.9024	1223.6292	1226.3567	1229.0851	1231.8142	1234.5442	1237.2750	1240.0066
540	1242.7390	1245.4722	1248.2062	1250.9410	1253.6766	1256.4130	1259.1501	1261.8881	1264.6269	1267.3665
550	1270.1069	1272.8480	1275.5899	1278.3327	1281.0762	1283.8205	1286.5655	1289.3114	1292.0580	1294.8054
560	1297.5536	1300.3026	1303.0523	1305.8028	1308.5541	1311.3062	1314.0590	1316.8126	1319.5669	1322.3220
570	1325.0779	1327.8345	1330.5919	1333.3501	1336.1090	1338.8687	1341.6291	1344.3903	1347.1522	1349.9149
580	1352.6783	1355.4425	1358.2074	1360.9731	1363.7395	1366.5066	1369.2745	1372.0432	1374.8126	1377.5827
590	1380.3535	1383.1251	1385.8974	1388.6705	1391.4443	1394.2188	1396.9940	1399.7700	1402.5467	1405.3241

	0	1	2	3	4	5	6	7	8	9
600	1408.1023	1410.8812	1413.6608	1416.4411	1419.2221	1422.0039	1424.7863	1427.5695	1430.3534	1433.1380
610	1435.9234	1438.7094	1441.4962	1444.2836	1447.0718	1449.8607	1452.6503	1455.4405	1458.2315	1461.0232
620	1463.8156	1466.6087	1469.4025	1472.1970	1474.9922	1477.7880	1480.5846	1483.3819	1486.1798	1488.9785
630	1491.7778	1494.5779	1497.3786	1500.1800	1502.9821	1505.7849	1508.5883	1511.3924	1514.1973	1517.0028
640	1519.8090	1522.6158	1525.4233	1528.2316	1531.0404	1533.8500	1536.6602	1539.4711	1542.2827	1545.0950
650	1547.9079	1550.7215	1553.5357	1556.3506	1559.1662	1561.9824	1564.7993	1567.6169	1570.4351	1573.2540
660	1576.0736	1578.8938	1581.7146	1584.5361	1587.3583	1590.1811	1593.0046	1595.8287	1598.6535	1601.4789
670	1604.3050	1607.1317	1609.9591	1612.7871	1615.6158	1618.4451	1621.2750	1624.1056	1626.9368	1629.7687
680	1632.6012	1635.4344	1638.2681	1641.1026	1643.9376	1646.7733	1649.6096	1652.4466	1655.2842	1658.1224
690	1660.9612	1663.8007	1666.6408	1669.4816	1672.3229	1675.1649	1678.0075	1680.8508	1683.6946	1686.5391
700	1689.3842	1692.2299	1695.0762	1697.9232	1700.7708	1703.6190	1706.4678	1709.3172	1712.1672	1715.0179
710	1717.8691	1720.7210	1723.5735	1726.4266	1729.2803	1732.1346	1734.9895	1737.8450	1740.7011	1743.5578
720	1746.4152	1749.2731	1752.1316	1754.9908	1757.8505	1760.7109	1763.5718	1766.4333	1769.2955	1772.1582
730	1775.0215	1777.8854	1780.7499	1783.6150	1786.4807	1789.3470	1792.2139	1795.0814	1797.9494	1800.8181
740	1803.6873	1806.5571	1809.4275	1812.2985	1815.1701	1818.0423	1820.9150	1823.7883	1826.6622	1829.5367
750	1832.4118	1835.2874	1838.1636	1841.0404	1843.9178	1846.7957	1849.6742	1852.5533	1855.4330	1858.3133
760	1861.1941	1864.0755	1866.9574	1869.8399	1872.7230	1875.6067	1878.4909	1881.3757	1884.2611	1887.1470
770	1890.0335	1892.9205	1895.8082	1898.6963	1901.5851	1904.4744	1907.3642	1910.2547	1913.1456	1916.0372
780	1918.9293	1921.8219	1924.7151	1927.6089	1930.5032	1933.3981	1936.2935	1939.1895	1942.0860	1944.9831
790	1947.8807	1950.7789	1953.6776	1956.5769	1959.4767	1962.3771	1965.2780	1968.1794	1971.0814	1973.9840

TABLE IX (continued)

	0	1	2	3	4	5	6	7	8	9
800	1976.8871	1979.7907	1982.6949	1985.5996	1988.5049	1991.4107	1994.3170	1997.2239	2000.1313	2003.0392
810	2005.9477	2008.8567	2011.7663	2014.6764	2017.5870	2020.4982	2023.4099	2026.3221	2029.2348	2032.1481
820	2035.0619	2037.9763	2040.8911	2043.8065	2046.7225	2049.6389	2052.5559	2055.4734	2058.3914	2061.3100
830	2064.2291	2067.1487	2070.0688	2072.9894	2075.9106	2078.8323	2081.7545	2084.6772	2087.6005	2090.5242
840	2093.4485	2096.3733	2099.2986	2102.2244	2105.1508	2108.0776	2111.0050	2113.9329	2116.8613	2119.7902
850	2122.7196	2125.6495	2128.5800	2131.5109	2134.4424	2137.3744	2140.3068	2143.2398	2146.1733	2149.1073
860	2152.0418	2154.9768	2157.9123	2160.8483	2163.7848	2166.7218	2169.6594	2172.5974	2175.5359	2178.4749
870	2181.4144	2184.3545	2187.2950	2190.2360	2193.1775	2196.1195	2199.0620	2202.0050	2204.9485	2207.8925
880	2210.8370	2213.7820	2216.7274	2219.6734	2222.6198	2225.5668	2228.5142	2231.4621	2234.4106	2237.3595
890	2240.3088	2243.2587	2246.2091	2249.1599	2252.1113	2255.0631	2258.0154	2260.9682	2263.9215	2266.8752
900	2269.8295	2272.7842	2275.7394	2278.6951	2281.6513	2284.6079	2287.5650	2290.5226	2293.4807	2296.4393
910	2299.3983	2302.3579	2305.3179	2308.2783	2311.2393	2314.2007	2317.1626	2320.1250	2323.0878	2326.0511
920	2329.0149	2331.9792	2334.9439	2337.9091	2340.8748	2343.8409	2346.8075	2349.7746	2352.7421	2355.7102
930	2358.6786	2361.6476	2364.6170	2367.5869	2370.5572	2373.5281	2376.4993	2379.4711	2382.4433	2385.4159
940	2388.3891	2391.3627	2394.3367	2397.3112	2400.2862	2403.2616	2406.2375	2409.2139	2412.1907	2415.1679
950	2418.1457	2421.1238	2424.1025	2427.0816	2430.0611	2433.0411	2436.0216	2439.0025	2441.9839	2444.9657
960	2447.9479	2450.9307	2453.9138	2456.8975	2459.8815	2462.8661	2465.8511	2468.8365	2471.8224	2474.8087
970	2477.7954	2480.7827	2483.7703	2486.7584	2489.7470	2492.7360	2495.7255	2498.7154	2501.7057	2504.6965
980	2507.6877	2510.6794	2513.6715	2516.6640	2519.6570	2522.6505	2525.6443	2528.6387	2531.6334	2534.6286
990	2537.6242	2540.6203	2543.6168	2546.6138	2549.6112	2552.6090	2555.6073	2558.6059	2561.6051	2564.6046
1,000	2567.6046	2570.6051	2573.6059	2576.6072	2579.6090	2582.6111	2585.6137	2588.6168	2591.6202	2594.6241

TABLE X

Common Logarithms of Numbers to Four Decimal Places.

N	L. 0	1	2	3	4	5	6	7	8	9	Proportional Parts 1	2	3	4	5
10	0000	0043	0086	0128	0170	0212	0253	0294	0334	0374	4	8	12	17	21
11	0414	0453	0492	0531	0569	0607	0645	0682	0719	0755	4	8	11	15	19
12	0792	0828	0864	0899	0934	0969	1004	1038	1072	1106	3	7	10	14	17
13	1139	1173	1206	1239	1271	1303	1335	1367	1399	1430	3	6	10	13	16
14	1461	1492	1523	1553	1584	1614	1644	1673	1703	1732	3	6	9	12	15
15	1761	1790	1818	1847	1875	1903	1931	1959	1987	2014	3	6	8	11	14
16	2041	2068	2095	2122	2148	2175	2201	2227	2253	2279	3	5	8	11	13
17	2304	2330	2355	2380	2405	2430	2455	2480	2504	2529	2	5	7	10	12
18	2553	2577	2601	2625	2648	2672	2695	2718	2742	2765	2	5	7	9	12
19	2788	2810	2833	2856	2878	2900	2923	2945	2967	2989	2	4	7	9	11
20	3010	3032	3054	3075	3096	3118	3139	3160	3181	3201	2	4	6	8	11
21	3222	3243	3263	3284	3304	3324	3345	3365	3385	3404	2	4	6	8	10
22	3424	3444	3464	3483	3502	3522	3541	3560	3579	3598	2	4	6	8	10
23	3617	3636	3655	3674	3692	3711	3729	3747	3766	3784	2	4	6	7	9
24	3802	3820	3838	3856	3874	3892	3909	3927	3945	3962	2	4	5	7	9
25	3979	3997	4014	4031	4048	4065	4082	4099	4116	4133	2	4	5	7	9
26	4150	4166	4183	4200	4216	4232	4249	4265	4281	4298	2	3	5	7	8
27	4314	4330	4346	4362	4378	4393	4409	4425	4440	4456	2	3	5	6	8
28	4472	4487	4502	4518	4533	4548	4564	4579	4594	4609	2	3	5	6	8
29	4624	4639	4654	4669	4683	4698	4713	4728	4742	4757	1	3	4	6	7
30	4771	4786	4800	4814	4829	4843	4857	4871	4886	4900	1	3	4	6	7
31	4914	4928	4942	4955	4969	4983	4997	5011	5024	5038	1	3	4	5	7
32	5051	5065	5079	5092	5105	5119	5132	5145	5159	5172	1	3	4	5	7
33	5185	5198	5211	5224	5237	5250	5263	5276	5289	5302	1	3	4	5	7
34	5315	5328	5340	5353	5366	5378	5391	5403	5416	5428	1	2	4	5	6
35	5441	5453	5465	5478	5490	5502	5514	5527	5539	5551	1	2	4	5	6
36	5563	5575	5587	5599	5611	5623	5635	5647	5658	5670	1	2	4	5	6
37	5682	5694	5705	5717	5729	5740	5752	5763	5775	5786	1	2	4	5	6
38	5798	5809	5821	5832	5843	5855	5866	5877	5888	5899	1	2	3	5	6
39	5911	5922	5933	5944	5955	5966	5977	5988	5999	6010	1	2	3	4	5
40	6021	6031	6042	6053	6064	6075	6085	6096	6107	6117	1	2	3	4	5
41	6128	6138	6149	6160	6170	6180	6191	6201	6212	6222	1	2	3	4	5
42	6232	6243	6253	6263	6274	6284	6294	6304	6314	6325	1	2	3	4	5
43	6335	6345	6355	6365	6375	6385	6395	6405	6415	6425	1	2	3	4	5
44	6435	6444	6454	6464	6474	6484	6493	6503	6513	6522	1	2	3	4	5
45	6532	6542	6551	6561	6571	6580	6590	6599	6609	6618	1	2	3	4	5
46	6628	6637	6646	6656	6665	6675	6684	6693	6702	6712	1	2	3	4	5
47	6721	6730	6739	6749	6758	6767	6776	6785	6794	6803	1	2	3	4	5
48	6812	6821	6830	6839	6848	6857	6866	6875	6884	6893	1	2	3	4	5
49	6902	6911	6920	6928	6937	6946	6955	6964	6972	6981	1	2	3	4	4
50	6990	6998	7007	7016	7024	7033	7042	7050	7059	7067	1	2	3	3	4
51	7076	7084	7093	7101	7110	7118	7126	7135	7143	7152	1	2	3	3	4
52	7160	7168	7177	7185	7193	7202	7210	7218	7226	7235	1	2	3	3	4
53	7243	7251	7259	7267	7275	7284	7292	7300	7308	7316	1	2	2	3	4
54	7324	7332	7340	7348	7356	7364	7372	7380	7388	7396	1	2	2	3	4
N	L. 0	1	2	3	4	5	6	7	8	9	1	2	3	4	5

TABLE X (continued)

N	L. 0	1	2	3	4	5	6	7	8	9	Proportional parts 1	2	3	4	5
55	7404	7412	7419	7427	7435	7443	7451	7459	7466	7474	1	2	2	3	4
56	7482	7490	7497	7505	7513	7520	7528	7536	7543	7551	1	2	2	3	4
57	7559	7566	7574	7582	7589	7597	7604	7612	7619	7627	1	1	2	3	4
58	7634	7642	7649	7657	7664	7672	7679	7686	7694	7701	1	1	2	3	4
59	7709	7716	7723	7731	7738	7745	7752	7760	7767	7774	1	1	2	3	4
60	7782	7789	7796	7803	7810	7818	7825	7832	7839	7846	1	1	2	3	4
61	7853	7860	7868	7875	7882	7889	7896	7903	7910	7917	1	1	2	3	3
62	7924	7931	7938	7945	7952	7959	7966	7973	7980	7987	1	1	2	3	3
63	7993	8000	8007	8014	8021	8028	8035	8041	8048	8055	1	1	2	3	3
64	8062	8069	8075	8082	8089	8096	8102	8109	8116	8122	1	1	2	3	3
65	8129	8136	8142	8149	8156	8162	8169	8176	8182	8189	1	1	2	3	3
66	8195	8202	8209	8215	8222	8228	8235	8241	8248	8254	1	1	2	3	3
67	8261	8267	8274	8280	8287	8293	8299	8306	8312	8319	1	1	2	3	3
68	8325	8331	8338	8344	8351	8357	8363	8370	8376	8382	1	1	2	3	3
69	8388	8395	8401	8407	8414	8420	8426	8432	8439	8445	1	1	2	3	3
70	8451	8457	8463	8470	8476	8482	8488	8494	8500	8506	1	1	2	3	3
71	8513	8519	8525	8531	8537	8543	8549	8555	8561	8567	1	1	2	3	3
72	8573	8579	8585	8591	8597	8603	8609	8615	8621	8627	1	1	2	3	3
73	8633	8639	8645	8651	8657	8663	8669	8675	8681	8686	1	1	2	2	3
74	8692	8698	8704	8710	8716	8722	8727	8733	8739	8745	1	1	2	2	3
75	8751	8756	8762	8768	8774	8779	8785	8791	8797	8802	1	1	2	2	3
76	8808	8814	8820	8825	8831	8837	8842	8848	8854	8859	1	1	2	2	3
77	8865	8871	8876	8882	8887	8893	8899	8904	8910	8915	1	1	2	2	3
78	8921	8927	8932	8938	8943	8949	8954	8960	8965	8971	1	1	2	2	3
79	8976	8982	8987	8993	8998	9004	9009	9015	9020	9025	1	1	2	2	3
80	9031	9036	9042	9047	9053	9058	9063	9069	9074	9079	1	1	2	2	3
81	9085	9090	9096	9101	9106	9112	9117	9122	9128	9133	1	1	2	2	3
82	9138	9143	9149	9154	9159	9165	9170	9175	9180	9186	1	1	2	2	3
83	9191	9196	9201	9206	9212	9217	9222	9227	9232	9238	1	1	2	2	3
84	9243	9248	9253	9258	9263	9269	9274	9279	9284	9289	1	1	2	2	3
85	9294	9299	9304	9309	9315	9320	9325	9330	9335	9340	1	1	2	2	3
86	9345	9350	9355	9360	9365	9370	9375	9380	9385	9390	1	1	2	2	3
87	9395	9400	9405	9410	9415	9420	9425	9430	9435	9440	1	1	2	2	3
88	9445	9450	9455	9460	9465	9469	9474	9479	9484	9489	0	1	1	2	2
89	9494	9499	9504	9509	9513	9518	9523	9528	9533	9538	0	1	1	2	2
90	9542	9547	9552	9557	9562	9566	9571	9576	9581	9586	0	1	1	2	2
91	9590	9595	9600	9605	9609	9614	9619	9624	9628	9633	0	1	1	2	2
92	9638	9643	9647	9652	9657	9661	9666	9671	9675	9680	0	1	1	2	2
93	9685	9689	9694	9699	9703	9708	9713	9717	9722	9727	0	1	1	2	2
94	9731	9736	9741	9745	9750	9754	9759	9763	9768	9773	0	1	1	2	2
95	9777	9782	9786	9791	9795	9800	9805	9809	9814	9818	0	1	1	2	2
96	9823	9827	9832	9836	9841	9845	9850	9854	9859	9863	0	1	1	2	2
97	9868	9872	9877	9881	9886	9890	9894	9899	9903	9908	0	1	1	2	2
98	9912	9917	9921	9926	9930	9934	9939	9943	9948	9952	0	1	1	2	2
99	9956	9961	9965	9969	9974	9978	9983	9987	9991	9996	0	1	1	2	2
N	L. 0	1	2	3	4	5	6	7	8	9	1	2	3	4	5

TABLE XI

Sequential Constants a and b from α and β

Base 10 logarithms

α if finding a, β if finding b

		.001	.01	.02	.03	.05	.10
	.001	3.000	2.000	1.699	1.522	1.301	1.000
	.01	2.996	1.996	1.695	1.519	1.297	.996
β if finding a	.02	2.991	1.991	1.690	1.514	1.292	.991
α if finding b	.03	2.987	1.987	1.686	1.510	1.288	.987
	.05	2.978	1.978	1.677	1.501	1.279	.978
	.10	2.954	1.954	1.653	1.477	1.255	.954

$a = \log\left(\frac{1-\beta}{\alpha}\right)$ $b = \log\left(\frac{1-\alpha}{\beta}\right)$

Base e logarithms

α if finding a, β if finding b

		.001	.01	.02	.03	.05	.10
	.001	6.907	4.604	3.911	3.506	2.995	2.302
	.01	6.898	4.595	3.902	3.497	2.986	2.293
β if finding a	.02	6.888	4.585	3.892	3.486	2.976	2.282
α if finding b	.03	6.877	4.575	3.882	3.476	2.965	2.272
	.05	6.856	4.554	3.861	3.455	2.944	2.251
	.10	6.802	4.500	3.807	3.401	2.890	2.197

$a = \ln\left(\frac{1-\beta}{\alpha}\right)$ $b = \ln\left(\frac{1-\alpha}{\beta}\right)$

TABLE XII

Sequential Range Test for Process Capability. To Meet Total Tolerance of T. Acceptable $\sigma = T/10$, Rejectable $\sigma = T/6$, with Corresponding Risks α and β of Wrong Decisions. Ranges for Samples of $n = 8$. Multiply Given Acceptance and Rejection Numbers by Specified T.*

No. of Ranges in Sum	Cumulated Sample Size	Multiples of Tolerance T for Acc.-Rej. Criteria			
		Risks α = .05		Risks α = .01	
		Acc. No.	Rej. No.	Acc. No.	Rej. No.
1	8	0.19	0.54	0.09	0.63
2	16	.55	.90	.46	1.00
3	24	.92	1.26	.82	1.36
4	32	1.28	1.63	1.18	1.73
5	40	1.64	1.99	1.55	2.09
6	48	2.01	2.36	1.91	2.45
7	56	2.37	2.72	2.27	2.82
8	64	2.91	2.91	2.64	3.18
9	72			3.00	3.54
10	38			3.36	3.91
11	88			3.73	4.27
12	96			4.36	4.36

* From H. M. Wies and I. W. Burr, Simple capability acceptance test, a sequential test on ranges. Indust. Quality Control 21 (No. 5), 266-268 (1964), with kind permission of American Society for Quality Control.

TABLE XIII

Acceptance and Rejection Numbers for <u>One-way</u> Sequential Check of Process Setting. Maximum Risks of Wrong Decisions at μ_1 and μ_2, $\alpha = \beta = .10$. Given for Classes of σ_x. Code x's So that $1 \leq \sigma_x \leq 10$.

Protection Against Proc.Aver.High by:		1.21		1.46		1.77		2.15		2.61		3.16	
Proc.Std.Dev.σ_x		1.00-1.21		1.22-1.46		1.47-1.77		1.78-2.15		2.16-2.61		2.62-3.16	
		Acc.	Rej.	Acc.	Rej.	Acc.	Rej.	Acc.	Rej.	Acc.	Rej.	Acc.	Rej.
Sample Size	1	-2.1	+3.3	-2.5	+3.9	-3.1	+4.8	-3.7	+5.8	-4.4	+7.1	-5.4	+8.5
	2	-1.5	3.9	-1.8	4.7	-2.1	5.7	-2.6	6.9	-3.1	8.3	-3.8	10.1
	3	- .8	4.5	-1.1	5.4	-1.2	6.6	-1.5	7.9	-1.8	9.7	-2.2	11.7
	4	- .2	5.1	- .3	6.2	- .4	7.5	- .4	9.1	- .5	10.9	- .6	13.3
	5	+ .4	5.7	+ .4	6.9	+ .5	8.4	+ .7	10.1	+ .8	12.3	+ .9	14.9
	6	.9	6.3	1.2	7.6	1.4	9.2	1.7	11.2	2.1	13.6	2.5	16.4
	7	1.6	6.9	1.9	8.4	2.3	10.1	2.8	12.3	3.4	14.9	4.1	18.1
	8	2.2	7.5	2.6	9.1	3.2	11.1	3.9	13.4	4.7	16.2	5.7	19.6
	9	2.8	8.1	3.4	9.8	4.1	11.9	4.9	14.4	6.1	17.5	7.3	21.2
	10	3.4	8.7	4.1	10.6	4.9	12.8	6.1	15.5	7.3	18.8	8.9	22.8
	11	6.0	6.1	8.0	8.1	9.8	9.9	11.8	11.9	14.3	14.4	17.3	17.4

TABLE XIII (continued)

Protection Margin, $1\sigma_x$		3.83		4.64		5.62		6.81		8.25		10.00	
Std. Dev. σ_x		3.17-3.83		3.84-4.64		4.65-5.62		5.63-6.81		6.82-8.25		8.26-10.0	
		Acc.	Rej.	Acc.	Rej.	Acc.	Rej.	Acc.	Rej.	Acc.	Rej.	Acc.	Rej.
Sample Size	1	- 6.5	+10.3	- 7.9	+12.5	- 9.5	+15.2	-11.6	+18.4	-14.1	+22.3	-16.9	+26.9
	2	- 4.6	12.2	- 5.6	14.8	- 6.7	17.9	- 8.2	21.8	- 9.9	26.4	-11.9	31.9
	3	- 2.7	14.2	- 3.2	17.2	- 3.9	20.8	- 4.8	25.2	- 5.8	30.5	- 6.9	36.9
	4	- .8	16.1	- .9	19.5	- 1.1	23.6	- 1.3	28.6	- 1.6	34.6	- 1.9	41.9
	5	+ 1.2	17.9	+ 1.4	21.8	+ 1.7	26.4	+ 2.1	32.1	+ 2.5	38.8	+ 3.1	46.9
	6	3.1	19.9	3.7	24.1	4.5	29.2	5.5	35.4	6.6	42.9	8.1	51.9
	7	4.9	21.8	6.1	26.4	7.3	32.1	8.9	38.8	10.8	47.1	13.1	56.9
	8	6.9	23.7	8.4	28.8	10.1	34.8	12.3	42.2	14.9	51.2	18.1	61.9
	9	8.8	25.7	10.7	31.1	12.9	37.7	15.7	45.6	19.1	55.3	23.1	66.9
	10	10.7	27.6	13.1	33.4	15.8	40.5	19.1	49.1	23.1	59.4	28.1	71.9
	11	21.1	21.2	25.5	25.6	30.9	31.0	37.4	37.5	45.4	45.5	55.0	55.1

TABLE XIV

Acceptance and Rejection Numbers for Two-way Sequential Check of Process Setting. Maximum Risks of Wrong Decisions at μ, and at $\mu \pm \sigma_x$ are $\alpha = .10$, $\beta = .10$. Given for Classes of σ_x. Code x's so that $1 \leq \sigma_x \leq 10$.

Protection Against Proc.Aver.Off by:	1.21		1.46		1.77		2.15		2.61		3.16	
Class, Std.Dev. σ_x	1.00-1.21		1.22-1.46		1.47-1.77		1.78-2.15		2.16-2.61		2.62-3.16	
	Acc.	Rej.	Acc.	Rej.	Acc.	Rej.	Acc.	Rej.	Acc.	Rej.	Acc.	Rej.
Sample Size 1	-	4.1	-	4.9	-	6.1	-	7.3	-	8.8	-	10.7
2	-	4.7	-	5.7	-	6.9	-	8.4	-	10.2	-	12.3
3	-	5.3	-	6.4	-	7.8	-	9.5	-	11.5	-	13.9
4	-	5.9	-	7.2	-	8.7	-	10.5	-	12.8	-	15.5
5	.9	6.5	1.2	7.9	1.5	9.6	1.8	11.6	2.1	14.1	2.6	17.1
6	1.7	7.1	2.1	8.6	2.6	10.5	3.1	12.7	3.8	15.4	4.6	18.6
7	2.4	7.7	2.9	9.4	3.5	11.4	4.3	13.8	5.2	16.7	6.3	20.2
8	3.1	8.3	3.7	10.1	4.4	12.3	5.4	14.8	6.5	17.9	7.9	21.8
9	3.6	8.9	4.4	10.8	5.3	13.1	6.4	15.9	7.8	19.3	9.5	23 4
10	4.2	9.6	5.1	11.6	6.2	14.1	7.5	16.9	9.1	20.6	11.1	24.9
11	4.8	10.2	5.9	12.3	7.1	14.9	8.6	18.1	10.4	21.9	12.6	26.5
12	5.4	10.8	6.6	13.1	7.9	15.8	9.7	19.1	11.7	23.2	14.2	28.1
13	6.1	11.4	7.3	13.8	8.9	16.7	10.8	20.2	13.1	24.5	15.8	29.7
14	6.7	11.9	8.1	14.5	9.8	17.6	11.8	21.3	14.3	25.8	17.4	31.3
15	7.3	12.6	8.8	15.3	10.7	18.5	12.9	22.4	15.7	27.1	18.9	32.9
16	7.9	13.2	9.5	15.9	11.6	19.4	13.9	23.5	16.9	28.4	20.5	34.4
17	8.5	13.8	10.3	16.7	12.4	20.3	15.1	24.5	18.3	29.7	22.1	36.1
18	9.1	14.2	11.1	17.2	13.3	20.9	16.1	25.3	19.6	30.6	23.7	37.1
19	9.7	14.2	11.7	17.2	14.2	20.9	17.2	25.3	20.9	30.6	25.3	37.1
20	10.3	14.2	12.5	17.2	15.1	20.9	18.3	25.3	22.2	30.6	26.9	37.1
21	10.9	14.2	13.2	17.2	15.9	20.9	19.4	25.3	23.5	30.6	28.4	37.1
22	14.1	14.2	17.1	17.2	20.8	20.9	25.2	25.3	30.5	30.6	37.0	37.1

Protection Margin, $1\sigma_x$	3.83		4.64		5.62		6.81		8.25		10.00	
Class, Std. Dev. σ_x	3.17-3.83		3.84-4.64		4.65-5.62		5.63-6.81		6.82-8.25		8.26-10.00	
	Acc.	Rej.	Acc.	Rej.	Acc.	Rej.	Acc.	Rej.	Acc.	Rej.	Acc.	Rej.
1	-	12.9	-	15.7	-	19.1	-	23.1	-	27.9	-	33.9
2	-	14.9	-	18.1	-	21.9	-	26.5	-	32.1	-	38.9
3	-	16.8	-	20.4	-	24.7	-	29.9	-	36.2	-	43.9
4	-	18.7	-	22.7	-	27.5	-	33.3	-	40.4	-	48.9
5	3.1	20.7	3.8	25.1	4.6	30.3	5.6	36.7	6.8	44.5	8.2	53.9
6	5.5	22.6	6.7	27.3	8.1	33.1	9.8	40.1	11.9	48.6	14.4	58.9
7	7.6	24.5	9.2	29.7	11.1	35.9	13.5	43.5	16.3	52.7	19.8	63.9
8	9.5	26.4	11.6	31.9	13.9	38.7	16.9	46.9	20.5	56.9	24.9	68.9
9	11.5	28.3	13.9	34.3	16.8	41.6	20.4	50.4	24.7	61.1	29.9	73.9
10	13.4	30.2	16.2	36.6	19.7	44.4	23.8	53.8	28.8	65.1	34.9	78.9
11	15.3	32.1	18.5	38.9	22.5	47.2	27.2	57.2	32.9	69.3	39.9	83.9
12	17.2	34.1	20.9	41.3	25.3	49.9	30.6	60.6	37.1	73.4	44.9	88.9
13	19.1	35.9	23.2	43.6	28.1	52.8	34.1	63.9	41.2	77.5	49.9	93.9
14	21.1	37.9	25.5	45.9	30.9	55.6	37.4	67.4	45.4	81.6	54.9	98.9
15	22.9	39.8	27.8	48.2	33.7	58.4	40.8	70.8	49.5	85.8	59.9	103.9
16	24.9	41.7	30.2	50.5	36.5	61.2	44.3	74.2	53.6	89.9	64.9	108.9
17	26.8	43.6	32.5	52.9	39.3	64.1	47.7	77.6	57.7	94.1	69.9	113.9
18	28.7	44.9	34.8	54.4	42.2	65.8	51.1	79.7	61.9	96.6	74.9	117.0
19	30.6	44.9	37.1	54.4	44.9	65.8	54.5	79.7	65.9	96.6	79.9	117.0
20	32.6	44.9	39.4	54.4	47.8	65.8	57.9	79.7	70.1	96.6	84.9	117.0
21	34.5	44.9	41.8	54.4	50.6	65.8	61.3	79.7	74.3	96.6	89.9	117.0
22	44.8	44.9	54.3	54.4	65.7	65.8	79.6	79.7	96.5	96.6	116.9	117.0

TABLE XV

Parameters for a General System of Distributions (15.62), for Selected Curve Shape Characteristics α_3 and α_4.

α_3	α_4	c	k	μ	σ
.00	2.8	3.939	19.865	.4275	.1243
	3.0	4.874	6.158	.6447	.1620
	3.2	6.065	3.745	.7673	.1644
	3.4	7.696	2.701	.8507	.1519
	3.6	10.182	2.090	.9111	.1300
.10	2.8	3.520	19.606	.3901	.1257
	3.0	4.297	6.283	.6077	.1714
	3.3	5.832	3.235	.7830	.1776
	3.6	8.298	2.164	.8896	.1539
	3.9	13.716	1.549	.9595	.1097
.20	2.8	3.071	30.547	.2957	.1069
	3.0	3.705	7.244	.5420	.1737
	3.3	4.895	3.548	.7347	.1944
	3.6	6.640	2.353	.8538	.1803
	3.9	9.858	1.694	.9352	.1447
.30	3.0	3.140	9.872	.4411	.1617
	3.3	4.058	4.168	.6636	.2054
	3.6	5.289	2.684	.8000	.2054
	3.9	7.227	1.933	.8951	.1818
	4.2	11.492	1.410	.9648	.1353
.40	3.0	2.625	19.800	.2887	.1214
	3.3	3.341	5.369	.5638	.2039
	3.6	4.226	3.223	.7240	.2235
	3.9	5.449	2.289	.8353	.2152
	4.2	7.492	1.706	.9190	.1863
.50	3.2	2.538	12.523	.3353	.1477
	3.5	3.162	4.927	.5653	.2172
	3.8	3.917	3.156	.7134	.2378
	4.1	4.925	2.306	.8211	.2332
	4.4	6.514	1.752	.9054	.2093
.60	3.2	2.066	190.007	.0700	.0356
	3.6	2.738	5.983	.4836	.2086
	4.0	3.549	3.253	.6859	.2503
	4.4	4.673	2.219	.8232	.2484
	4.8	6.748	1.574	.9287	.2141
.70	3.5	2.065	20.670	.2080	.1087
	3.9	2.656	5.246	.5017	.2258
	4.3	3.354	3.155	.6815	.2641
	4.7	4.292	2.238	.8105	.2656
	5.1	5.896	1.635	.9141	.2386
.80	3.8	2.005	14.088	.2434	.1325
	4.3	2.653	4.406	.5413	.2494
	4.8	3.448	2.717	.7260	.2823
	5.3	4.651	1.897	.8616	.2727
	5.8	8.262	1.204	.9858	.2044

GLOSSARY OF SYMBOLS

A = multiplier of σ for $\pm$ limits for $\bar{x}$, around μ

A = an event (in probability)

A_2 = multiplier of $\bar{R}$ for $\pm$ limits for $\bar{x}$, around $\bar{\bar{x}}$

A_3 = multiplier of $\bar{s}$ for $\pm$ limits for $\bar{x}$, around $\bar{\bar{x}}$

Ac = acceptance number for variable, single sampling

Ac_1 = acceptance number on first sample for defectives or defects

Ac_2 = acceptance number after two samples for total number of defectives or defects

AOQ = average outgoing quality

= average fraction defective of all lots, after screening of rejected lots

$AOQ_{p'}$ = AOQ if incoming quality is at p'

AOQL = average outgoing quality limit

= maximum of AOQ's for all p''s for a plan

AQL = acceptable quality level

= a nominal lot or process level considered satisfactory by consumer

ARL = average run length

= average number of trials to a decision

ASN = average sample number

= average number of inspections or tests till a decision is reached

ATI = average total inspection

= average number of inspections per lot, including accepted and rejected lots

B = an event (in probability)

B_3 = multiplier of $\bar{s}$ to give lower limit for s's

B_4 = multiplier of $\bar{s}$ to give upper limit for s's

B_5 = multiplier of σ to give lower limit for s's

B_6 = multiplier of σ to give upper limit for s's

℄ = symbol for center line

c = number of defects on a sample of product

c = acceptable number of defectives, in single sampling

c_1 = acceptance number on first sample for defectives or defects (like Ac_1)

c_2 = acceptance number after two samples for total number of defectives or defects (like Ac_2)

c_4 = multiplier of σ to give theoretical average of s's

c_5 = multiplier of σ to give theoretical standard deviation of s's

c' = true or theoretical average number of defects per sample of specified size

$\bar{c}$ = average number of defects per sample

C' = true or theoretical average number of defects per 100 units or pieces

CR = consumer's risk, or probability of erroneous acceptance of a lot at some given unacceptable quality, if offered

ChSP = chain sampling plan

CSP = continuous sampling plan

C(n,d) = number of different possible combinations of d objects taken from n distinct ones

d = number of defective or non-conforming pieces in a sample of n pieces

d_1 = number of defective or non-conforming pieces in the first sample

d_2 = multiplier of σ to give theoretical average of R's

d_2 = number of defective or non-conforming pieces in a second sample

d_3 = multiplier of σ to give theoretical standard deviation of R's

D = number of defective or non-conforming pieces in lot

D = total number of demerits on a sample or samples

D_1 = multiplier of σ to give lower limit for R's

D_2 = multiplier of σ to give upper limit for R's

D_3 = multiplier of $\bar{R}$ to give lower limit for R's

D_4 = multiplier of $\bar{R}$ to give upper limit for R's

e = natural logarithmic base, 2.71828 ···

E = expected value or theoretical average of whatever variable is written in parentheses after E

f = proportion of units sampled in continuous plans

f_i = observed frequency of cases in i'th class

f(x) = probability density function of x

F_i = theoretical or calculated frequency of cases in i'th class

F(y) = probability that variable $x \leq y$

H_1 = basic or null hypothesis, often that quality level is satisfactory

H_2 = alternate hypothesis, often that the quality characteristic is at some unsatisfactory level

i = number of a frequency class or variable

i = number of units, all good in succession, to qualify for sampling in continuous sampling plans CSP

i = number of previous lot-samples to be free of defectives to tolerate a defective in present lot in ChSP

i = required number of consecutive lots, all accepted, to qualify for skipping in SkSP

k = number of samples

k = an amount of departure from μ_0 considered rejectable, for μ

k = a multiple for s, so that if $\bar{x} + ks \leq U$, accept

K = acceptance-rejection limit for $\bar{x}$ or for s^2

K = number of units inspected per 1000 if always on sampling, in CSP

L = lower tolerance or specification limit for individual x's

LCL = lower control limit

LQ_{10} = that fraction defective having Pa of .10

LTPD = lot tolerance proportion (or percent) defective

= a nominal fraction defective to be accepted only a small proportion of times offered

n = sample size, number of observations in sample

n_1 = number of pieces in first sample

n_2 = number of pieces in second sample

n-1 = number of degrees of freedom in a sample of n

$\bar{n}$ = average of a set of sample sizes n

np = number of defectives in a sample of n

N = number of pieces or numbers in a lot or population

N(A) = number of distinct ways event A can occur

N(A and B) = number of distinct ways in which both A and B can simultaneously occur

$N(A|B)$ = number of distinct ways in which event A can occur, given that event B did occur

p = sample fraction defective (= d/n)

$\bar{p}$ = average fraction defective over several samples (= $\Sigma d/\Sigma n$)

p' = true or theoretical fraction defective

= D/N in finite lot

p'_1 = an AQL fraction defective (often p'_{95})

p'_2 = an LTPD fraction defective (often p'_{10})

p'_{10} = that fraction defective having a 10% chance of acceptance whenever offered to a plan

p'_{50} = that fraction defective having a 50% chance of acceptance whenever offered to a plan

p'_{95} = that fraction defective having a 95% chance of acceptance whenever offered to a plan

p_L = estimated fraction defective below L

p_t = lot tolerance per cent defective in Dodge-Romig (= p'_{10})

p_U = estimated fraction defective above U

P' = theoretical average of defectives per 100 units

Pa = probability of acceptance for a given plan on some given quality level

PR = producer's risk, or probability of erroneous rejection of a lot at acceptable quality, if offered

P(A) = probability of event A occurring

P(A|B) = probability that event A occur, given that event B did occur

P(A and B) = probability that both of events A and B occur on the trial

q = sample fraction good or conforming (= 1-p)

q = a quality characteristic in general

q' = lot or process fraction good or conforming (= 1-p')

Q_L = quality index at L in MIL-STD 414 (like -z)

Q_U = quality index at U in MIL-STD 414 (like z)

R = range, maximum minus minimum values in sample

$\bar{R}$ = average of a set of ranges

Re = rejection number for a single sample

Re_1 = rejection number for first sample

Re_2 = rejection number for total defectives or defects on first two samples

RQL = rejectable quality level on some characteristic

s = sample standard deviation, square root of variance

s_x = sample standard deviation of x, ($\sqrt{s_x^2}$

$\bar{s}$ = average of a set of s's

s^2 = sample variance

s_x^2 = sample variance, $\sum_1^n (x-\bar{x})^2/(n-1)$

S = the space of all possible outcomes of an experiment

SkSP = skip lot sampling plan

t = a standardized variable

= $(\bar{x}-\mu)/(s/\sqrt{n})$

T = tolerance (= U-L)

u = defects per unit

= (total defects on n units)/n

u' = true or theoretical average of u's for a process

$\bar{u}$ = average of a set of u's

$\bar{U}$ = average demerits per unit

U = upper tolerance or specification limit for individual x's

UCL = upper control limit

V_x = coefficient of variation (σ_x/μ_x)

w_A = demerit points for each class A defect, w_B, w_C, w_D similar

x = observed measurement

x_i = the i'th x in a sample

$\bar{x}$ = an average of a sample of x's = $\sum_1^n x/n$

$\bar{\bar{x}}$ = average of a set of $\bar{x}$'s = $\sum_1^k \bar{x}/k$

y = a measurable characteristic of an assembly

z = standardized variable = $(x-\mu)/\sigma_x$

z_α = normal curve value with α probability above $\int_{z_\alpha}^{\infty} \phi(u)du = \alpha$

Z = random variable

α = probability of rejecting hypothesis H_1 when true

= probability of rejecting lot or process when it is "acceptable"

= producer's risk

α_3 = a measure of lack of symmetry or skewness of a frequency distribution

= $E[(\alpha-\mu)^3]/\sigma^3$

α_4 = similar, measuring kurtosis or contact

β = probability of accepting hypothesis H_1 when false by some amount

= probability of accepting lot or process when it is rejectable

= consumer's risk

$\Gamma(k)$ = gamma function of $k = \int_0^\infty x^k e^{-x} dx$

$\Delta x = x-\mu_x$

μ = population mean or average of some random variable

μ_x = population or process mean of x's

μ_0 = a desired or nominal mean of x's

μ_1 = a particular value of μ, often an acceptable value

μ_2 = a particular value of μ, often a rejectable value

$\mu_2 = E[(x-\mu)^2]$

$\mu_k = E[(x-\mu)^k]$, special cases μ_3, μ_4

$\mu'_k = E(x^k)$, μ'_3, μ'_4 special cases

$\hat{\mu}$ = estimate of μ, often $\bar{\bar{x}}$

ν = degrees of freedom

$\phi(z) = e^{-z^2/2}/\sqrt{2\pi}$ = normal curve density function

$\Phi(z) = \int_{-\infty}^{z} (e^{u^2/2}/\sqrt{2\pi})\, du$ = cumulative normal function

= P(z or less)

σ = population, lot or process standard deviation

$\sqrt{E[(x-\mu)^2]}$, for x's

$\sigma_x, \sigma_{\bar{x}}, \sigma_{s^2}, \sigma_s, \sigma_R$ = population standard deviation of $x, \bar{x}, s^2, s, R$

σ^2 = population, lot or process variance

= $E[(x-\mu)^2]$, for x's

$\hat{\sigma}_x$ = an estimate of σ_x (might use s, $\bar{s}/c_4$ or $\bar{R}/d_2$)

$\sum$ = summation of whatever follows the sigma

χ^2 = chi square random variable

= $(n-1)s^2/\sigma^2$ for normal curve samples

= $\sum_{i=1}^{k} \frac{(f_i - F_i)^2}{F_i}$ for testing goodness of fit

AUTHOR INDEX

SUBJECT INDEX